Sustainable Agriculture Reviews

Volume 32

Series Editor
Eric Lichtfouse
CEREGE, Aix Marseille Univ
CNRS, IRD, INRA, Coll France
Aix-en-Provence, France

Other Publications by Dr. Eric Lichtfouse

Books
Scientific Writing for Impact Factor Journals
https://www.novapublishers.com/catalog/product_info.php?products_id=42242

Environmental Chemistry
http://www.springer.com/978-3-540-22860-8

Sustainable Agriculture
Volume 1: http://www.springer.com/978-90-481-2665-1
Volume 2: http://www.springer.com/978-94-007-0393-3

Book series
Environmental Chemistry for a Sustainable World
http://www.springer.com/series/11480

Sustainable Agriculture Reviews
http://www.springer.com/series/8380

Journal
Environmental Chemistry Letters
http://www.springer.com/10311

Sustainable agriculture is a rapidly growing field aiming at producing food and energy in a sustainable way for humans and their children. Sustainable agriculture is a discipline that addresses current issues such as climate change, increasing food and fuel prices, poor-nation starvation, rich-nation obesity, water pollution, soil erosion, fertility loss, pest control, and biodiversity depletion.

Novel, environmentally-friendly solutions are proposed based on integrated knowledge from sciences as diverse as agronomy, soil science, molecular biology, chemistry, toxicology, ecology, economy, and social sciences. Indeed, sustainable agriculture decipher mechanisms of processes that occur from the molecular level to the farming system to the global level at time scales ranging from seconds to centuries. For that, scientists use the system approach that involves studying components and interactions of a whole system to address scientific, economic and social issues. In that respect, sustainable agriculture is not a classical, narrow science. Instead of solving problems using the classical painkiller approach that treats only negative impacts, sustainable agriculture treats problem sources.

Because most actual society issues are now intertwined, global, and fast-developing, sustainable agriculture will bring solutions to build a safer world. This book series gathers review articles that analyze current agricultural issues and knowledge, then propose alternative solutions. It will therefore help all scientists, decision-makers, professors, farmers and politicians who wish to build a safe agriculture, energy and food system for future generations.

More information about this series at http://www.springer.com/series/8380

Eric Lichtfouse
Editor

Sustainable Agriculture Reviews 32

Waste Recycling and Fertilisation

Springer

Editor
Eric Lichtfouse
CEREGE, Aix Marseille Univ
CNRS, IRD, INRA, Coll France
Aix-en-Provence, France

ISSN 2210-4410 ISSN 2210-4429 (electronic)
Sustainable Agriculture Reviews
ISBN 978-3-319-98913-6 ISBN 978-3-319-98914-3 (eBook)
https://doi.org/10.1007/978-3-319-98914-3

Library of Congress Control Number: 2018957274

This Springer imprint is published by the registered company Springer Nature Switzerland AG
The registered company address is: Gewerbestrasse 11, 6330 Cham, Switzerland

Preface

In the near future, waste recycling will no longer be an option because natural resources become rare and costly, urbanisation is blooming and population is growing. In theory, most waste could be recycled efficiently. In practice, most waste is wasted, notably in rich countries where most people have somehow forgotten that food production by agriculture is simply vital. In other words, without food we die, to put it bluntly. For food security we need both more funds for agricultural research, and more ideas and inventions to produce food using waste. This book presents advanced research in fertilisation and recycling.

Spring pea field in Burgundy, France. Cernay et al. Chap. 4

In the first chapter, Drangert applies systems thinking to develop the concept of waste hierarchy, which is at the basis of improving waste recycling in smart cities and eco-houses. He shows that more than 50% of mined phosphorus (P) actually used for fertilisation can be replaced by phosphorus from waste. Liwei et al. review food losses and waste in the Chinese food systems, and found that the loss ratio during harvest could be reduced by 62%, in Chap. 2. Ipsilantis et al. review the role of mycorrhizal fungi and P-mobilising bacteria to improve plant nutrition, in Chap. 3. A meta-analysis of the yield of world grain legumes shows that soybean, narrowleaf lupin and faba bean are interesting alternatives to pea in Europe, as explained in Chap. 4 by Cernay et al. In the same vein, in Chap. 5, Mahmoud et al. recommend to foster legume cultivation in Europe because grain legumes occupy only 1.8% of arable lands.

Yu et al. explain that less than 40% of applied nitrogen (N) fertiliser is used by crops ; they thus give management guidelines for fertilisation in rice-wheat systems in Chap. 6. Benefits and drawbacks of using oilseed rape residues for fertilisation are presented by Kriauciuniene et al. in Chap. 7. The production of biochar from organic wastes, and the use of biochar to fertilise and improve soils are reviewed by Singh et al. in Chap. 8. Mkonda and He discuss fertilisation and agropastoralism in semi-arid areas, and conclude that the use of organic manure and waste has increased crop yields from 0.8 to 18 tons per hectare, in Chap. 9. In Chap. 10, Raza et al. decribe the impact of climate change on agriculture in Pakistan, and the potential benefits of organic farming. Guleria and Kumar discuss the effect of transgenes and nanoparticles on plants and soil microbes in the last chapter.

Aix-en-Provence, France Eric Lichtfouse

Contents

About the Editor

Eric Lichtfouse, PhD, born in 1960, is an environmental chemist working at the University of Aix-Marseille, France. He has invented carbon-13 dating, a method allowing to measure the relative age and turnover of molecular organic compounds occurring in different temporal pools of any complex media. He is teaching scientific writing and communication and has published the book *Scientific Writing for Impact Factor Journals*, which includes a new tool – the Micro-Article – to identify the novelty of research results. He is founder and chief editor of scientific journals and series in environmental chemistry and agriculture. He got the Analytical Chemistry Prize by the French Chemical Society, the Grand Prize of the Universities of Nancy and Metz, and a Journal Citation Award by the Essential Indicators.

Chapter 1
Nutrient Recycling: Waste Hierarchy, Recycling Cities and Eco-houses

Jan-Olof Drangert

Abstract Food security presupposes access to sunshine, nutrients and water. With an increase in population to 10–11 billion in this century, the Malthusian issue of resources boundaries is still on the global agenda. Urban flows of nutrient-rich waste from the food chain and excreta need to be redesigned. This chapter elaborates on measures to ensure a sustainable supply of plant nutrients for future food production.

An extended waste hierarchy is employed here to structure the analysis of nutrient waste recovery. Reduction, reuse and recycling measures show that recovered P from the waste flows in Europe can substitute 50–70% of mined phosphorus in fertilizers. The rate of losses between the mine and plate control the degree of substitution. A practical city-level example of improved design of nutrient flows indicates increases in recovery of both P and N of 90% and 80% respectively. Examples of eco-houses built to recover and reuse/recycle nutrient-rich liquid and solid waste displays required piping.

Keywords Nutrient recovery · Food loss · N · P · K · Planetary resources boundaries · Waste hierarchy · Reuse · Recycling · Urban infrastructure · Food security · Urban agriculture

1.1 Introduction

The Malthusian issue whether food production can cope with population increase is still on the global agenda and is likely to continue to be as 85% of the world population of 10–11 billion is expected to be urbanites at the end of this century (Malthus 1798; OECD 2013). Meanwhile, the rural population will be halved to 1.5 billion. Thanks to mechanization and other productivity improvements, each farmer can feed more people (Krausmann et al. 2008). Specialization and international trade support this trend. But, will enough plant nutrients be available in the future?

J.-O. Drangert (✉)
Linköping University, Linköping, Sweden

E. Lichtfouse (ed.), *Sustainable Agriculture Reviews 32*, Sustainable Agriculture Reviews 32, https://doi.org/10.1007/978-3-319-98914-3_1

Over the centuries, agricultural production has gradually been geographically separated from food consumption. An extreme case is where active cereal farmers in Canada have moved to nearby towns with their families and started commuting daily to their farms. A result of the disconnection between food production and consumption is that nutrient-rich urban food waste and human excreta, which were traditionally returned as a local fertiliser resource, has turned into a disposal problem, while imported food, feed and mineral fertilisers fill the nutrient gap (Senthilkumar et al. 2012).

Does the disconnection mean that residents in cities become unwilling to return nutrient-rich waste to agriculture? Not necessarily, as shown by the fact that garden cities or suburbs were built already a century ago where families produced flowers, fruits, berries and vegetables (Smit et al. 1996). Today, there is an emerging trend of greening the cities, and more concerted efforts go into urban agriculture with local roof-top production of some food items (Stringer 2010). But, recycling of nutrient-rich household waste in order to replace mineral fertilisers is still poorly developed (Schoumans et al. 2015).

The subject of this chapter deals with options that urbanites have to end present wastage of valuable nutrient resources and instead direct societies' organic waste back to agriculture. Such a step would support a sustainable supply of plant nutrients to food production in an era when easy access to potassium and phosphorus is diminishing (USGS 2015; van Dijk et al. 2015). A hierarchy of actions to manage both solid and liquid wastes is employed to guide the modification of global nutrient flows (Drangert et al. 2018), followed by a practical city-level approach to bend nutrient flows. Lastly, some options to achieve a change in a single house are presented.

1.2 Flows and Sinks of Nutrient Resources

There is a growing concern about crossing planetary boundaries to access natural resources (Rockström et al. 2009). Nine global resources have been identified to be vulnerable to transgression of set boundaries. All such estimates build on more or less certain data, and both exploitable resources and reserves may differ over time depending on technological advances, newly found deposits, or simple estimate errors (USGS 2015).

Steffen et al. (2015) revised the boundaries and found that the nitrogen and phosphorus flows have already transgressed the boundary. Next, land-system change and potassium are imminent candidates (USGS 2016). All these resources are vital inputs to agriculture and food production. The challenge is to manipulate present resource flows to avoid crossing planetary boundaries. A systems approach is applied to display some options.

The production of food requires sun, water and nutrients, and in conventional agriculture these resources are drawn directly or indirectly from rainfall and mainly mined mineral nutrients. Figure 1.1 visualizes the two main options to access

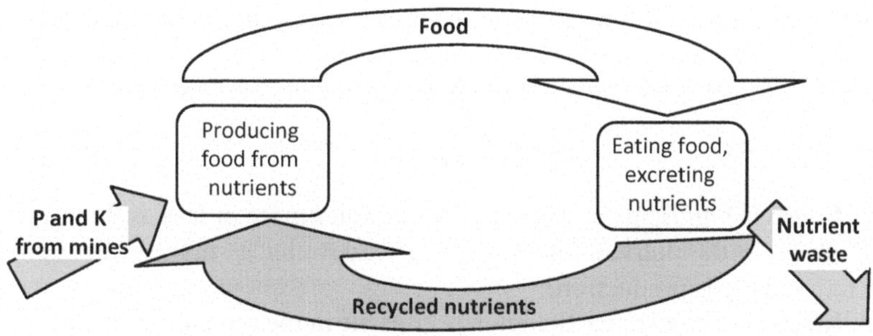

Fig. 1.1 Instead of wasting nutrients in e.g. excreta (red arrow), they can be recycled (green arrow) and replace nutrients in mineral fertilisers such as P (phosphorus) and K (potassium)

nutrients for food production. Commonly, the disposal of nutrient-rich human excreta and biowaste takes place via untreated wastewater causing eutrophication, or field application of sludge containing e.g. heavy metals, or dumping on landfills (EUP 2011). The main alternative is to recycle treated nutrient-rich wastes back to fields and food production – the ultimate sink and resource (Drangert 2000).

In the following the focus is on the potential to bend excreta and biowaste flows from urban settlements to become part of the flow of nutrients to agriculture, and thereby replacing nutrients that are presently being mined (P and K) or converted nitrogen (N) from the air.

1.3 Systems Thinking – The 'Extended Waste Hierarchy'

The guiding principle is to turn nutrients in urban liquid and solid waste into an agricultural resource by transforming the urban sanitation systems. Life-cycle thinking is applied and looks at environmental impacts throughout the entire life cycle of a product, from extraction of the resource to – and including – its disposal phase. Actions begin where waste originates, rather than where it ends up. Previous focus on "end-of-pipe" treatment is thus avoided, and the initial attention goes to controlling the substances used for the making of products (ECHA 2007).

The food sector has been singled out because it contains most of the nutrients in urban wastes (Zeeman 2012). The macro-nutrients N, P and K are in focus since there is no substitute for P or K in agriculture and mineral N requires a high energy input when converted from N_2 in the air. Some 15% of the globally mined K is used in non-food industries and only some 7% of the mined P, mainly for making detergents (Cordell et al. 2009; USGS 2016). The agricultural sector plays an important role as a potential recipient of urban nutrients, but is not analyzed in its own right as provider of food.

A hierarchy, originally developed for handling solid waste (EU 2008), is productively applied to handle both solid and liquid waste. This *extended waste hierarchy* has five steps, starting with how to produce more while generating less waste, and how to reduce unwanted substances in the products in order to facilitate later reuse and recycling. The measures to recover nutrients in each step are exhausted before entering the next step, and ideally little nutrients remains in step 4 and 5.

Step 1. Reduce (a) waste generation, and **(b)** harmful contents in products and flows;
Step 2. Reuse the nutrients in waste and wastewater more or less as they are;
Step 3. Recycle the nutrients in waste and wastewater as input to new products (including biogas production);
Step 4. Incinerate to extract the energy content in the remaining waste;
Step 5. Safely **landfill** residues remaining after exhausting the previous steps.

Impacts of steps 1–3 of the "extended waste hierarchy" are studied from measurable and achievable results rather than being based on prescribed technical solutions. The selection of measures below reflects rough estimates that these will have a large enough impact to be considered in most local circumstances. Quantifications can be made more consistent by using national data bases with similar definitions and known local conditions than when using global data (Prud'homme 2011; Kabbe et al. 2014; van Dijk et al. 2015). The estimates presented next are therefore from one region, the European Union.

In a phosphorus context, the above steps are interpreted as follows. Step 1 is the most important step in the hierarchy. Step1a reduces the generation of solid and liquid wastes which contain nutrients, and thus the need to tap mineral nutrient reserves. For example, replacing phosphorus in detergents with potassium and minimizing food additives reduces the need for mined phosphate (EC 2012; Vallin et al. 2016). Reducing P-waste generation has substantial positive environmental benefits: avoids the toxic radioactive byproduct phospho-gypsum from processing phosphate rock (Ayres et al. 2001) and reduces eutrophication of water bodies. By replacing mineral fertilisers with non-processed nutrient-rich solid and liquid waste in agriculture the associated emissions are largely avoided (Tidåker et al. 2007). Equally important is Step 1b to minimize harmful and unwanted substances in products that otherwise end up in municipal waste flows together with the desired nutrients (ECHA 2007; Kümmerer 2007).

By not mixing various waste flows, it becomes both easier and safer to reuse nutrient-rich products right away (Step 2). For example, the almost sterile human urine may be applied on farmland and replace some amount of mineral fertilisers (WHO 2006). However, if the desired nutrients in the waste are not safe or not in a state that allows reuse, some kind of conversion into a new product is required (Step 3). For instance, organic waste such as faecal matter and toilet water sludge could be composted, hygienised and turned into a safe multi-nutrient fertiliser product. In addition, organic waste may be digested anaerobically to produce biogas while retaining the nutrients in the digestate for agricultural use. Such recycled nutrient inputs save on virgin mineral resources, energy, and transport.

Incineration of organic waste is the next step (Step 4). Incineration is mainly used to reduce the volume of solid waste and to recover some energy. Both ashes and smoke contain phosphorus and potassium but, when organic waste is incinerated

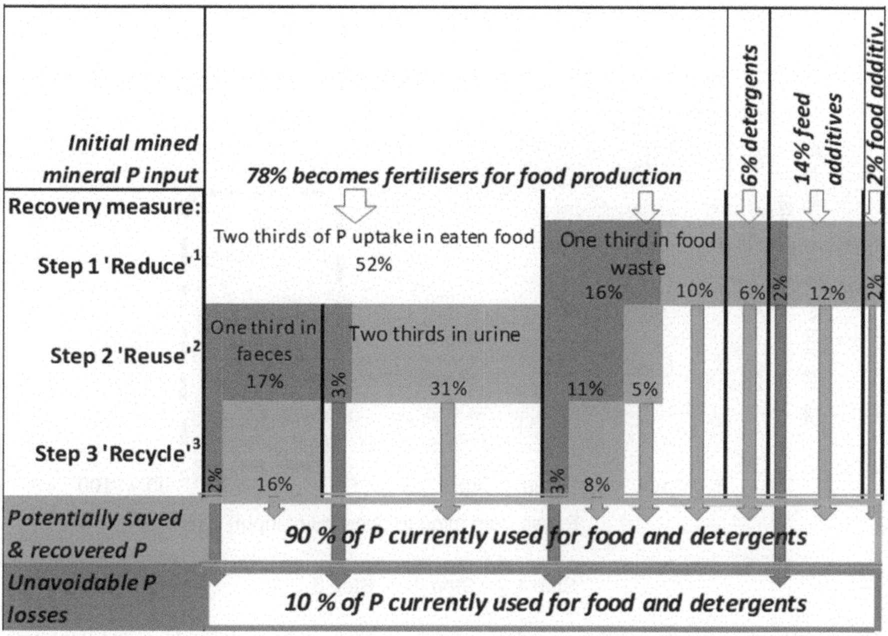

Fig. 1.2 Potential recovery of phosphorus (P) from the food chain, from human excreta, and banning P in detergents is guided by the first three steps of the 'extended waste hierarchy'. Green areas = recovered P and red areas = losses of P. (Source: Drangert et al. 2018)

at temperatures above 800 °C, the amount of plant-available phosphorus in the ashes decreases (cf. Zhang et al. 2001). Also, all carbon, nitrogen and sulphur are lost which makes the end products less valuable for agricultural use. Dumping waste on a landfill (Step 5) should be resorted to only after having exhausted the previous four steps. Currently, however, the most common practices employed in the world's solid waste management are Steps 5 and 4, whereas what is needed for food security is to shift the focus towards the first three steps applied to both solid and liquid nutrient waste.

The above steps for phosphorus recovery are brought together in a comprehensive format in Figs. 1.2 and 1.3 in order to estimate the potential for replacing mined P with saved and recovered P. Excreta are included since they contain most of the P in urban waste flows (Vinnerås et al. 2006). P in biodegradable paper, board and wood waste is not included since these flows are already recycled to a large extent for non-agricultural purposes. Garden waste is excluded due to a lack of reliable data, but it can easily be composted and recycled on site.

The main usages of mined P, as well as proportions food losses, and P in faeces and urine are presented in Fig. 1.2 together with estimates of the potential to recover these. The mined phosphate rock is used to manufacture fertilisers (78%), feed additives (14%), detergents (6%) and food additives (2%) (van Dijk et al. 2015). Non-P substances can substitute P in food and feed additives and detergents (Vallin et al. 2016;

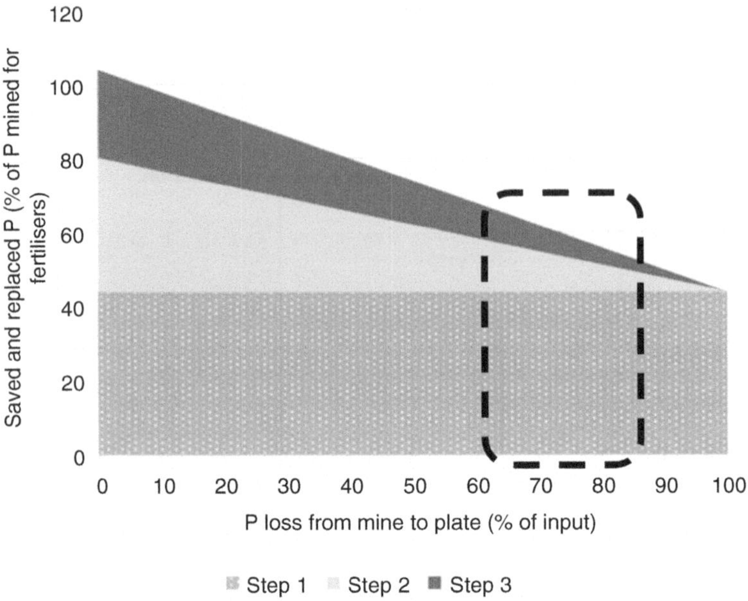

Fig. 1.3 Proportion of mined P for fertiliser production that can be replaced by saved P through reduced food waste and food/feed additives, and no use of P in detergents (Step 1), reused P in urine and food waste (Step 2), and recycled P in faeces and food waste (Step3) as a function of the percentage loss from mine to plate. The dashed box indicates the interval where most countries are likely to be. (Source: Drangert et al. 2018)

EC 2012). In the example given in Fig. 1.2, no P is used in detergents, while P in additives is reduced from a combined 16% to 2%, should there be a valid usage. It is also deemed possible to reduce food waste from one-third to 20% in Step 1, by buying less and eating more of the food that is bought and prepared. In this way, some 10% (0.33 – 0.2 = 0.13 of 78%) of the initial input of mined P for this hierherto food production is saved and can be left in the ground. Thus, the three measures could reduce P mining by 30% (6 + 14 + 10) and be saved and substitute 44% of the P-fertilisers needed for today's level of food production, or for increased food production, or be left in the ground. A change of P-related diets belongs to Step 1a and could save substantial amounts of mined P (Gustavsson et al. 2011; World Bank 2012). This measure is not proposed here, however, since it is deemed difficult to achieve, while arresting a further decrease of vegetarian food in e.g. China may be within reach.

Eaten food requires 54% of the total mined P (including losses from mine to table) and all eaten food is subsequently excreted (Cordell et al. 2009). Two-thirds of the excreted P is found in the urine, and one-third in the faeces. A well-designed city infrastructure can realistically recover 90% of the P in urine (Step 2) and faecal matter (Step 3), or from blackwater (Step 3). Some 30% of the food waste remaining after Step 1is suggested to be reused directly (Step 2), and 70% of the remaining food waste is recycled (Step 3).

The impact of Step 1 represents a direct saving of the currently mined P and is independent of the losses from mine to plate. Given the assumptions in Fig. 1.2, an amount that equals 44% of the P required for fertilisers to produce the current amount of eaten food (with 10% lower wastage than today) is saved. However, the proportion of mined P that can be replaced by measures taken in Step 2 and 3 is strongly impacted by losses as shown in Fig. 1.3. Losses vary between countries, diets, storage, collection methods, etc. and the dashed box indicates level of losses often cited. With an assumed average loss from mine to plate of X per cent, the amount of P recovered through reuse and recycling in Steps 2 and 3 are as follows:

Step 2: Reused P in urine, given the same food intake and a 90% recovery rate + Reused P in food waste (30% of what remains from Step 1) recovers:

$$31^{*}(100-X)/100+0.30^{*}(26-10)*(100-X)/100=(31+5)*(100-X)/100\text{units}$$

Step 3: Recycled P in faeces, given the same food intake and 90% recovery rate + Recycled P in food waste (70% of what remains from Step 2) recovers:

$$16^{*}(100-X)/100+0.70^{*}(26-10-5)*(100-X)/100=(16+8)*(100-X)/100\text{units}$$

In the unlikely case of no losses of P from mine to table, X = 0, the recovered P in Step 2 + 3 can replace 60% (36 + 24) of the mined P. Together with the 44% from Step 1, there is a surplus of mined P of 4%. If, instead, the loss from mine to table is 60%, the recovered P from Step 1–3 is 68% (44 + 14.4 + 9.6), and only 32% of present-day mining is needed for this purpose and the rest can be left in the ground for future needs. If the P-loss is 80%, about 43% of present-day mining is required. Therefore, the easily available global P resource will last two to three times longer in these cases, and the transgression of the planetary P resource boundary is delayed by several hundreds of years. This is a major reason for the European Union with only one phosphate mine, to engage in recovery of otherwise wasted nutrient resources and become a P-recycling society.

Spångberg et al. (2014) and Jönsson et al. (2012) calculated the theoretical economic value of the four nutrients N, P, K, and S (sulphur) for two Swedish scenarios: one with all toilet water (black water) being recovered, and another with all municipal mixed wastewater sludge being recovered. The total amount of N, P, and K from toilets was equivalent to 28%, 44%, and 55% respectively of the total amounts of these nutrients in chemical fertilisers sold in Sweden in the financial year 2010/2011. The annual monetary values in Fig. 1.4 are expressed as the value of the chemical fertilisers that were replaced by recovered nutrients. Since mixed wastewater contains not only toilet water but also detergents and food scraps, somewhat more P could be extracted from sludge than from toilet water, given a removal

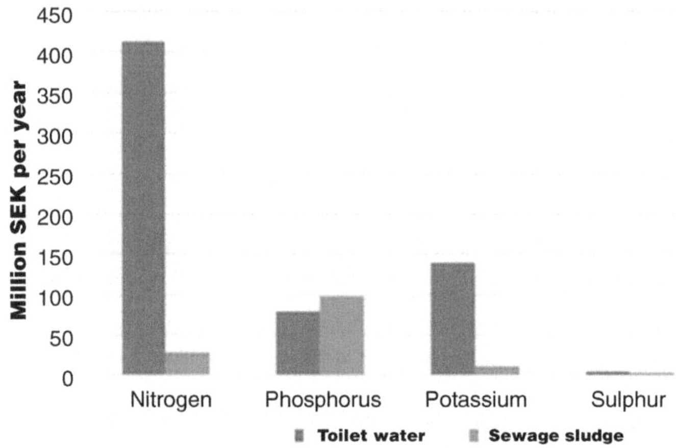

Fig. 1.4 Economic value of the plant nutrients N, P, K, and S in toilet water and in sewage sludge from Swedish households. SEK = Swedish Kronor. (Source: Jönsson et al. 2012)

rate of 100% in the wastewater treatment plant. P from the toilet water has the additional advantage of being more accessible for plants than the P in sewage sludge.

The economic value of nitrogen in toilet water stands out and is several times higher than that of the other nutrients. Also, the value of nitrogen and potassium in toilet water is considerable higher than in sludge. This reflects the fact that nitrogen disappears to air on its way from the toilet to sludge. This loss of nitrogen has to be replaced by the energy-intensive production of nitrogen from ammonia and hydrogen. Also, dissolved potassium K is not captured in the treatment plant and is therefore not found in the sludge.

In addition to the economic benefit of recycling, CO_2 emissions would be reduced in Sweden if chemical fertilisers were replaced by recovered nutrients from toilet water or sewage sludge. Jönsson et al. (2012) estimated the reduction to be 203,500 and 17,000 tons per year of CO_2 equivalents when replaced by N in toilet water and sludge respectively. Again, recovering nitrogen in the toilet water – but not in sludge – would contribute substantially to mitigate climate change.

1.4 Designing a Nutrient-Recycling City

Nutrient-smart cities are within reach at reasonable investments by keeping flows separate, treating each waste flow separately, and reuse/recycle the recovered nutrients for feasible purposes. Figure 1.5 represents a common urban situation with little nutrient recovery. The data is mainly from Fig. 1.2 and literature and differ from city to city. It illustrates how the total household outputs of nitrogen (N) and phosphorus (P) is distributed in the system. The theoretical flows indicate a modest one-fifth of the P and only 5% of the N that household discharge is being gainfully

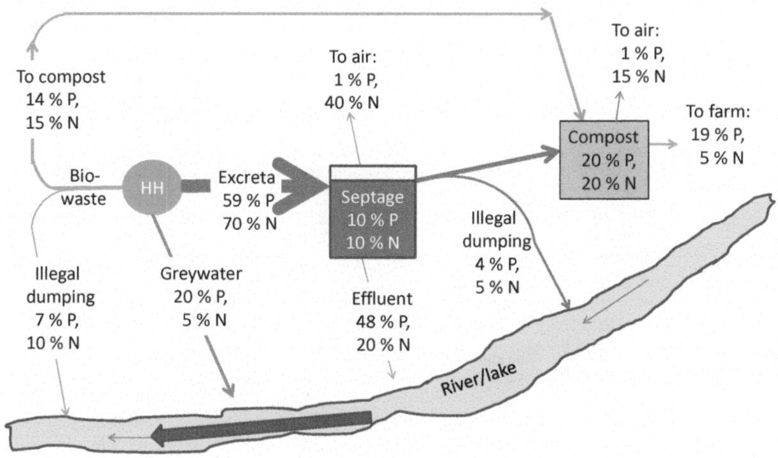

Fig. 1.5 Illustration of present-day nutrient flows from urban households (HH)

recovered and used. Most N is emitted to the atmosphere, while most other nutrients end up in water bodies (red arrow in Fig. 1.5) and causes eutrophication and algal blooms in receiving water bodies. This may, in turn, result in less aquatic flora and even dead zones on lake floors and reduced living space for fish (UNEP 2006).

The term "bio-waste" refers to such items as food waste, paper, and garden waste. Such solid waste is usually easier to manage than liquid organic waste which gets caught in sludge. Food remains, fat and grease on plates, pans and cutlery that is swept into the organic waste bin, is possible to compost or convert to biogas and apply the compost/slurry as fertiliser. Also, such a measure prevents fat, oil and grease to be washed away and clogging sewer pipes that requires costly cleaning.

In urban areas, the nutrient-rich excreta are commonly flushed to a septic tank for partial treatment. Ideally, settled sludge is cleaned out and brought to a compost facility but, due to infrequent emptying, much of the nutrients remain in the effluent. Illegal dumping is also commonplace in developing cities. Co-composted sludge and solid organic waste is made available for use in agriculture, although most of the nitrogen content gets lost to the atmosphere.

A modified sanitation systems in line with Steps 2 and 3 in the extended waste hierarchy can considerably improve the capacity to reuse and recycle nutrients. Figure 1.6 presents a hypothetical scenario for a typical city in the developing world that has taken four important measures. Urine-diverting toilets have been installed, which collect dehydrated urine separately (Senekal and Vinnerås 2017), while dewatered faecal matter is stored in line with the World Health Organisation recommendation, before being applied to soil (WHO 2006). The wastewater treatment plant has been upgraded to remove 90% of the P. Residents segregate household solid organic waste, and a waste-handling company composts it, and thereby reduces previous illegal dumping.

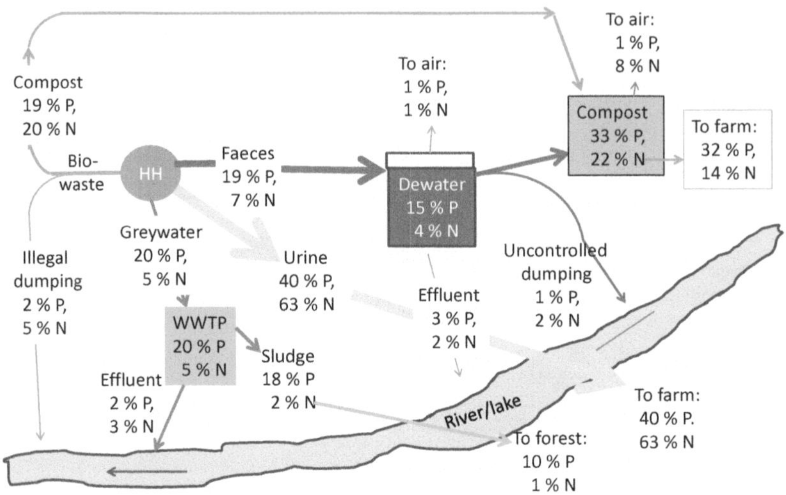

Fig. 1.6 A scenario for nutrient flows out of households (HH) in the year 2030

The P- and N deficient greywater and sludge contains polluting substances that may accumulate in soil (EC 2013). Therefore, this sludge is only applied on forest trees, after degrading organics to avoid clogging of soil pores.

The urine is safely applied on agricultural soil and it represents the least polluted fertiliser available on the market, and has a well-balanced nutrient composition (Spångberg 2014). The nutrient loss from well-managed urine is insignificant, even for nitrogen (Senekal and Vinnerås 2017). Likewise, the faecal matter is likely to be of good quality and, in addition, provides valuable organic material to the soil.

Such measures have the potential to increase the productive use of the P originating from households from 19% to 82%, while N increases from 6% to 78%. The accompanying reduction in wastage also implies that water bodies are less affected by nutrient pollution and eutrophication. This P recovery does not account for losses from mine to table and is therefore comparable to Step 2 and 3 the figure given in Fig. 1.3 for substitution of P fertilisers.

1.5 The "24/7 Eco-house" Concept and Sustainability

What possibilities are there to achieve recovery of nutrients in a single house? The perception of an eco-house commonly focuses on green plants, while smart houses often focus on electronic devices for managing various installations in the house. A 24/7 eco-house comprises all this and, in addition, handles the flows of water, nutrients and energy in a way to make the house function without relying on municipal services – in the middle of a city.

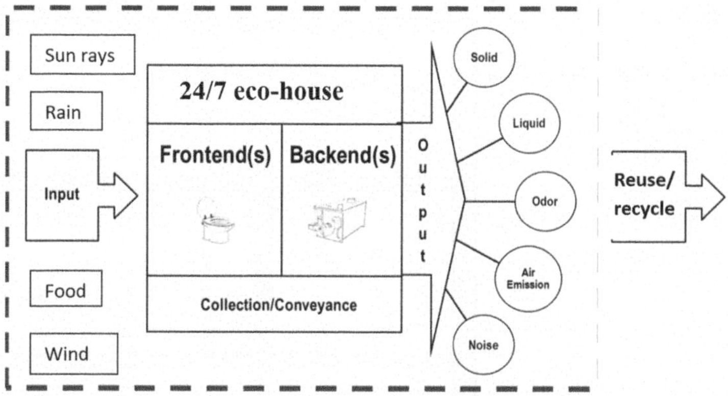

Fig. 1.7 Conceptual chart of flows through a 24/7 eco-house and back to specified uses

Figure 1.7 below conceptualizes liquid and solid nutrient flows to and from a 24/7 eco-house. Sustainability requirements for the output are set for solid and liquid matter in order for them to be safely recycled back to use in the house or compound. In addition, no odour is allowed and only low levels of noise and air emissions.

At the *frontend* residents interface with water-saving faucets, urinals, and pour flush urine-diverting toilets or water-less toilets. *Backend* treatment technologies and processes range from physical to biological methods, while restricting chemical use. Each treated flow is used for appropriate recycling purposes.

Safety concerns are related to pathogens in toilet water and bad odour, and to toxic chemicals in greywater from e.g. hygiene products and detergents used in washing machines. The crucial design idea is to fit the single house with pipes that keep separate each of the four differently composed wastewaters from all floors all the way to a treatment unit (Fig. 1.8). In warm climates the pipes may be attached to the outside wall to allow for easy inspection and repair and, also, allow for low-cost redesign of piping as need arises. Planners and builders can apply the same systems thinking as in the case of keeping industrial and hospital sewage separate from municipal sewers and stormwater drains.

The *greywater* from showers, hand-wash basins and washing machines can be treated *in situ* by letting harmful substances bond to filter particles in a resorption filter, and be diluted before recycled to fill washing machines, to water a garden, to wash a car, or flush the toilet.

So called *nutritious water* from the kitchen sink, urine, and leachate from excreta contains only background levels of chemical compounds and is feasible to use as a fertiliser in the garden after treatment. Health risks from pathogens are minimized by storage for long periods (WHO 2006). An extra safety measure is to avoid manual handling of this water by installing piped irrigation.

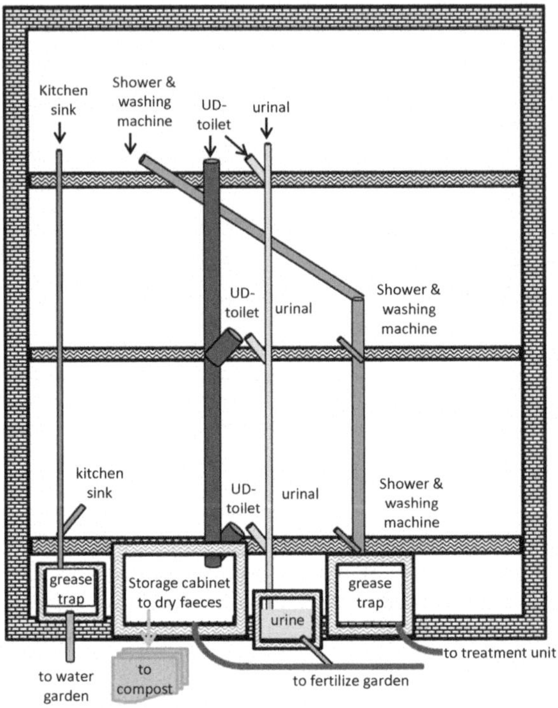

Fig. 1.8 Four separate wastewater flows in a 3-storey house (sink/green, flush or dry toilet/brown, urine/yellow, and shower and washing machine/grey)

Biowaste from kitchens and garden is collected separately and composted in a simple, insulated composter. Faecal material, after being stored for almost 2 years, can be added to the composter and the mix is used as a soil conditioner in the garden or nearby farm. With two to three yields per year, the urban food production could contribute a substantial part of urbanites' vegetable and fruit requirements. Depending on the area assigned for urban agriculture and distance to agricultural areas, part or all nutritious material is transported to farm land. This is becoming economically feasible as the liquid part in urine can be reduced (Senekal and Vinnerås 2017).

1.6 Food Security in Urban Areas

For long, the idea that food should not be produced in towns has been a feature of urban identity. However, this idea has been challenged during periods of societal stress, such as war and economic depression. Data gathered in the 1980s revealed that urban agriculture was strong in many capitals of the world: Lusaka, Dar es Salaam, Moscow, and other cities produced almost half of the consumed food within the city limits (Smit et al. 1996). Cofie et al. (2003) estimated that 800 million

Fig. 1.9 Restoring Nature in urban settings in El Bosco, Milan. Vertical forest on balconies. (https://www.stefanoboeriarchitetti.net/en/project/vertical-forest/)

people were involved worldwide in urban agriculture, and 150 million fully employed, while they contributed an estimated 15% of food production in 1993.

Gardening city dwellers may want to strengthen the family economy or enjoy fresh vegetables or just be fond of gardening. Home-grown vegetables, berries and fruits have a higher quality than when irrigated with untreated wastewater downstream of the city (Drechsel et al. 2010). An example of urban agriculture is Europe's allotment movement, which started in the late nineteen century. Initially, it was introduced to improve workers' well-being and complement their income by producing some food. In Ukraine, for instance, there are some 7 million allotment gardens on a population of about 40 million citizens (pers.com).

Today a renaissance for urban agriculture is ongoing in the western hemisphere driven by a healthy food movement. A novel view of the use of 'empty' space is emerging. At the Food and Climate Summit in New York 2009, an estimate was presented that New York City has 52,000 acres of backyard space that collectively could provide vegetables for 700,000 people (Stringer 2010).

The local situation determines what would suit the residents and their physical and economic status. An interesting option is to build and use balconies and roofs to grow plants and apply recovered nutrients. An ambitious example comprising twelve 27-storey apartment towers, called Bosco Vertical alluding to hundreds of full-size trees planted on the balconies, are built in Milan Italy (Fig. 1.9). Here, plants also provide shade, cooling and dust reduction in the summer, and allow light in the winter when the leaves have fallen (Financial Times 2011; INYT 2014). The corresponding planted area on the ground would require 5 ha of agricultural land and 1 ha of woodland. Such an area can potentially provide all vegetables required by the families in the high-rise building.

In most cases, only minor changes are required of resident water-use and waste-handling behaviours and routines. Perhaps the most important aspect of eco-houses is that residents are obliged to be more careful with what they mix into the water while using it, since they know that this water will come back to them in the tap after treatment. Therefore, the quality of the raw wastewater entering the mini-WWTP is likely to be of much better quality than that received by a municipal wastewater treatment plant. Residents are also encouraged to sort solid waste, including organic waste to be composted.

Eco-homes can be modified to suit local preferences and future options entering the market. Hydroponic technology is an emerging space-saving medium for growing e.g. salads faster than in soil and being fertilised by recovered nutrients in wastewater. Another novel method makes meat production more independent of mineral fertilisers and available land by letting earthworms or fly larvae process manure and organic waste into protein-rich animal feed (Lalander et al. 2013). This is in line with FAO's aim to increase insect-based food production in order to feed the growing global population (van Huis et al. 2013). By so doing, also the land area required for waste management would be reduced.

1.7 Conclusion

This chapter indicates that food security is within reach, if urban areas are designed to save, reuse and recycle nutrients in organic waste. Such systems create a win-win situation by also reducing health risks for humans and minimising polluting emissions to water bodies and greenhouse gases to the atmosphere. But, such a change is unlikely to come about by itself.

It is farfetched to hope for an international convention like the one on climate change for limiting global use of mined nutrients. Instead, a multi-pronged solid and liquid waste hierarchy emerges from the challenges posed by global resource constraints, food insecurity and environmental degradation. The measures comprise favourable building norms and environmental laws, product requirements, non-toxic production through substitution, etc. Five examples have been explored:

- Manufacture and use products that generate as little waste as possible.
- Produce non-toxic materials whenever possible to facilitate the creation of nutrient loops.
- Ban the use of P in detergents and minimize P in food and feed additives to delay planetary shortage of P.
- Keep flows of different wastewater qualities separated and segregate solid waste.
- Keep the nutrient-rich toilet water separate from other household wastewater in order to recover valuable nutrients.
- Enforce more stringent rules for agricultural use of municipal sludge and ban storage of sludge on landfills in order to create incentives to save, reuse and recycle nutrients.
- Encourage saving, reuse and recirculation of nutrients in order to minimize incineration and landfilling of organic materials.

The on-going urbanization provides a window of opportunity to build new recycling-friendly urban areas and infrastructure. At the end of this century, 8.5 billion will reside in cities – an increase from 3 billion the year 2000 – while only 1.5 billion people will reside in rural areas. This shift of people from rural to urban areas increases the magnitude of urban nutrient flows. At the same time, the shift means that twice as many homes and offices are to be built in the present century as the total building stock in the year 2000. The existing stock can be gradually upgraded when e.g. piping is worn out. Thus, city councils can select any infrastructure and building codes for these new urban areas without incurring extra investment for creating a nutrient-smart city.

References

Ayres RU, Holmberg J, Andersson B (2001) Materials and the global environment: waste mining in the 21st century. MRS Bull 26:477. https://doi.org/10.1557/mrs2001.119

Cofie O, Drechsel P, De Zeeuw H (2003) Improving agricultural productivity in the rural-urban Interface through recycling of urban waste. Abstract and presentation at the SCOPE workshop on the Peri-urban environmental change and urban waste management, commonwealth heads of government meeting. Dec. 2003, Abuja, Nigeria

Cordell D, Drangert J-O, White S (2009) The story of phosphorus: food security and food for thought. Glob Environ Chang 19(2009).292–305

Drangert J-O (2000) Reuse-the ultimate sink? Urine-diverting toilets to protect groundwater quality and to fertilize urban agriculture. In: Chorus I et al (eds) Water, sanitation and health: resolving conflicts between drinking water demands and pressures from society's wastes, World Health Organisation Series. IWA Publishing, London, pp 275–280

Drangert J-O, Tonderski K, McConville J (2018) Extending the European Union waste hierarchy to guide nutrient-effective urban sanitation toward global food security – opportunities for phosphorus recovery. Front Sustain Food Syst. February 2018 2:3. https://doi.org/10.3389/fsufs.2018.00003

Drechsel P, Scott CA, Raschid-Sally L, Redwood M, Bahri A (2010) Wastewater irrigation and health: assessing and mitigating risk in low-income countries. IDRC, Ottawa; 2010. Available from: http://idrc.ca/EN/Resources/Publications/Pages/IDRCBookDetails.aspx?PublicationID=93

EC (2012) Preparing a waste prevention programme. Guidance document. Drafted by BioIntelligence services, Paris, October, 2012. Accessed 18 Nov 2013 at http://ec.europa.eu/environment/waste/prevention/pdf/Waste%20prevention%20guidelines.pdf

EC (2013) Sewage sludge directive. EU commission home page. Accessed 15 Feb 2014 at http://ec.europa.eu/environment/waste/sludge/

ECHA (2007) REACH in brief. Environment directorate general. European commission. Brussels. http://ec.europa.eu/environment/chemicals/reach/pdf/publications/2007_02_reach_in_brief.pdf

EU (2008) Waste framework directive 2008/98/. Available at http://eur-lex.europa.eu/LexUriServ/LexUriServ.do?uri=CELEX:32008L0098:EN:NOT

EUP (2011) European parliament resolution of 6 July 2010 on the Commission green paper on the management of bio-waste in the European Union (2009/2153(INI)) (2011/C 351 E/07) In C 351716 E/48–55 Official Journal of the European Union 2.12.2011, pp 48–55

Financial Times (2011) The age of flower towers. Financial Times, October 8/9, 2011

Gustavsson J, Cederberg C, Sonesson U, van Otterdijk R, Meybeck A (2011) Global food losses and food waste – extent, causes and prevention. FAO and the International Congress SAVE FOOD at Interpack 2011. Dusseldorf, Germany

INYT (2014) For evidence of a rebound in Milan, just look up. Article by Eric Sylvers in International New York Times, Friday, March 14, 2014

Jönsson H, Hallin S, Bishop K, Gren I-M, Jensen ES, Rockström J, Vinnerås B, Bergkvist G, Strid I, Kvarnström E (2012) Många skäl att återvinna mer fosfor (several reasons to recover more phosphorus). Dagens Nyheter 2012-08-03

Kabbe C, Bäger D, Mancke R (2014) Phosphoruspotentiale in Land Berlin. Kompetenzzentrum. Wasser Berlin. Projektnummer 11400 UEPII/2

Krausmann F, Schandl H, Sieferle RP (2008) Socio-ecological regime transitions in Austria and the United Kingdom. Ecol Econ 65:187–201

Kümmerer K (2007) Sustainable from the very beginning: rational design of molecules by life-cycle engineering as an important approach for green pharmacy and green chemistry. Green Chem 9:899–907. https://doi.org/10.1039/b618298b

Lalander C, Diener S, Magri M, Zurbrügg C, Hellström A, Vinnerås B (2013) Faecal sludge management with the larvae of the black soldier fly (Hermetiaillucens) – from a hygiene aspect. Sci Total Environ 458-460:312–318

Malthus TR (1798) An essay on the principles of population. Penguin 1978. London

Mara D, Drangert J-O, Tonderski K, Gulyas H, Tonderski A (2007) Selection of sustainable sanitation arrangements. J Water Policy 9(2007):305–318

OECD (2013) Green growth in cities, OECD Green Growth Studies. OECD Publishing, Paris. https://doi.org/10.1787/9789264195325-en

Prud'homme M (2011) The global supply of nutrients. European commission workshop: NPK – will there be enough plant nutrients to feed a world of 9 billion in the year 2050? Brussels 5–6 December 2011. Available from: http://eusoils.jrc.ec.europa.eu/projects/NPK/

Rockström J, Steffen W, Noone K, Persson Å, Chapin FSC III, Lambin EF, Lenton TM, Scheffer M, Folke C, Schellnhuber HJ, Nykvist B, Wit CA, Hughes T, Leeuw S, Rodhe H, Sörlin S, Snyder PK, Costanza R, Svedin U, Falkenmark M, Karlberg L, Corell RW, Fabry VJ, Hansen J, Walker B, Liverman D, Richardson K, Crutzen P, Foley JA (2009) A safe operating space for humanity. Identifying and quantifying planetary boundaries that must not be transgressed could help prevent human activities from causing unacceptable environmental change. Nature 461:472–475

Schoumans OF, Bouraoui F, Kabbe C, Oenema O, van Dijk K (2015) Phosphorus management in Europe in a changing world. Ambio 44(Suppl. 2):S180–S192. https://doi.org/10.1007/s13280-014-0613-9

Senekal J, Vinnerås B (2017) Urea stabilization and concentration for urine-diverting dry toilets: urine dehydration in ash. Sci Total Environ 586(2017):650–657

Senthilkumar K, Nemse T, Mollier A, Pellerin S (2012) Conceptual design and quantification of phosphorus flows and balances at the country scale: the case of France. Glob Biogeochem Cycles 26:GB2008. https://doi.org/10.1029/2011GB004102

Smit J, Ratta A, Nasr J (1996) Urban agriculture: food, jobs and sustainable cities. UNDP, New York

Spångberg J, Tidåker P, Jönsson H (2014) Environmental impact of recycling nutrients in human excreta to agriculture compared with enhanced wastewater treatment. Sci Total Environ 493(2014):209–219. https://doi.org/10.1016/j.scitotenv.2014.05.123

Steffen W, Richardson K, Rockström J, Cornell SE, Fetzer I, Bennett EM, Biggs R, Carpenter SR, de Vries W, de Wit CA, Folke C, Gerten D, Heinke J, Mace GM, Persson LM, Ramanathan V, Reyers B, Sörlin S (2015) Planetary boundaries: guiding human development on a changing planet. Science (80-) 347:736–747. https://doi.org/10.1126/science.aaa9629

Stringer SM (2010) A blueprint for a sustainable food system. Presented at FoodNY, February 2010 by Manhattan Borrough President Singer

Tidåker P, Sjöberg C, Jönsson H (2007) Local recycling of plant nutrients from small-scale wastewater systems to farmland—a Swedish scenario study. Resour Conserv Recycl 49:388–405

UNEP (2006) Challenges to international waters – regional assessments in a global perspective. United Nation Environment programme, Nairoi

USGS (2015) Mineral commodity summaries. United States Geological Survey

USGS (2016) Mineral commodity summaries. United States Geological Survey

Vallin A, Grimvall A, Sundblad E-L, Djodjic F (2016) Changes in four societal drivers and their potential to reduce Swedish nutrient inputs into the sea. Swedish Institute for the Marine Environment, Report No. 2016:3

van Dijk KC, Lesschen JP, Oenema O (2015) Phosphorus flows and balances in the European Union member states. Sci Total Environ 542(Part B):1078–1093. https://doi.org/10.1016/j.scitotenv.2015.08.048 15 January 2016

van Huis A, Itterbeeck JV, Klunder H, Martens E, Halloran A, Muir G, Vantomme P (2013) Edible insects; future prospects for food and feed security. FAO forestry paper 171. FAO, Rome

Vinnerås B, Palmquist H, Balmer P, Jönsson H (2006) The composition of household wastewater and biodegradable solid waste – proposal for new norms for the flow of nutrients and heavy metals. Urban Water 3(1):3–11

WHO (2006) WHO guidelines for the safe use of wastewater, excreta and greywater. In: Excreta and greywater use in agriculture, vol 4. World Health Organisation, Geneva. Available at: http://www.who.int/water_sanitation_health/publications/gsuweg4/en/

World Bank (2012) Inclusive green growth. The pathway to sustainable development. Washington, DC

Zeeman G (2012) New sanitation: bridging cities and agriculture. Wageningen University, Wageningen. https://www.wur.nl/en/Publicationdetails.htm?publicationId=publication-way-343335303333

Zhang F-S, Yamasaki S, Nanzyo M (2001) Application of waste ashes to agricultural land – effect of incineration temperature on chemical characteristics. Sci Total Environ 264:205–214

Zhang X et al (2015) Managing nitrogen for sustainable development. Nature 528:51–58. https://doi.org/10.1038/nature15743

Chapter 2
Reducing Food Losses and Waste in the Food Supply Chain

Gao Liwei, Zhang Yongen, Xu Shiwei, Xu Zengrang, Cheng Shengkui, Wang Yu, and Muhammad Luqman

Abstract Globally around one-third of total food production is lost or wasted along the entire food chain, which is an issue for food security. Therefore, better understanding of food waste is needed for waste reduction. However, such knowledge is still vague and incomplete, particularly in developing countries and emerging countries such as China. Here we review food losses and waste in the Chinese food system. We found that food loss and waste occurred at each stage of food chain, each food department and each food item. Crop postharvest section and food post-consumer section were the two biggest sources of food losses and waste. The loss ratio of Chinese crop postharvest ranged from 7% to 11%, which is much higher than that of developed counties, below 3%, though the Chinese ratio is decreasing with improvement of postharvest technologies. The loss ratio at the food post-consumer stage ranged from 3.8% to 11.1%, which is much lower than that of developed counties, around 10%, but the Chinese ratio is increasing with urbanization blooming and residential income rising. Food losses and waste still has not been investigated in several processing steps including food processing, cool chain logistics and retail.
To investigate losses reduction potential, a meta-analysis was used here to explore reduction potential of Chinese food losses and waste, mainly focussing on crop postharvest section and food post-consumer section. The results show that the loss ratio during harvest can be reduced by 62.2% compared with the level in 2010. Here, on-farmer traditional storage and harvest had the highest reduction potential,

G. Liwei (✉) · Z. Yongen · X. Shiwei · W. Yu
Agricultural Information Institute, Chinese Academy of Agricultural Sciences, Beijing, China

Key Laboratory of Agricultural Information Service Technology, Ministry of Agriculture, Beijing, China
e-mail: gaoliwei@caas.cn; zhangyongen@caas.cn; xushiwei@caas.cn

X. Zengrang · C. Shengkui
Institute of Geographic Sciences and Natural Resources Research, CAS, Beijing, China

M. Luqman
Agricultural Information Institute, Chinese Academy of Agricultural Sciences, Beijing, China

University of the Punjab, Lahore, Pakistan

E. Lichtfouse (ed.), *Sustainable Agriculture Reviews 32*, Sustainable Agriculture Reviews 32, https://doi.org/10.1007/978-3-319-98914-3_2

and drying and transport had the second highest potential. We also assessed the impacts of policy on food waste in the restaurant industry. Results indicate that there has been significantly waste declines in Chinese restaurants, particularly in large restaurants and medium-sized restaurants. But food waste in households was still not given detail evaluations because of data deficiency.

Keywords Food security · Food losses and waste · Reduction of food losses and waste · Food supply chain

2.1 Introduction

It is estimated that food production will have to increase by 70% worldwide to be able to meet the demand of increasing population and diet changes by the year of 2050 (Tilman et al. 2011). However, each year we still lost or waste about 30% to 50% of the edible parts of food that is produced and intended for human consumption (Godfray et al. 2010; Gustavsson et al. 2011). The staggering food waste aggravated global food security burden (Gustavsson et al. 2011), and the need to feed an ever-increasing world population makes it obligatory to reduce the millions of tons of avoidable food waste along the food supply chain.

The quantity and proportion of food losses or waste along the entire food supply chain is staggering, but the underlying reasons differ between developed and developing countries. In developing nations, more than 40% of losses occur at the post-harvest and processing stages due to the absence of infrastructure in food chain and lack of knowledge or investment related to storage technologies at farmer levels (Nellemann et al. 2009; Gustavsson et al. 2011). But, in developed nations, more than 40% of losses occur at the retail and consumer stages for a variety of reasons (Godfray et al. 2010; Gustavsson et al. 2011). For example, food wastage per capita by consumers in Europe and North-America amounted to 95–115 kg a^{-1}, while the figures in Sub-Saharan Africa and South/Southeast Asia were only 6–11 kg a^{-1} (Gustavsson et al. 2011).

Food losses and waste not only threatens world food security but also negatively effects resources, environment and human health (Hall et al. 2009; Cuellar and Webber 2010), which has been substantial implications for sustainable development (Godfray et al. 2010; Kummu et al. 2012).

In addition to the actual food wasted, resource inputs, eg. arable land, irrigated water, fertilizer, oil, coal, natural gas, and environmental emissions, eg. CO_2, N_XO, CH_4, embedded in the whole food supply chain are also wasted (Gustavsson et al. 2011; FAO 2013), even so those losses have accumulative effects (Garnett 2008; Porter et al. 2016). Furthermore, the methane gas, extra generated from food waste landfill, has a 20–25 times more potent than carbon dioxide (Garnett 2011).

A report from Food and Agriculture Organization showed that global wasted food consumed around 250 cubic kilometers of water, and it was equivalent to annual water discharge of the Volga River, or three times the volume of Lake

Geneva. Produced but uneaten food also relied on almost 1.4 billion ha of land, which means about 30% of the world's agricultural land area exploited in vain (FAO 2013). Kummu et al. (2012) calculated the resource costs associated with food losses and waste within food supply chain, and found that around one quarter of global produced food was lost and wasted, representing 24% of total freshwater resources used in food crop production, 23% of total global cropland area, and 23% of total global fertilizer use. The USA is the most concerned area where considerable food is wasted. It was estimated that food wasted by each American had increased by 50% since 1974, accounting for more than one quarter of the total freshwater consumption and 300 million barrels of oil per year (Kantor et al. 1997). In China, Liu et al. (2013a, b) investigated food loss and waste across the food supply chain, and found that 19% of grain produced was lost and wasted, and the consumer segment contributed the most (7.3%). In addition, the water and arable land costs from Chinese food loss and waste were 135 billion cubic meters and 26 million ha (Liu et al. 2013a, b), respectively.

Besides resources impacts of food waste, food production also contributes to 19–29% of worldwide greenhouse gas emissions (Vermeulen et al. 2012), including emissions from the decomposition of food waste after disposal in landfills and from the embedded emissions associated with its production, processing, transport and retailing. The later impact requires a life-cycle view of wasted food (Garnett 2008). It was estimated that the global footprint of wasted food was equivalent to 3.3 billion tons of carbon dioxide annually, ranking as the third largest source of emissions after USA and China (FAO 2013), but that was not included greenhouse gas emissions from land use change. Venkat (2011) calculated the emissions from wasted food using life cycle assessment from production to disposal for each food commodity in the USA, data showed that avoidable food waste produced approximately 113 million metric tons of CO_2 e annually, equivalent to 2% of national emissions. Figures in the UK from the year of 2010, indicated that the total avoidable household food and drink waste to be 4.4 million tons, and it was equivalent to 17 million tons of CO_2 emissions (WRAP 2011). Within the grocery supply chain, the 3.6 million tons of food waste each year was estimated to generate 8.4 million tons of CO_2 emissions (WRAP 2010). In summary, reducing food losses and waste may be one of the best ways to cut down the emissions and mitigate anthropogenic climate change (Garnett 2011).

According to this increasingly serious problem, there is a growing global consensus that curbing food loss and waste has been increasingly becoming another way to enhance food supply, ensure food security and reduce environmental emissions. In the year of 2011, identifying food as a key sector where resource efficiency should be improved, the European Commission set targets to halve the disposal of edible food waste by 2020 (EC 2011). Meanwhile, in the year of 2012, the European Parliament also issued a resolution to halve food waste by 2025 and designated 2014 as the "European Year against Food Waste" (EP 2012). Some other governments have started to define specific targets for reduction of food losses and waste including the United Kingdom, Republic of Korea, Japan, the Netherlands, France,

Spain and Austria (HLPE 2014). Reduction of food losses and waste has been becoming a worldwide campaign.

Moreover, any rapid and urgent actions to reduce food waste would have help saving resources and reducing emissions. PBL (2009) estimated that effective measures for food waste reduction can reduce global land claim for agriculture by approximate 5 million km^2 to 2050. At least 40% of the food waste produced in Britain was estimated to be disposed in landfill (Defra 2011), large volumes of which are biodegradable. Decomposition of this waste results in the production of the greenhouse gas methane, which can contribute to climate change if not properly managed. According to the UK Waste and Resource Action Programme, preventing one ton of food waste had the potential to save an estimated 4.2 tons of CO_2e emissions in the UK (accounting for the lifecycle emissions and including emissions from landfill) (WRAP 2009).

However, knowledge of food loss and waste along the food supply chain is still inadequate worldwide (Schneider 2013), especially in developing countries and regions with rapid economic transition (Song et al. 2015a, b; Gao et al. 2015; Rembold et al. 2011). Therefore, the objective of this chapter, taking China as an example, from crop postharvest section to food post-consumer section, would try to give a whole review of food losses and waste along the food supply chain, and to reveal the characteristics of food losses and waste, and then to discuss and explore some approaches of reducing food losses and waste including improved technology, management strategy, changing consumer behavior, policy guidance and so on. At the end of this paper, to achieve sustainable food production and consumption in whole food system, we presented and summarized some problems urgently needed to be resolved and some policy suggestions urgently needed to be put into effects.

2.2 Food Production in China

Before discussion on food loss and waste in China, we firstly elaborated the evolution of food production and its resource inputs and environmental emissions in order to clarify the importance of curbing food loss or waste along food supply chain in China.

China is a country not only with huge food production but also with huge food consumption. Hence, both sustainable food production and consumption are mostly critical for food security in China (UNEP 2012). For more than six decades, in order to ensure national food security, China has spared no effort to enhance food production through implementing reforms, open-policy, technology and material inputs. And the increasing amounts of agricultural products were produced from limited land and water resources, and satisfied the huge food demands of a doubling population and growing economy, creating a miracle of using only 7% of the world's cultivated land to feeding 22% of the world's population (FAOSTAT 2013).

For example, beef production increased by a factor of more than 20 from 0.3 million tons in 1980 to 6.5 million tons in 2010, and contemporaneously, mutton, poultry, pork and aquatic products increased by factors of 9, 10, 3.5 and 11 respectively. Fresh milk and egg production have increased by a factor of 26.5 and 8.8, respectively, during the same period. However, total grain production only had a growth rate of 28.5% during the three decades (FAOSTAT 2013) (Fig. 2.1). However, improving crop yield will be more and more difficult within limited cultivated land resources in the future, particularly because of challenges such as land competition from industrialization, urbanization, and infrastructure development, and ongoing soil erosion and desertification.

Although many works has been done to ensure food security in China, the growth in China's food production has made the resource and environmental cost prominent (Guo et al. 2010; Beman et al. 2005; Jane 2011; Fang 2009) (Fig. 2.2), and made people more aware of the true state of the national agricultural development. Since the year of 2002, China has become the largest country of synthetic fertilizers production and consumption in the world (Li et al. 2013). Data showed that, agricultural capital goods such as chemical nitrogen and phosphors inputs in food production, increased from 8.3 and 2.2 million tons to 23.5 and 8.1 million tons between 1978 and 2010 (Fig. 2.2), respectively. And during that time, agricultural machinery inputs increased to 90 million KW from 10 million KW and irrigation water utilization in agriculture was more than 300 billion cubic meters, which consumed 60% of domestic water consumption (FAOSTAT 2013).

There is no doubt that overused agricultural resources, especially excessive applied chemical nutrients discharging, caused many environmental problems including eutrophication of water body, soil acidification, and greenhouse gas emissions including CO_2, CH_4, N_2O (Guo et al. 2010; Zhang et al. 2013; Liu et al. 2013a, b), which has the potential to intensify global climate change. Taking agricultural activity for example, CH_4, N_2O emission from agricultural sources accounted for 50.2%, 92.5% of total emissions of CH_4, N_2O, respectively, and greenhouse gas emissions shared 17% of the total emission in China (Dong et al. 2008).

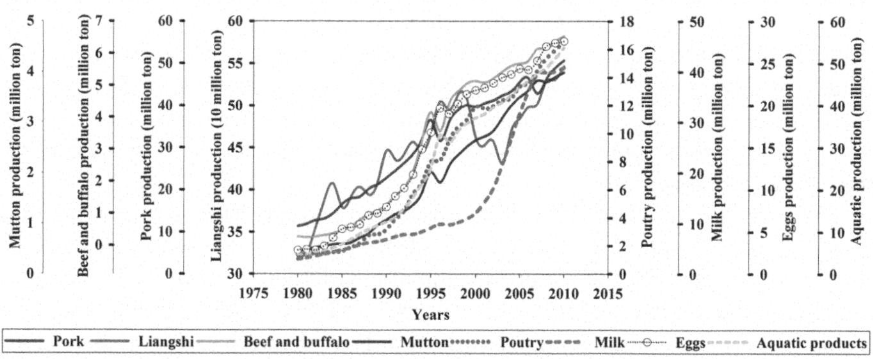

Fig. 2.1 Total outputs of food production between 1980 and 2010 in China. Note the increase outputs of agricultural production. Data was from FAOSTAT, 2013 and CNBS, 2017; Outputs including liangshi, pork, beef and buffalo, mutton, poultry, milk, eggs and aquatic products. And here liangshi included grain, tuber and beans

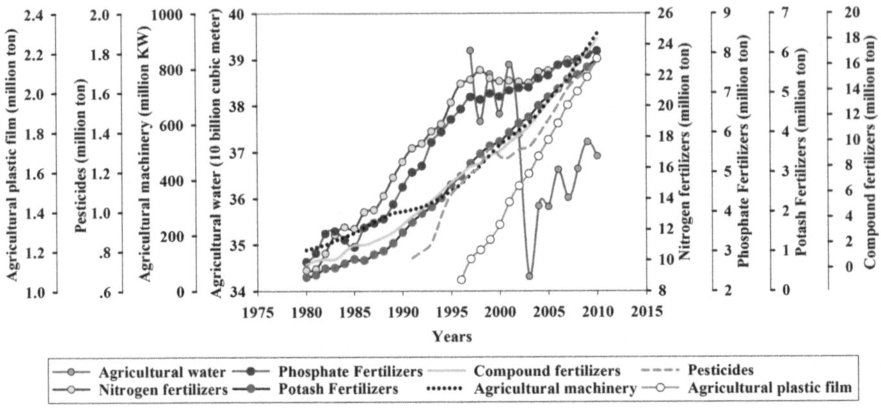

Fig. 2.2 Total inputs of crop production between 1980 and 2010 in China. Note the increase inputs of agricultural production. Data was from FAOSTAT, 2013 and CNBS, 2017; Inputs including water use, nutrients application of chemical fertilizer, inputs of plastic mulch and machinery power

However, when challenge facing by resources and environmental cost of food production, food losses and waste has been long ignored. Since the early 1990s, the proportion of food loss and waste from 'farm to fork' was 18.1% in China, and the post-consumer segment (consumer segment) accounted for almost one third of total waste (5.4%), followed by postharvest stage (4.9%) (Zhan 1995). However, it seemed that the situation had not been changed yet. In recent years, the ratio of food loss for grains across the total supply chain was 19% in China, with the consumer segment responsible for the single largest portion of food waste of 7.3% (Liu et al. 2013a, b). Nearly 20% of grain produced along food supply chain was loss and wasted in China, which is equivalent to more than 1200 million tons in the year of 2014, which was nearly equal to half of total grain production in Africa. In other words, if we can reduce by half of the loss or waste, more than 15 million people will be fed (calculation based on 400 kilogram grain per capita per annum), more than 32 million ha of arableland and 65 billion cubic meter of water would be saved (Liu et al. 2013a, b). Therefore, in addition to maximizing crop yields, reducing food loss or waste along food supply chain, especially in the consumer segment, will be more significant to ensure food security in China because of its spending on the least cost (Garnett 2011).

The rapid development of Chinese urbanization has been driving the diversification of food consumption patterns and changes in food consumption behavior. Statistics indicated that the ratio of animal-based food to total food consumed increased from 10.7% to 22.1% between 1985 and 2010 (CNBS 2011). The transformation of food consumption patterns towards to animal-based food drives more greenhouse gas and resource-intensive food types (Garnett 2011). Therefore, the magnitude of greenhouse gas and resource-intensive food wasted would incur greater resources and environmental costs than the same plant-based food by weight (Garnett 2008, 2011; Hamerschlag and Venkat 2011). Beef, for example, accounting for 16% of total emissions, was the single largest contributor to emissions from wasted food in USA, even though the quantity of beef wasted amounts to less than 2% of total waste (Hamerschlag and Venkat 2011). This is because of the high emissions intensity of beef and its low feed conversion efficiency (Hamerschlag and Venkat 2011). However, with 65% of China's population expected to be urbanized by 2030, the volume of food wasted from food post-consumer segment in urban China will be most likely to increase dramatically unless long-term and effective measures are adopted by government and policy-makers (Cheng et al. 2012).

2.3 Food Losses and Waste in the Food Supply Chain

Since at least the 1970s, reducing post-harvest losses of food has been identified as an element integral to supporting a growing population, particularly in developing countries (Hall 1970; Bourne 1977; Gao 1977). However, the problem has not been resolved to date, and even more serious. And besides post-harvest stage, other stages including food processing, distribution and consumption also increasingly contributed to food losses.

China is taking responsibility of ensuring food security for 22% of the world's population, but almost 20 percent of grains produced each year was lost or wasted in human food supply chain (Liu et al. 2013a, b), threatening to undermine future food and resources security (Liu et al. 2013a, b), and intensifying climate change (Cuéllar and Webber 2010). Therefore, prioritizing methods of reducing food wastage must be another try to increase food supply. However, the characteristics and scales of food losses and waste from all the stages of food supply chain still has not been understood systematically in China (Cheng et al. 2012), impeding the loss reduction campaign, and the primarily reason was the unknown information hampered by fragmented and outdated data (Parfitt et al. 2010; Liu et al. 2013a, b).

Food loss and waste can occur at each stage of food supply chain (Fig. 2.3). In this part, a framework for food waste and losses was built along Chinese food supply chain (Fig. 2.3), and we will try our best to give a detail review around Chinese food system, mainly focus on crop postharvest section and food consumed stage, despite the uncertainties due to data limitation and lacking of literature presented.

2.3.1 Crop Losses in Postharvest Section

Crop postharvest stage in China contains major four segments including crop harvesting, crop transported from field to farmer household, crop drying and crop storage, and each segment of crop postharvest exists food losses (Fig. 2.3). It was estimated that each year there was a loss ratio of 7% to 11% in Chinese crop postharvest, and the figure was much higher than that of developed counties (below 3%), such as America and Europe (Godfray et al. 2010; Gustavsson et al. 2011).

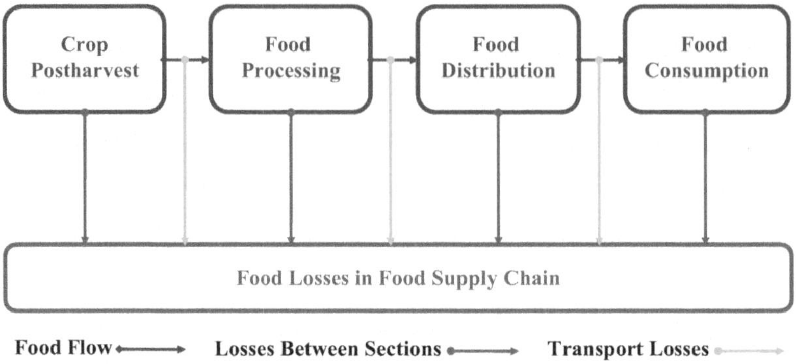

Fig. 2.3 Food losses along human food supply chain. (Note: the blue arrows mean food products flow from production to consumption, and the red arrows show food losses and waste within each section of food system, and the orange arrows indicate food losses in transportation. However, food losses mainly happened in crop postharvest and food waste mainly in food consumption)

Besides climate factors, the mainly reasons were because of primitive storage method, simple and crude facility, poor technology of grain crop (SAG 2011).

Losses in grain storage were serious in China, particularly in farmer storage. Based on a report from China Administration of Grain, we can found that the percentage of grain loss in farmer storage has reached 8% during "the Eleventh Fifth-Year" (the average value between the year of 2006 and 2010). However, because of scientific methods of grain storage, there was a much lower loss ratio in intensive grain storage including storages of governments and enterprises, and the losses ratio was between 0.5% and 1.0% (SAG 2011).

Among Chinese major losses in grain storage at farmer level, the highest loss value was maize, with an average of 11%. Paddy and wheat was about 6.5% and 4.7%, respectively. And the distributions of grain loss were, damage caused by rats accounted for 49% of the total losses, fungi and insect pests accounted for 30% and 21%, respectively (SAG 2011). Compared with other countries, the loss values were only slightly smaller than undeveloped countries (Gustavsson et al. 2011), such as Nigeria (Thylmann et al. 2013) and Sri Lanka (NSC 1980), but that was much bigger than the developed countries (Gustavsson et al. 2011).

However, because almost half of grain was stored at farmer level (Fig. 2.4), so grain losses in storage is still an inconvenient truth. And calculation based on 400 kilogram grain per capita per annum, if half of its losses were saved, nearly 25 million people would be fed in China.

Following grain storage, crop harvesting had the second highest loss ratio in China. Research showed that grain loss ratio varied in harvesting approaches (Gao et al. 2016). There are two different approaches of crop harvesting in China, and one is called combined harvesting using combine harvester to finish cutting, threshing and cleaning of grain at one time, with the loss value between 0.2% and 6.0% (Chen et al. 2011); while the other is called two-stage harvesting, with grain cutting, threshing and cleaning by labor or machine, and the maximal value can achieve

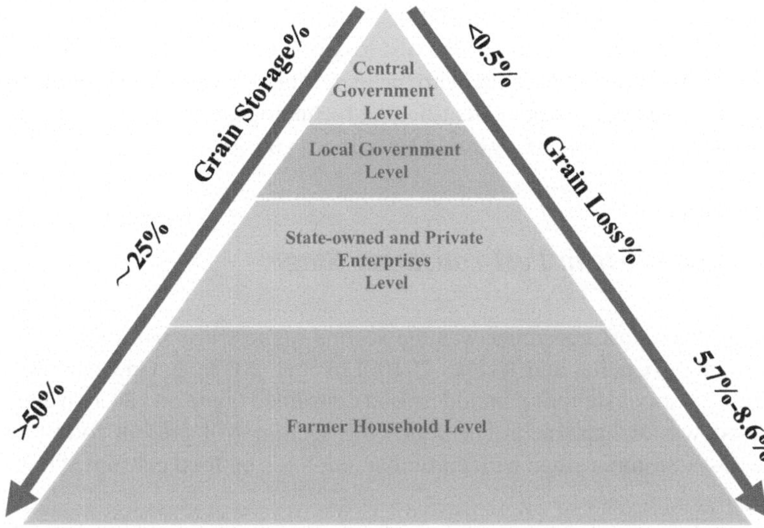

Fig. 2.4 Pyramid structure of grain storage and its associated loss rate in China. Note: losses rates happened in different grain storage level, and modified after Liu 2014

more than 10% (Zhan 1995), and the minimum can also more than 1.0% (Song et al. 2015a, b). Therefore, enhancing the level of crop mechanical harvesting will be beneficial to reduce loss from crop postharvest (Gao et al. 2016).

In addition to crop loss in storage and harvest, grain transport and drying also had grain loss, although there was a small portion of grain loss. It was estimated that grain transported loss in the packaging bags was on an average of 1.0%, and loading in bulk about 0.3% (Gao et al. 2016). In grain drying, about 0.5% of losses occurred by dehumidifying equipment and 1.5% by natural withering, however, the later had still much great proportion, with a percentage of more than 50. Actually, in the last few decades, grain often got poorly packaged for transport in Chinese countryside. Some transporters use sacks, or polythene bags or simply load the "naked" products directly onto the trucks, leading to compression damage during transport, adding the poor state of roads, especially worsens during the rainy season when it was common to see trucks ferrying grain products breaking down or getting stuck in the mud, which further aggravated the losses during transportation.

In summary, compared with Europe and the United States (with 5% to 6% of grain postharvest loss) (Gustavsson et al. 2011), China still has a higher grain loss ratio in postharvest, especially for the on-farmer stage of grain storage, which is far from the level of 5% loss percentage, proposed by the United Nations food and agriculture organization (Gustavsson et al. 2011). Besides grain, perishable food also has higher loss percentage in postharvest stage in China. It is estimated that postharvest loss of vegetables and fruits are between 25% and 30%, which is 5–6 times as high as western developed countries (Gustavsson et al. 2011). Meanwhile because of China lacking of adequate cold storage facilities, meat and aquatic products can reach between 10% and 15% (Zhao 2008). Therefore, based on the discussion above, the reduction potential of food postharvest loss was mainly in food storage, but we also need to know that in the future with the increasingly rising of agricultural mechanization level and agriculture scientific and technological progress, percentage of crop postharvest loss most likely could be reduced considerably.

2.3.2 Food Waste in Post-consumer Stage

Food waste from post-consumer was the second highest loss worldwide based on previous studies (Cuéllar and Webber 2010; Liu et al. 2013a, b; Gao et al. 2015). In China, food post-consumption includes food consumed home and away from home, where food wastes happened. The following section will present food waste in Chinese post-consumer stage with particular emphasis on food consumed home and outside.

2.3.2.1 Food Waste at Home

With the increasing income level of Chinese residents, food-purchasing power enhanced, and coupled with the food sales promotion in supermarket, all of which intensified food surplus, so food wasted in home occurred and has been becoming more and more serious, particularly in urban China. If nothing to do, the situation would be towards to the extent of developed countries just as the America and Europe (Cuéllar and Webber 2010; HLPE 2014).

Based on the China Health and Nutrition Survey data including nine provinces from 1991 to 2009 in China, the ratio of food wasted in home was calculated by Song et al. (2015a, b), and the results showed that the average ratio of total food wasted per capita per year was on average of 3.8% in Chinese resident household (Fig. 2.5), of which vegetables contributed 54% to the total food waste by weight, followed by rice with 13%. And the overall of pork, legumes and fruits represented 15% of the total consumed but 13% of the total discarded, respectively. Moreover, the results for the food items also presented a significant dependence of the generation of food waste on consumption, with a correlation coefficient of 0.87 (Song et al. 2015a, b), implying that commonly consumed more foods generated more waste.

And with the data, spatial variation of food wasted in different provincial families was analyzed by Ding (2015), and the results indicated that an average of 16 kg of food consumed per capita per year was wasted, and the most amount of food wasted per capita was from Hubei province with an annual median value of 28.9 kg in the 7 years, but the least was from Heilongjiang province with an median value of 12.0 kg (Fig. 2.6).

In addition to regional household diet leading to food waste, the spatial diversity of climate may be another more important factor effects food waste, and compared

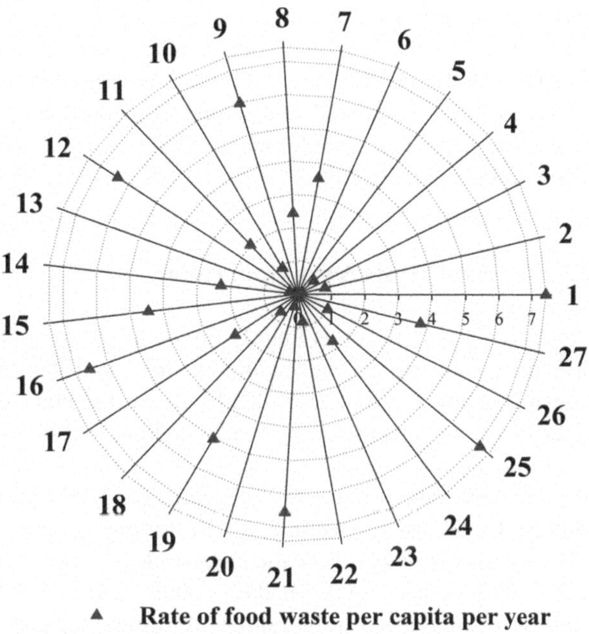

Fig. 2.5 Food waste in Chinese resident household. (Note: waste rate per capita per year in Chinese household by food items, and modified after Song et al. 2015a, b. The numbers showed different food items, and NO. 1–27 represented aquatic products, beef, biscuits, bread, butter, cheese, dried fruit, eggs, fruits, lamb, legumes, maize, milk, pork, potatoes, poultry meat, rice, snacks, sugar, sweets, vegetables, wheat, yogurt, other cereals, other meats, others and the total food products, respectively)

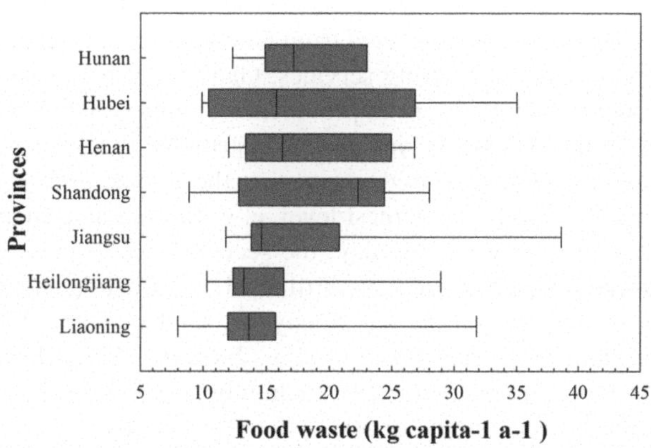

Fig. 2.6 Food waste in different provinces' household in China. (Note: box-plot for household food waste per capita per year by provinces in China, and modified by Ding 2015)

with North China, South China would be more likely to spoil food because of the wet weather (Xu 2005; SAG 2011). Food culture could be another reason caused the difference.

But there still existed large difference between researches. A wastage of 43 g food per person per day was higher than a previous estimate of 11 g (Song et al. 2015a, b), but far less than the average of 490 g of food wasted in Beijing households (Zhang and Fu 2010). A person in China discards an average of 16 kg of food per year according to Song et al. (2015a, b), slightly higher than their counterparts in sub-Saharan Africa and regions of southern and Southeast Asia, which have annual per capita food wastages of 6–11 kg (Gustavsson et al. 2011), and far less than the USA, the UK and Turkey, with annual per capita wastages of 124 kg (Buzby and Hyman 2012), 138 kg (WRAP 2012) and 116 kg (FAO 2006), respectively. Besides the spatial and temporal disparities, an inconsistent standard for measuring food waste would be the main factor that led to the differences between results (Gao et al. 2015).

2.3.2.2 Food Waste Away from Home

Consumer waste, which is mainly linked to restaurants and canteens, is increasing driven by growing affluence, urbanization, and the growth of the restaurant and catering sector in China (Liu 2014). It is estimated there are 3.5 million catering enterprises in China including large, medium and small restaurants, snack and fast-food outlets, and cafeterias. Food wasted away from home in Chengdu, Sichuan province was investigated by Wang and Xu (2012), and results showed that 26.7% of served food was wasted in 2011. In Beijing, compared with food wasted at home (0.07 kg per capita per day), the magnitude of food wasted in restaurants was much higher (0.3 kg per capita per day) (Zhang and Fu. 2010). Xu (2005) investigated food waste of Beijing restaurants with different scales (large, medium and small), and found that wasted food accounted for 11.1% of total food consumed, of which animal-based food served was 14.7% and plant-based food was 15.6%.

Based on data collected from news and reports related to food waste by consumers in Chinese catering services, Gao et al. (2013) estimated that food waste was roughly 6 million tons in provincial capitals with the year of 2008 (Fig. 2.7).

Total food waste was mainly distributed in the economically developed eastern regions of China. Based on city size, the amount of food wasted in cities was divided into five levels: the largest level, e. g. Beijing and Shanghai, produced 1000–1600 tons of food waste per day; the second large level, e. g. Changsha, Nanjing, produced 600–1000 tons; the third level, e.g. Fuzhou and Taiyuan, produced 360–600 tons; the fourth level, e.g. Shenyang, produced 150–360 tons; and the lowest level produced less than 100 tons (Fig. 2.7 left). Taking Hefei as an example, there are 3350 restaurants in this city, including hotels and collective canteens, and there was a food waste amount of 500–700 tons each day.

Following the above, we also divided food waste per meal per capita into five levels; and data was showed in Fig. 2.7 right. But different from data of total food waste in food 7 left, data of food waste per meal per capita showed that most of food waste produced in East China (Fig. 2.7 right), mainly distributed in cities with low population density and non-tourist cities, which caused the higher food waste per capita. But Lhasa in Tibet, as one of the famous tourist cities in China, also had the most food waste, and that was mainly because food away from home was wasted mostly by tourist, and the climate of hypoxia contributed to the most of food wastage (Gao et al. 2017).

Organic components embedded in food waste were high in catering services. The proportion of fat and protein were between 16.9% to 38.9% and 6.6% to 15.9%,

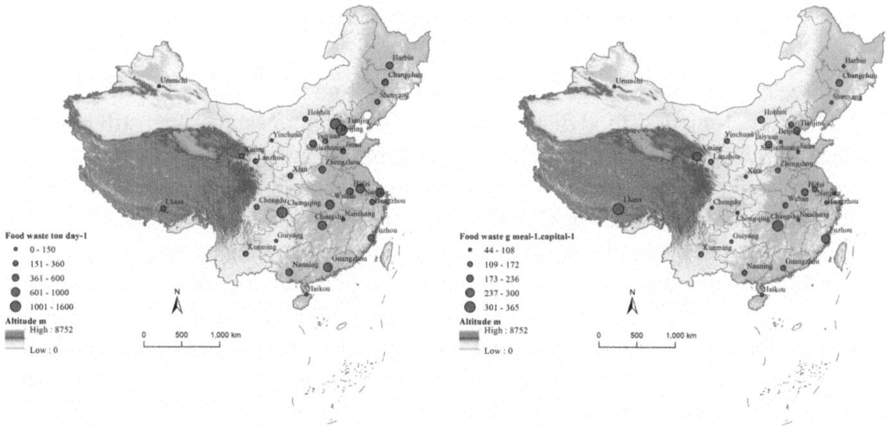

Fig. 2.7 Food waste per day (left) and per capita per meal (right) in restaurants in Chinese provincial capitals. (Note: variations of food waste in Chinese restaurants, and calculated using news reports and the population of provincial cities, and tourist populations considered in Beijing and Lhasa, and modified by Gao et al. 2013)

respectively (Xu et al. 2011; Wu et al. 2006). Based on data from China Agriculture University, protein and fat contained in catering food waste reached eight and 3 million tons per year, respectively, equivalent to the amount of nutrients consumed by 200 million people a year, and when adding food wasted in food consumed home, the total food waste can feed 200–300 million people each year (Xu 2007), which may be probably overestimated. Another study estimated that the magnitude of wasted food was equal to 5 million tons of grain yield, nearly equal to the amount of total grain imported by China (CNBS 2010). It was estimated that based on a ratio of 10% food wasted away from home (the actual value is greater than that), it was rough estimated that food wastage led to a financial loss of 150 billion yuan, accounting for 8.4% of the gross domestic product (1780 billion yuan) of Beijing in 2012.

Now food waste has been becoming a pervasive problem, especially in urban catering (Cheng et al. 2012). Research on catering food waste in Beijing indicated that 81% of interviewed consumers ever had wasted food, and 28% of consumers did not consider packing up leftovers when dining out, and 53% of consumers would pack up leftovers only when too much food was wasted (Zhang and Fu 2010). Official business, weddings, funerals and dinner parties are the major occasions where food was wasted (Xu 2005), and these situations were even common in rural regions of China few years ago.

In short, no matter how wasted food occurred in post-consumer stage, and it was more on related to human customer behavior, but studies on behavioral interventions for food consumption is seldom applied in reduction of food waste (Whitehair et al. 2013), so it is very important to carry out researches around behavior from food wasted by human.

2.3.3 Food Loss and Waste in Other Stages

Now there are still many stages existing food loss and waste, including food industry, cold-chain logistics, food distribution and so on, but the stages had still not been given systemic analysis in China, because of limited data presented. However, documents were listed here as much as possible to reflect and reveal the seriousness of the food loss in other stages in Chinese food supply chain.

Firstly, in grain primarily processing stage, residents in China excessively pursue the heavily processed grain, not only reducing milling yield of paddy and wheat but also resulting in great losses of vitamins and essential micronutrients of grains, and if long-time exposure to this environment, human dietary nutritional balance could be lost (Fan et al. 2015), which will probably bring another pressure on food security. Currently, data showed that the edible portion of grain was only between 65% and 70% in China, which has a large gap between 20% and 30% compared with developed country. Data also showed that milled rice ratio of the third degree decreased by 2–4% compared with the second degree, but polished rice reduced by about 15% than the third degree (Fan et al. 2015). Now, nearly half of the rice consumed is from intensive processing. As for wheat, 50 kg wheat could produced wheat flour about 42.5 kg twenty years ago, but now the data has changed already and decreased to 36.5 kg, because of domestic market demand to the intensive wheat flour. And all the by-product of grain processing is used for feed, which may be another kind of food waste.

Secondly, cold-chain logistics, important to keep perishable food fresh and avoid food wastage, has been a booming area in developing countries like China. However, compared with developed countries, the cold-chain logistics transportation of fresh food has taken more than 50% in many developed countries, and the ratio even reaches 80% in America and Japan, but in China the cold-chain logistics transportation of fresh food merely has taken 15% (Zhou and Sun 2015). Wang et al. (2013) reported that as a result of inadequate refrigerated facilities and poor cold-chain systems, up to 90% of meat products, 80% of aquatic products, and the majority of dairy and bean products were transported and sold without using any refrigerated equipment and outside the cold chain system, which has led up to 20% to 30% of fruits and vegetables, 12% of meat, and 15% of aquatic products to be lost. But losses in most developed countries were about 5% and in the USA, this figure was even less than 2% (Bolton and Liu 2006).

Furthermore, lacking of cold-chain logistics was not only producing food loss but also aggravating healthy potential. It was estimated that more than 94 million people became ill due to bacterial food borne disease, which was attributed to the lack of cold chain facilities and improper handling of food products in 2011 (Mao et al. 2011), which has severely affected consumer confidence.

Thirdly, as one of the most important parts of food distribution, the retail sector of the food supply chain is not the largest contributor to food waste, but the amounts are still high and the share of unnecessary waste is also high (Eriksson 2015), which has been considered as an important issue (Table 2.1), but in China, especially in urban retail markets, food losses and waste has still not been investigated.

Table 2.1 Studies quantifying losses and waste by food items in supermarkets in developed countries

Reference	Country	Data collection method	Reference base	Product group	Relative waste (%)
Buzby et al. (2009)	USA	Supplier records	Supplier	Fruit	8.4–10.7
			Shipment data	Vegetables	8.4–10.3
Buzby and Hyman (2012)	USA	Analysis of national statistics	Food supply value	Fresh fruit and vegetables	9
Buzby and Hyman (2012)	USA	Analysis of national statistics	Food supply value	Dairy products	9
Göbel et al. (2012)	Germany	Analysis of national statistics	Delivered mass	Retail sector	1
Katajajuuri et al. (2014)	Finland	Interviews	Not specified	Retail sector	1–2
Stensgård and Hanssen (2015)	Norway	Store records	Sales value	Fruit	4.5
				Vegetables	4.3
Stensgård and Hanssen (2015)	Norway	Store records	Sales value	Milk products	0.8
				Cheese	0.9
Lebersorger and Schneider (2014)	Austria	Store records	Sales in cost price	Fresh fruit and vegetables	4.3
Lebersorger and Schneider (2014)	Austria	Store records	Sales in cost price	Dairy products	1.3
Beretta et al. (2013)	Switzerland	Estimate from store records	Volumes of sales	Fresh fruit and vegetables	8–9
Fehr et al. (2002)	Brazil	Quantification at retailer	Delivered mass	Fresh fruit and vegetables	8.8
Mattsson and Williams (2015)	Sweden	Store records	Sold mass	Fresh fruit and vegetables (only in-store waste)	1.9

Supermarkets, as a typical component of food retail, produce much food waste. Food loss happened when food over shelf life (Kantor et al. 1997), when food spoiled in storage (Buzby et al. 2014), even when food stolen (Bamfield 2011). Data showed that considerable losses happened in food retail, although lacking data from Chinese retail. Taking Netherlands for example, since 2011 the largest retailer-Ahold, published data on food lost or waste in its Corporate Social Responsibility report. In 2012, the volume of food loss and waste was between 1% and 2% of total food sales, with fresh food loss and waste between 2% and 3% and dry food between 0% and 1% (HLPE 2014). High losses at the retail stage occurred in perishable commodities such as fruits and vegetables, fish and seafood, meat, dairy products, baked foods and cooked foods. In the United States of America alone, it was estimated that the in-store food losses were 10% of the total food supply (Buzby et al. 2014).

Supermarket giant Tesco had also revealed it generated 28,500 tons of food waste in the first 6 months of 2013. Of the total waste, 21% was made up of fruit and vegetables and 41% of bakery items (Tesco 2014).

From all of the above, we can see that food loss and waste occurred at each stage of food chain, each food department and each food item. Varieties of food loss and waste along food supply chain we presented here is just to explain to what extent food was lost or wasted at stages of food system. But we do not give the detail reasons for each kind of food loss or waste, which is influenced by many factors, and even there are many causes of food loss or waste that they are often linked and that they are also often very specific to the nature of different products and to local conditions.

To be summarized, we can conclude, firstly, including demarcation, status, reasons, and reduction potentials and so on, food losses and waste in food supply chain still has not been systematically investigated and analyzed yet, particular in China. Secondly, with the increasing investment of agricultural mechanization, food losses in crop postharvest will be expected to be decreasing, but with rising incomes and an anticipated shift towards more animal-product based diets, if nothing to do, the balance of food losses will be most likely skewed towards to food consumption side, and food waste levels may even reach those found in developed countries (Buzby et al. 2009; Cuéllar and Webber 2010). Thirdly, the antiquated data discouraged researches of food losses and waste, and except crop postharvest and post-consumer stage, the data of other sections including food transportation, processing and retail hardly was blank in Chinese food supply chain.

2.4 Reducing Food Losses Along Food Supply Chain

In the previous discourse, based on exiting researches about food losses and waste, we had given a systematic review of the characteristics and some causes of food loss or waste along food supply chain in China, and a wide range of causes, organized in different levels, called for a wide range of solutions. But the knowledge of reduction potential of food losses and waste, especially focus on crop postharvest and post-consumers, still had not been given. Hence, the following will present some domestic research cases about reducing food loss and waste in crop postharvest section and food consumption, in order to reveal the potential of cutting food loss and waste in China. And solutions used here contained approaches from technique, management, policy and the compound.

2.4.1 Losses Reduction in Grain Post-harvest Section

From the above we can see that widely discussions and considerable researches has been carried out around crop post-harvest loss and its reduction in China, but the focus was mainly on grain storage due to its most serious losses, and the situation of Chinese crop postharvest loss has not been explored systematically yet, and in addition, the reduction potential still has not been known. So, based on the deepening and development of China's agricultural science and technology, a method of calculating grain post-harvest loss was built and its reduction potential also was indentified the following.

Here, a study was carried out to quantify the status of grain postharvest losses and calculate the reduction potential in China. Here, based on agricultural products flow footprint in China, losses of three major grains, including paddy, wheat and maize, were analyzed in each segment of postharvest section, which was showed in Fig. 2.8.

Grain postharvest segments were divided into four sections, including grain harvest, transport, drying and storage, and each was also divided into several different loss ways related to technologies and agricultural machines. Grain harvest was divided into combine harvesting and two-stage harvesting, and transport was divided into package and bulk transporting, and drying was divided into air and mechanical drying, and storage was divided into household and depot storage, based on which the data of various documents were collected and classified (Fig. 2.8).

Based on above, loss partition coefficient of each crop in each segment of postharvest section was identified in the year of 2010 (Figs. 2.9, 2.10 and Table 2.2), which was as the baseline of scenario analysis to explore the loss reduction potential of crop postharvest. And scenarios setting based on changes and improvements of different technical conditions in different phases were built, which was showed in Fig. 2.11.

There was a greater loss in grain harvest section. Data showed that 31.4% of total losses in grain postharvest were from the harvest in 2010 (Gao et al. 2016), and major grains of paddy, wheat and maize had different loss ratios because of their different harvesting approaches (Fig. 2.9). Two major grain harvesting approaches were chosen and compared. The results in Fig. 2.9 showed that, compared with

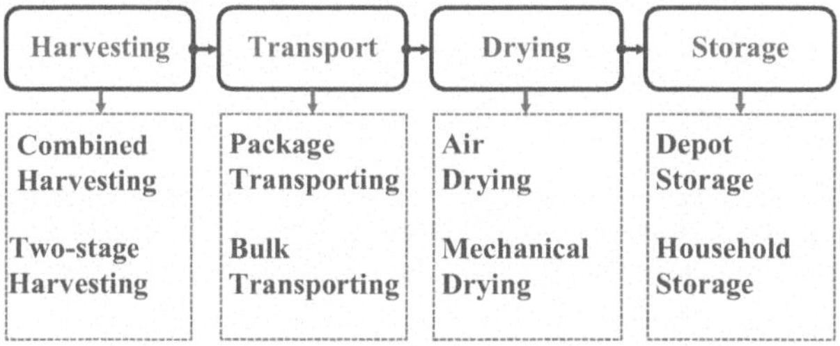

Fig. 2.8 Postharvest losses of major grain in China. Note pathways of postharvest losses of major grains including wheat, maize and paddy rice, and different pathways have different loss rate. In harvest, combine harvesting means grain from reaping to threshing at one time, and two-stage harvesting means first reaping with manpower or machinery, and then threshing with machinery. In transport, grain loading after harvest from field to farmer household, package transporting means grain transported in bags, and bulk transporting means by bulk-grain truck. In drying, air drying means grain dried by nature wind roof or ground; in storage, depot storage indicates grain stored by governments and enterprises while household storage by farmer households

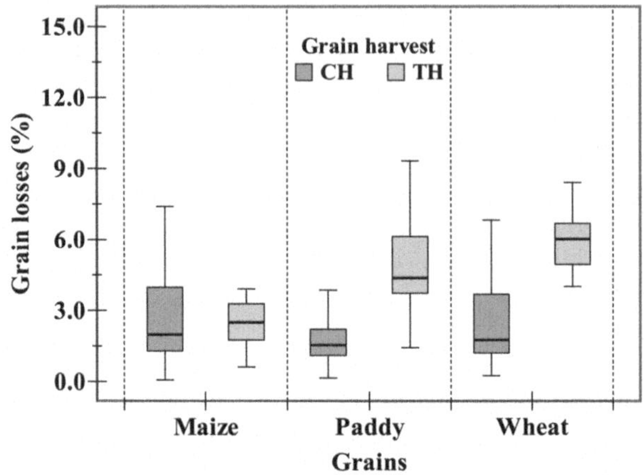

Fig. 2.9 Major grain losses of wheat, paddy and maize in different harvesting approaches in China in 2010. (Note: box-plot for grain losses in harvest. CH, Combined Harvest, represented grains harvested mainly with combine-harvester, and TH, Two-Stage Harvest, represented grains harvested firstly cut down by labor or machine, and then threshed by labor or machine. Modified after Gao et al. 2016)

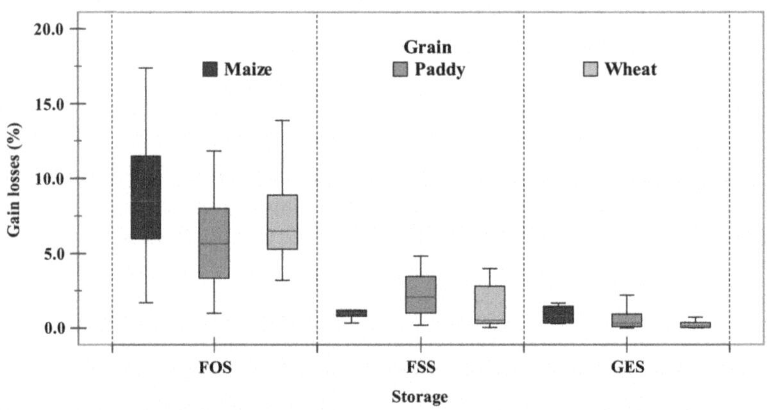

Fig. 2.10 Major grain losses of maize, paddy and wheat in different storage patterns in China. (Note: box-plot for grain losses in storage. FOS, FSS, GES are Famer Ordinary Storage, Farmer Scientific Storage, Government and Enterprise Storage, respectively. Modified after Gao et al. 2016)

combined harvest, two-stage harvest had higher loss ratios between grains, so the enhanced harvesting percentage with combine-harvester can considerably reduce grain harvesting loss.

Storage may be the greatest loss in grain postharvest section. Data showed that more than 40% of total losses happened in grain storage in 2010 (Gao et al. 2016). Variations of the loss ratio differed by grain storage patterns. Famer ordinary storage, farmer scientific storage and government and enterprise storage were the

Table 2.2 Partition coefficient of each crop in each segment of postharvest section

	Harvesting/%		Transport/%		Drying/%		Storage/%		
							Farmer storage		
Crop	CH	TH	PT	BT	ED	ND	TS	SS	DS
Paddy	60.0	40.0	85.0	15.0	10.0	90.0	30.0	10.0	60.0
Wheat	86.0	14.0	85.0	15.0	10.0	90.0	50.0	10.0	40.0
Maize	27.5	72.5	85.0	15.0	10.0	90.0	50.0	10.0	40.0

Note: *CH* grain harvested by combine harvester, *TH* grain harvested by labor and machine, *PT* grain transported in package, *BT* grain transported by closed bulk truck, *ED* dried by grain drying equipment, *ND* grain dried naturally, *TS* Traditional grain storage, *SS* Scientific grain storage, *DS* grain stored in depot of government and enterprises

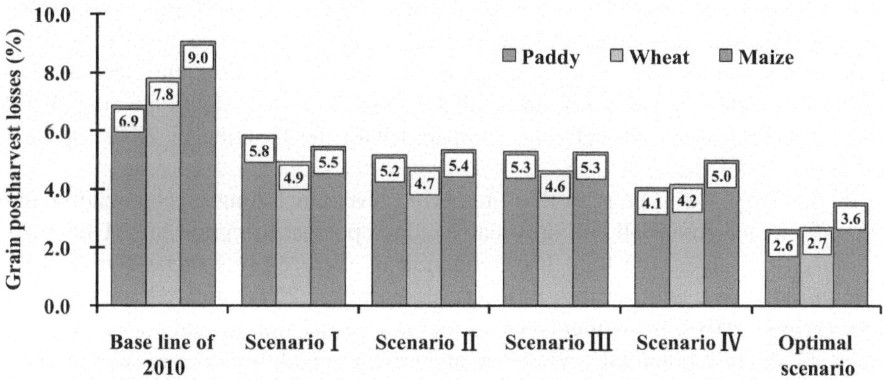

Fig. 2.11 Potentials of postharvest losses reduction for paddy, wheat and maize in different scenarios in China. (Note: Base line of 2010 means postharvest losses of paddy, wheat and maize in the year of 2010; Scenario I, means based on base line of 2010, FOS was all replaced by FSS; Scenario II means, based on scenario I, FSS all became GES; Scenario III means, based on scenario I, percentage of TH decreased by fifty percent for paddy, wheat and maize, respectively; Scenario IV means, based on scenario II, percentage of CH increased to one hundred. Optimal scenario means, based on scenario IV, grain bulk transporting and mechanical drying for paddy, wheat and maize achieved one hundred percentages, respectively; Modified after Gao et al. 2016)

mainly patterns in Chinese grain storage patterns (Fig. 2.10). And among all the patterns, famer ordinary storage had the largest loss ratios between grains, and the less was farmer scientific storage, and the least was the government and enterprise storage. Therefore, the increasing grain intensive storage had a good beneficial to the reduction of grain storage losses.

Based on the data collected above, the reducing potential of major crops loss was identified and quantified (Fig. 2.11). The results showed that compared with the baseline (the year of 2010), when all farmer traditional grain storage was replaced by farmer scientific grain storage, there was a considerable loss decline in post harvest for the three crops, in which the ratio of maize post-harvest loss decreased the most with a percentage of 3.5 (Scenario I).

Based on scenario I, when government and enterprise storage substituted all farmer scientific grain storage, that means all grains stored in depots of government and enterprise. Because there was a similar losses ratio between farmer scientific grain storage and government and enterprise storage, there produced a slightly losses descent (Scenario II) compared with scenario I.

Based on scenario I, when the percentage of the two-stage harvest of all the crops decreased by 50%, and that means the percentage of combined harvesting of the three crops increased to 80.0%, 93.0%, 63.8%, respectively. There were also a slightly losses descent (Scenario III) compared with scenario I, because of a minor discrepancies between two-stage harvest and combined harvesting.

Based on scenario II, when the percentage of combined harvesting of the three crops all increased to 100%, and that means grain harvest and storage all achieved mechanization and scientific management. Compared with scenario I arioIII, there was also a slightly losses descent (scenario IV).

But based on scenario IV, when all of grain loaded with packaging was substituted by that loaded in bulk, and mechanical drying of grain was instead of all of grain air dried, loss ratio of three crops above had a sharp drop, and decreased to 2.6%, 2.7%, 3.6%, respectively, and with the amount of loss reduction were 8.3, 5.9 and 9.7 million tons, respectively, compared with the baseline of 2010 (optimal scenario).

In summary, it can be seen that improving levels including mechanization and scientific management all can drastically reduce postpartum grain loss. Under the optimized measures, the total loss reduction of three crops can be achieved by 23.9 million tons, and the loss ratio decreased by 62.2% compared with present level (year of 2010). In segments, on-farmer traditional storage and harvest had the greatest reduction potential, and drying and transport had the second greater potential. Therefore, the formation of scientific management consciousness for farmers may be one of the most important factors to crop loss reduction in China.

2.4.2 Influence of Policy on Catering Food Waste

China may be the first country in the world that the central government had implemented the most drastic measures to crack down food waste in catering food services. Facing that food waste from catering had become an important social and political issue in China to both government officials and the public, in early 2013, China's Central government commented on an article titled "netizen's call upon restaurants to restrict food waste" and called for rigorous measures to stop the waste of resources. All mainstream media in China immediately followed and reported on the issue of food waste and the anti-waste campaigns have flourished.

The effect of the government's and the public campaign against food waste has been immediate and impressive, and the topic quickly became a priority for both government and civil society. The authorities tried their best to put an end to extravagant feasts and reduce expenses on receptions and banquets, because they have to take the lead in saving food, especially when taxpayers' money was used to create waste. The campaign was initiated by nongovernmental organizations and activists, which urged people to save food by not wasting anything on the dining table. The campaign, launched on weibo, was soon joined by millions of netizens across China in a bid to curb food wastage and appreciate the virtue of being thrifty even in times of plenty.

Several years have been past already until the policy has been released, and the continuing effects of policy on food wasted in restaurant have not been given an evaluation. Here, with the survey data from different scales of restaurants in urban China, intervention effects of policy released on restaurant food wasted were shown in this section.

Based on 3 years' survey data of food waste from catering between 2001 and 2015 in urban China (2011, 2013 and 2015), the characteristics of food waste per meal per capital was analyzed. The results showed that policy, to a great extent, curbing catering food waste, had occurred significantly intervening effect (Fig. 2.12 and Table 2.3).

Food waste per meal per capita decreased from 181.0 gram in 2010 to 88.7 gram in 2013 and to 65.7 gram in 2015, in which the absolute decreased magnitude of pant-based food wasted was higher than that of animal-based food wasted, but the absolute magnitude of pant-based food wasted was still higher than that of animal-based food wasted. In plant-based food wasted, the absolute magnitude of vegetable

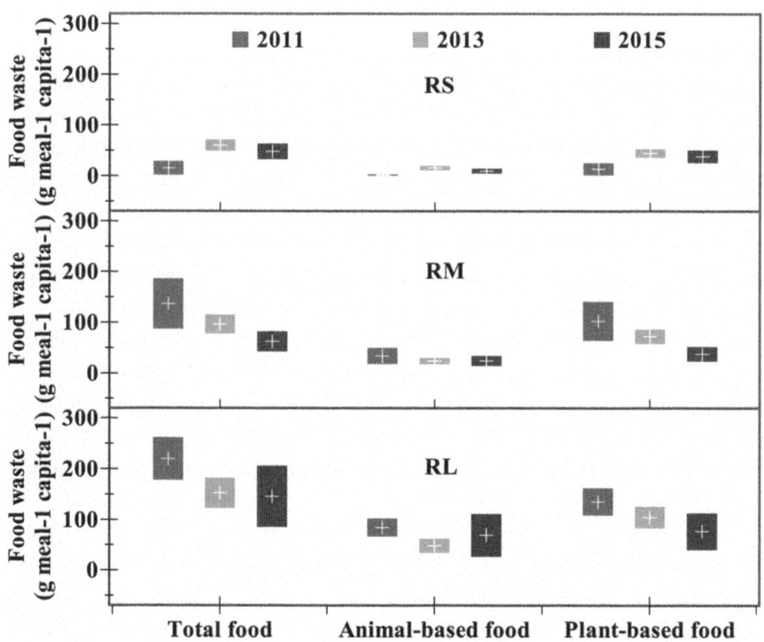

Fig. 2.12 Food waste in catering by different restaurant types in urban China. (Note: error bar of catering food waste per meal per capita by food categories and years. 95% confidence interval for food waste mean. RS, RM, RL mean food waste in Small Restaurants, Middle Restaurant, Large Restaurant, respectively. Modified after Gao et al. 2017)

Table 2.3 Impacts of policy on food waste (g meal^{-1} cap^{-1})

Foot Item	Year	2013	2015	2013 vs. 2015
Pork	2011	1.85 ± 2.20	2.84 ± 2.44	0.99 ± 1.92
Beef	2011	13.17 ± 1.44**	14.66 ± 1.59**	1.49 ± 1.25a
Mutton	2011	5.39 ± 0.88**	4.50 ± 0.98**	−0.89 ± 0.77a
Poetry	2011	8.38 ± 1.16**	8.07 ± 1.29**	−0.32 ± 1.02
Aquatic product	2011	10.50 ± 1.46**	13.82 ± 1.62**	3.32 ± 1.27**
Egg	2011	0.83 ± 1.25	−1.81 ± 1.39	−2.63 ± 1.09**
Vegetable	2011	35.33 ± 5.07**	51.65 ± 5.63**	16.32 ± 4.43**
Rice	2011	11.11 ± 2.65**	16.52 ± 2.94**	5.41 ± 2.31**
Flour	2011	5.74 ± 2.55**	5.05 ± 2.83	−0.69 ± 2.23
Animal-based food	2011	40.12 ± 4.36**	42.08 ± 4.84**	1.96 ± 3.81
Plant-based food	2011	52.18 ± 6.79**	73.22 ± 7.54**	21.04 ± 5.94**
Total waste	2011	92.30 ± 9.60**	115.30 ± 10.67**	23.00 ± 8.40**

Note: Using the single factor analysis of variance (One Way ANOVA analysis) t test method (LSD); ** means average difference between groups have significant difference at P < 0.05 level; Average differences±SD, P < 0.05

wasted was the highest, however had significant reduction, but the total waste was still the greatest among all food items.

With one-way analysis of variance (one-way ANOVA analysis) and t test (LSD), significant analysis of food wasted was carried out during policy released before (2011) and after (2013 and 2015). Table 2.3 showed that compared with the policy approved before the year of 2011, except eggs and pork, the total amount of food waste, animal-based food waste and plant-based food waste all had significantly decreased in the year of 2013 and 2015 (P < 0.05) (Table 2.3), which indicated that the policy intervention of reducing food wasted in restaurants had significant effects. Moreover, compared with the year of 2013, the total amount of food waste and plant-based food waste had significantly decreased, and animal-based food waste had not significantly decreased, but still gave a downward trend in quantity (Table 2.3).

Food waste from different scales of restaurants was also analyzed, and the results indicated that significantly intervening effect of policy on food waste mainly concentrated on large and medium-sized restaurants. The total amount of food waste, animal-based food waste and plant-based food waste all had significantly declined in large restaurants (P < 0.05), but for medium-sized restaurants, only total amount of food waste and plant-based food waste occurred significantly declined (P < 0.05). However, small restaurant had not significantly declined among all food in 2013 and 2015, compared with 2011, even increased which may be affected by other important factors and should be given a further research in the future (Table 2.4).

From the above, the results indicated that food waste from large and medium-sized restaurants still has the potential to be cut down, particularly for animal-based food in medium-sized restaurants such as pork. And the researches to be carried out in the future should be strengthened to focus on large and medium-sized restaurants, and to explore the influence factors. Finally, inducting sustainable development of catering industry should be the ultimate goal of reducing food waste in urban China.

Table 2.4 Impacts of policy on food waste with different scales restaurants (g $meal^{-1}$ cap^{-1})

Scales of restaurant	Food item	Years	2013	2015
Large	Animal-based food	2011	67.34 ± 21.14**	73.94 ± 28.75**
	Plant-based food	2011	36.45 ± 10.44**	15.19 ± 14.20a
	Animal-based food	2011	30.89 ± 13.92**	58.74 ± 18.94**
Middle	Animal-based food	2011	40.34 ± 16.14**	74.07 ± 17.91**
	Plant-based food	2011	9.95 ± 6.28a	9.51 ± 6.97a
	Animal-based food	2011	30.39 ± 12.41**	64.56 ± 13.77**
Small	Animal-based food	2011	−44.87 ± 19.59**	−32.81 ± 19.95a
	Plant-based food	2011	−13.87 ± 8.18a	−7.94 ± 8.33a
	Animal-based food	2011	−31.00 ± 15.50**	−24.86 ± 15.79a

Note: Using the single factor analysis of variance (One Way ANOVA analysis) t test method (LSD); ** means average difference between groups have significant difference at $P < 0.05$ level; Average differences±SD, $P < 0.05$

2.5 Outlook

Producing enough food for the world's population in 2050 will be easy. But doing it at an acceptable cost to the planet will depend on research into everything from high-tech seeds to low-tech farming practices (Nature 2010). This challenge requires changes in the way food is produced, stored, processed, distributed and consumed. Godfray et al. (2010) suggested five major strategies to meet these challenges: closing the yield gap, increasing production limits by genetic modification, expanding aquaculture, dietary changes, and reducing waste. These all involve utilizing the full potential of the production system so that more food can be consumed without increased resource demand at the same rate.

Reducing waste is unique in this context, since it focuses on food that is already produced, but not consumed for various reasons. However, reducing waste of edible food is also one of the least controversial ways to make the food supply chain more productive, it has the potential to be used immediately to decrease the competition for natural resources that could be saved for future production to avoid a future food crisis (Nellemann et al. 2009). So in this sense, reducing food losses and waste towards sustainable food production and consumption appears to be particularly important, particularly in China. However, it is still a hard work, and many works need to be done.

2.5.1 Definition to Be Unified Urgently Worldwide

Although lots of researches have been done around food loss and waste along supply chain, the definition has not been clearly and explicitly formed, particularly for food waste, which existed considerable difference between countries.

Food and Agriculture Organization firstly gave the definition of food loss and waste (Gustavsson et al. 2011), which believed that food loss and waste in food system all can be called food waste. But the different was that food loss mainly focus on early stage of food supply chain, including before food entering into terminal products or food distributed, presenting food decreased in quantity or food decline in quality, which caused the reduction of food supply. Food loss mainly happened in undeveloped countries, e.g. South Africa and India, and regions because of the lack of proper storage facilities, lagged agricultural technology and unsuitable climate factor, and food loss mainly occurred in stages including production, harvest, postharvest and processing of early food supply chain, which caused edible food to be lost in vain (Lundqvist et al. 2008). While food waste mainly referred to foodstuff that should had been eaten or processed was wasted because of various reasons, which happened in later supply chain including food consumption and retail (e.g. supermarket, retail market, wholesale market, household, restaurant and canteen) (Gustavsson et al. 2011).

Based on different purposes, the terms of food loss and waste were also given by different scholars and research institutions, which mainly located in stages of food distribution, consumption and post consumption. Waste and Resources Action Programme, a non-governmental organization in Britain, divided kitchen waste into unavoidable food waste, avoidable food waste and possibly avoidable food waste (WRAP 2009, 2011a, b). Food wastage from supermarket was also investigated by WARP (Lyndhurst and WRAP 2012). In addition, based on characteristics of food waste generated, food waste was also called swill, waste cooking oil, food scrap, food leftover and plate food waste (Hayes and Kendrick 1995; Wu and Xing 2003; Zhang et al. 2012; Tai et al. 2011). And the different definitions led the different estimations of food wasted, which are difficult to contrast and comparison, hence it is very urgent to build a global unified estimate standard for food loss and waste along supply chain.

It is important to note that food loss and waste referred to properties of food for eating, that is to say food that should had been eaten was thrown away. But the deep insight of different dimensions of food was also got worthy of attention. Firstly, from the economic dimension, part of food produced cannot enter into market, because of food harvested not match the market access standards or food laws and regulations in quality, size and appearance, which caused lots of food loss, especially in developed countries (Lundqvist et al. 2008). Secondly, from the culture dimension, regional disparity in life habits, food diet patterns and religions, all of which caused the difference of food waste in statistics (Parfitt et al. 2010), for example, most countries and regions worldwide do not take animal internal organs as food but feed except in China, which needed to be additional consideration when standard of food wasted in preparation. Thirdly, from the healthy dimension, energy from food intake by quite a number of persons worldwide, particularly in developed countries, has been far exceeded the demand standard recommended by international organization, and excessive food intake led to increasingly overweight and obese population, which seriously affected country's public health and safety. Data showed that since 1980 the obesity rate in the United States has doubled, and two-thirds of the population was overweight (Flegal et al. 2002, 2010), which seems to be happening in China (Yu et al. 2012). Hence, reducing over consumed food may be another way to cut food waste.

2.5.2 Several Issues to Be Studied Further

Globally, more than 20% of food loss and waste along supply chain was from consumer stage, even in North America, European and industrialized Asian, which has achieved 30% above (Gustavsson et al. 2011). It was also indicated that 24% of energy embedded in food production was loss and waste in global food system, and 35% of the energy was from consumer level, in North America and European even reaching more than 50% (Lipinski et al. 2013). Increasing researches should focus on consumer food waste, understanding the pattern and scale of daily food waste is thus vital for each consumer, especially in China, so that they can prioritize methods of reducing their food waste and the embedded footprints; however, doing so is hampered by fragmented and outdated data (Parfitt et al. 2010). The following tries to introduce several issues need to be studied further around consumer food waste.

Section of food consumption contains chilling storage, ingredients processing, cooking and post consumer at home and away from home, of which post-consumer food waste has presented considerable researches, but other segments of food waste was lack of systematic investigations (Dryerre and Andross 1946; Chappell 2007), particularly in regions with rapidly increasing urbanization and emerging family miniaturization trend.

Firstly, the magnitude of food loss in cold-storage equipments cannot be neglected. WRAP (2009) estimated that 4.4 million tons of avoidable food was thrown away from refrigerator in Britain households, of which more than 50% was wasted because of food stored exceeding the shelf life, losing freshness and deterioration, and unsuitable refrigerator temperatures. James et al. (2008) investigated that compared with the optimal food cold storage temperature (5 °C), the temperature of operating refrigerator was higher, with the average value of 7 °C, and further studied by Brown et al. (2014) showed that by decreased the temperature to 4 °C, a value of 160 million pounds of food could be saved but another 270 thousand tons of emissions of greenhouse gases (CO_2 e) would be produced, and trade-off between impacts of food wasted on environment and benefits of reducing food waste should be given in order to achieve the optimal solution for resources and environment.

Secondly, in dietary nutrition investigation, the actual data of food nutrition intake is difficult to access so that data of food consumed or bought was instead to evaluate residents dietary nutrition, but this neglected food wastage in preparation and plate (Chappell 2007), which is also an important factor and reference on adjusting and assessing residents dietary nutrition (Muth et al. 2011; Putnam et al. 2000). Waste and Resources Action Programme proposed unavoidable and avoidable food waste within Britain households, but food wastage in preparation was not listed alone (WRAP 2009). And previous studies proved that food wastage in food preparation cannot be looked down upon, variations of food wastage between different food ingredients was large from 1.2% to 80% (Chappell 2007), especially for vegetable, which is influenced by many factors (e.g. storage time). Taking zucchini (*Cucurbita pepo L.*) for example, fresh zucchini wastage in preparation can be reach to 45.3%, but after a period of storage, it would decrease to 30% (Dryerre and Andross 1946). In China, data from catering survey indicated that vegetables in preparation contributed to the most part of wastage, and wastage ratio with same kind of vegetable in different scales of restaurants (e.g. top-grade Chinese restaurant compared with small cookshop) can be a differential of 60% to 70%, which is also related to Chinese food culture.

Thirdly, quality and quantity of food in cooking changed in various degrees (Garnett 2011; Stewart 1946), so the conversion ratio of raw material to food cooked is another useful parameter for residential dietary nutrition and accounting of resources and environmental cost (Dryerre and Andross 1946; Chappell 2007; Matthews and Garrison 1975; Putnam et al. 2000). The evaluation of human body nutrients intake was based on uncooked food consumed, but when cooking, the quality and quantity of food changed, and some nutrients in food lost during cooking, and the ratio can help to more accurately calculate human nutrients intake (Putnam et al. 2000). Moreover, when cooked, plant-based food such as rice and flour, the weight increased to more than twice, respectively, while for animal-based food, such as pork and beef, the weight lost a little bit (Matthews and Garrison 1975), and when we calculate impacts of food waste on resources and environment, the ratio would be even more critical (Zhang et al. 2016; Monier et al. 2011).

Finally, the relationship between food waste and customer behavior is vital for reducing food wasted on consumer level (Harrison et al. 1975; Wechsler et al. 2000; Baranowski et al. 2003). The behavior of plate food waste was analyzed between elementary school students with the age of 6–9 by Baik and Lee (2009), and the results showed that food waste had a lot to do with food preference, and students with more serious food preference had more food wasted, but a further study of how to reduce food wasted by changing of bad eating habits between students was not carried out. By information intervention, Whitehair et al. (2013) conducted a comparison test of food waste among college students, and the intervening effects indicated that compared with blank control (without any information presented when the sampling student eating in dining hall), the total amount of food wasted by students decreased by 15% in group with information intervention, but information intervention between different forms had not been presented significant differences. Moreover, besides food consumption, impacts of behavior on food waste also exists many section of food supply chain (e.g. food distributed section) (WRAP 2013). However, existing researches still has been few until now, and it is time to step up research on food waste reduction.

2.5.3 Systematic Engineering Need to Be Strengthen

It is estimated that there still had more than 70 countries around the world in a state of severe food shortage, and nearly 800 million people are undernourished (FAO et al. 2015). Moreover, by the middle of the twenty-first century, the world's population will exceed 9 billion, more than one third higher than today, and nearly all population increase will happen in developing countries (FAO 2009). Simultaneously, urbanization will keep at an accelerated pace with about 70% of the world's population to be in city by 2050 (compared to 49% today) (United Nations et al. 2014). All of those presented above would have been continuing to challenge global food system for a long time. To resolve the challenge, global food production must increase by 60% at least, but for the developing countries, food production must enhance by 80% to meet growing food demand of a rapidly increasing population (FAO et al. 2015). Annual cereal production will need to rise to about 3 billion tons compared to 2.1 billion today and annual meat production will need to rise by over 200 million tons to reach 470 million tons (FAO 2009). However, with rising incomes and an anticipated shift towards to more animal-based diets in developing countries and emerging countries, food waste levels and its resources and environmental cost may even reach to the extent presented in developed countries if nothing to do (Porter et al. 2016).

It is estimated that if 50% of food loss and waste happened in food system can be saved presently (supposing that the total proportion of food waste is 30%), based on the analysis above (global food production must increase by 60% in 2050), another 24% of food would be produced in 2050, that is to say meeting the food demand of increasing population in 2050, global food production only increase by 36%, let alone the embedded footprints.

But to achieve the goal needs system planning and design for food system. First, on a global scale, establishing unified evaluation criteria of food loss and waste and forming a complete set of theoretical method for enhancing efficiency of global food system are urgent. Second, case studies need to be urgently carried out, particularly developing countries and emerging countries, so that deficiency of data of food wastage can be resolved and influencing factors related to food wastage can be understood, and effective measures can be worked out by policy makers in order to increase food supply. Ultimately, integration of food system management must be adopted to provide a sustainable solution to cutting down food losses and waste along food supply chain. Governments, corporations, research institutions and individuals all must be involved towards to global sustainable food system.

References

Baik JY, Lee H (2009) Habitual plate-waste of 6- to 9-year-olds may not be associated with lower nutritional needs or taste acuity, but undesirable dietary factors. Nutr Res 29(12):831–838. https://doi.org/10.1016/j.nutres.2009.10.009

Bamfield J (2011) The global retail theft barometer 2011. Centre for Retail Research, Nottingham

Baranowski T, Cullen KW, Nicklas T, Thompson D, Baranowski J (2003) Are current health behavioral change models helpful in guiding prevention of weight gain efforts? Obes Res 11(S10):23S–43S. https://doi.org/10.1038/oby.2003.222

Beman JM, Arrigo KR, Matson PA (2005) Agricultural runoff fuels large phytoplankton blooms in vulnerable areas of the ocean. Nature 434(7030):211–214. https://doi.org/10.1038/nature03370

Beretta C, Stoessl F, Baier U, Hellweg S (2013) Quantifying food losses and the potential for reduction in Switzerland. Waste Manag 33:764–773. https://doi.org/10.1016/j.wasman.2012.11.007

Bolton JM, Liu WW (2006) Creating an effective China cold supply chain-current status, challenges and implementation considerations. Accenture report

Bourne M (1977) Cornell international agriculture mimeograph 53, post harvest food losses-the neglected dimension in increasing the world food supply. New York State College of Agriculture and Life Sciences, Cornell University, Ithaca

Brown T, Hipps NA, Easteal S, Parry A, Evans JA (2014) Reducing domestic food waste by lowering home refrigerator temperatures. Int J Refrig 40:246–253. https://doi.org/10.1016/j.ijrefrig.2013.11.021

Buzby JC, Hyman J (2012) Total and per capita value of food loss in the United States. Food Policy 37(5):561–570. https://doi.org/10.1016/j.foodpol.2012.06.002

Buzby JC, Farah WH, Axtman B, Mickey J (2009) Supermarket loss estimates for fresh fruits, vegetables, meat, poultry, and seafood and their use in the ERS loss-adjusted food availability data, Economic Information Bulletin Number 44. United States Department of Agriculture, Economic Research Service, Washington, DC

Buzby JC, Wells HF, Hyman J (2014) The estimated amount, value, and calories of postharvest food losses at the retail and consumer levels in the United States. EIB-121, Economic Research Service, US Department of Agriculture, USA

Chappell GM (2007) Food waste and loss of weight in cooking. Brit J Nutr 8(04):325–240. https://doi.org/10.1079/BJN19540050

Chen D, Wang S, Kang F, Zhu Q, Li X (2011) Mathematical model of feeding rate and processing loss for combine harvester. Trans of the CSAE 27(9):18–21. https://doi.org/10.3969/j.issn.1002-6819.2011.09.004

Cheng SK, Gao LW, Xu ZR, Tang CC, Wang LE, Dhruba BGC (2012) Food waste in catering industry and its impacts on resources and environment in China. China Soft Sci 7:106–114. https://doi.org/10.3969/j.issn.1002-9753.2012.07.012

CNBS (Chinese National Bureau of Statistics) (2010) China compendium of statistics 1949–2008. China Statistics Press, Beijing (in Chinese)

CNBS (Chinese National Bureau of Statistics), the National Commerce Department, China General Chamber of Commerce (2011) The retail and chain restaurant statistical yearbook 2011. China Statistics Press, Beijing (in Chinese)

CNBS (Chinese National Bureau of Statistics) (2017) China statistics book 2017. China Statistics Press, Beijing (in Chinese)

Cuéllar AD, Webber ME (2010) Wasted food, wasted energy: the embedded energy in food waste in the United States. Environ Sci Technol 44:6464–6469. https://doi.org/10.1021/es100310d

Defra (2011) Government review of waste policy in England 2011. Department for Environment, Food and Rural Affairs, London

Ding S (2015) Study on the household food waste and its ecological footprint, carbon footprint, water footprint and methane emission. Master thesis, Dalian University of Technology, Dalian (in Chinese)

Dong H, Li Y, Tao X, Peng X, Li N, Zhu Z (2008) China greenhouse gas emissions from agricultural activities and its mitigation strategy. Trans of the CSAE 24(10):269–273 (in Chinese with English abstract). https://doi.org/10.3969/j.issn.1002-6819.2008.10.054

Dryerre H, Andross M (1946) Losses of nutrients in the preparation of foodstuffs. Proc Nutr Soc 4(02):155–164. https://doi.org/10.1079/PNS19460030

EC (2011) Roadmap to a resource efficient Europe. European Commission, Brussels

EP (European Parliament) (2012) Parliament calls for urgent measures to halve food wastage in the EU – plenary sessions. European Commission, Brussels Available at http://www.europarl.europa.eu/pdfs/news/expert/infopress/20120118IPR35648/20120118IPR35648_en.pdf

Eriksson M (2015) Supermarket food waste-prevention and management with the focus on reduced waste for reduced carbon footprint. Doctoral thesis, Swedish University of Agricultural Sciences, Uppsala

Fan Q, Liu MY, Qi HQ (2015) Research on loss in rice processing and governance measures. Sci Technol Cereals, Oils Foods 23(05):117–120 (in Chinese with English abstract). https://doi.org/10.3969/j.issn.1007-7561.2015.05.027

Fang CL (2009) Issues of resources and environment protection in China's rapid urbanization process and suggestions on countermeasures. Bull Chin Acad Sci 5:468–474 (in Chinese with English abstract). https://doi.org/10.3969/j.issn.1000-3045.2009.05.002

FAO (2006) Household food wastage in Turkey. Food and Agriculture Organization of the United Nations (FAO), Rome

FAO (2009) How to feed the world in 2050. FAO, Rome Available at http://www.fao.org/fileadmin/templates/wsfs/docs/expert_paper/How_to_Feed_the_World_in_2050.pdf

FAO (2013) Food wastage footprint, impacts on natural resources-summary report. Food and Agriculture Organization of the United Nations, Rome

FAO, IFAD, WFP (2015) The state of food insecurity in the world 2015. Meeting the 2015 international hunger targets: taking stock of uneven progress. FAO, Rome

FAOSTAT (2013) FAOSTAT statistics database. Available at http://faostat.fao.org/site/291/default.aspx

Fehr M, Calcado MDR, Romão DC (2002) The basis of a policy for minimizing and recycling food waste. Environ Sci Pol 5:247–253. https://doi.org/10.1016/S1462-9011(02)00036-9

Flegal KM, Carroll MD, Ogden CL, Johnson CL (2002) Prevalence and trends in obesity among US adults, 1999-2000. JAMA 288(14):1723–1727. https://doi.org/10.1001/jama.288.14.1723

Flegal KM, Carroll MD, Ogden CL, Curtin LR (2010) Prevalence and trends in obesity among US adults, 1999-2008. JAMA 303(3):235–241. https://doi.org/10.1001/jama.2009.2014

Gao (1977) Food waste: an opportunity to improve resource use. U.S. General Accounting Office, CED-77-118, Washington, DC. Available at: http://www.gao.gov/assets/120/119733.pdf

Gao LW, Cheng SK, Cao XC, Zhang D, Liu XJ, Qin Q, Liu Y (2013) An overview of the resources and environmental issues from wasted food in urban catering across China. J Resour Ecol 4(4):337–343. https://doi.org/10.5814/j.issn.1674-764x.2013.04.006

Gao LW, Cheng SK, Cao XC, Zhang D, Liu XJ, Qin Q, Liu Y (2015) Review of food loss and waste research and its prospects. J Nat Resour 30(3):523–536. https://doi.org/10.11849/zrzyxb.2015.03.015

Gao LW, Xu SW, Li ZM, Cheng SK, Yu W, Zhang YE, Li DH, Wang Y, Wu C (2016) Main grain crop postharvest losses and its reducing potential in China. Trans CSAE 32(23):1–11. https://doi.org/10.11975/j.issn.1002-6819.2016.23.001

Gao LW, Cheng SK, Xu SW, Li ZM, Liu XJ, Zhang D, Yu W, Cao XC, Wang LE, Liu Y, Wang Y, Qin Q (2017) Impacts of national policy on catering food waste in urban China-a case study of Lhasa city. Food Nutr China 23(03):44–48. https://doi.org/10.3969/j.issn.1006-9577.2017.03.011

Garnett T (2008) Cooking up a storm: food, greenhouse gas emissions and our changing climate. Food Climate Research Network, Center for Environmental Strategy, Surrey, UK. http://www.fcrn.org.uk/sites/default/files/CuaS_web.pdf

Garnett T (2011) Where are the best opportunities for reducing greenhouse gas emissions in the food system (including the food chain)? Food Policy 36:S23–S32. https://doi.org/10.1016/j.foodpol.2010.10.010

Göbel C, Teitscheid P, Ritter G, Blumenthal A, Friedrich S, Frick T, Grotstollen L, Möllenbeck C, Rottstegge L, Pfeiffer C, Baumkötter D, Wetter C, Uekötter B, Burdick B, Langen N, Lettenmeier M, Rohn H (2012) Reducing food waste – identification of causes and courses of action in North Rhine-Westphalia. Abridged version, University of Applied Sciences Münster, Institute for Sustainable Nutrition and Food Production – iSuN, Münster

Godfray HCJ, Beddington JR, Crute IR, Haddad L, Lawrence D, Muir JF, Pretty J, Robinson S, Thomas SM, Toulmin C (2010) Food security: the challenge of feeding 9 billion people. Science 327(5967):812–818. https://doi.org/10.1126/science.1185383

Guo J, Liu X, Zhang Y, Shen J, Han W, Zhang W, Christie P, Goulding K, Vitousek P, Zhang F (2010) Significant acidification in major Chinese croplands. Science 327(5968):1008–1010. https://doi.org/10.1126/science.1182570

Gustavsson J, Cederberg C, Sonesson U, Van Otterdijk R, Meybeck A (2011) Global food losses and food waste – extent, causes and prevention. Food and Agriculture Organization of the United Nations, Rome

Hall DW (1970) Handling and storage of food grains in tropical and subtropical areas. FAO, Rome

Hall KD, Guo J, Dore M, Chow CC (2009) The progressive increase of food waste in America and its environmental impact. PLoS One 4(11):e7940. https://doi.org/10.1371/journal.pone.0007940

Hamerschlag K, Venkat K (2011) Meat eater's guide to climate change and health-life-cycle assessments: methodology and results. Environmental Working Group, Washington, DC

Harrison GG, Rathje WL, Hughes WW (1975) Food waste behavior in an urban population. J Nutr Educ 7(1):13–16. https://doi.org/10.1016/S0022-3182(75)80062-8

Hayes J, Kendrick OW (1995) Plate waste and perception of quality of food prepared in conventional VS commissary systems in the nutrition program for the elderly. J Am Diet Assoc 95(5):585–586. https://doi.org/10.1016/S0002-8223(95)00159-X

HLPE (2014) Food losses and waste in the context of sustainable food systems. A report by the high level panel of experts on food security and nutrition of the committee on world food security, Rome

James SJ, Evans J, James C (2008) A review of the performance of domestic refrigerators. J Food Eng 87(1):2–10. https://doi.org/10.1016/j.jfoodeng.2007.03.032

Jane Q (2011) China to spend billions cleaning up groundwater. Science 334(11):745. https://doi.org/10.1126/science.334.6057.745

Kantor LS, Lipton K, Manchester A, Oliveira V (1997) Estimating and addressing America's food losses. Food Rev 20(1):2–12

Katajajuuri JM, Silvennoinen K, Hartikainen H, Heikkilä L, Reinikainen A (2014) Food waste in the Finnish food chain. J Clean Prod 73:322–329. https://doi.org/10.1016/j.jclepro.2013.12.057

Kummu M, De Moel H, Porkka M, Siebert S, Varis O, Ward PJ (2012) Lost food, wasted resources: global food supply chain losses and their impacts on freshwater, cropland, and fertiliser use. Sci Total Environ 438:477–489. https://doi.org/10.1016/j.scitotenv.2012.08.092

Lebersorger S, Schneider F (2014) Food loss rates at the food retail, influencing factors and reasons as a basis for waste prevention measures. Waste Manag 34:1911–1919. https://doi.org/10.1016/j.wasman.2014.06.013

Li Y, Zhang W, Ma L, Huang G, Oenema O, Zhang F, Dou Z (2013) An analysis of China's fertilizer policies: impacts on the industry, food security, and the environment. J Environ Qual 42:972–981. https://doi.org/10.2134/jeq2012.0465

Lipinski B, Hanson C, Lomax J, Kitinoja L, Waite R, Searchinger T (2013) Reducing food loss and waste- working paper, installment 2 of creating a sustainable food future. World Resources Institute, Washington, DC Available online at http://www.worldresourcesreport.org

Liu G (2014) Food losses and food waste in China: a first estimate. OECD Food, Agriculture And Fisheries Papers, No. 66, OECD Publishing, Paris

Liu J, Lundqvist J, Weinberg J, Gustafsson J (2013a) Food losses and waste in China and their implication for water and land. Environ Sci Technol 47(18):10137–10144. https://doi.org/10.1021/es401426b

Liu X, Zhang Y, Han W, Tang A, Shen J, Cui Z, Vitousek P, Erisman JW, Goulding K, Christie P (2013b) Enhanced nitrogen deposition over China. Nature 494(7438):459–462. https://doi.org/10.1038/nature11917

Lundqvist J, Fraiture C, Molden D (2008) Saving water: from field to fork-curbing losses and wastage in the food chain. SIWI Policy Brief, Stockholm International Water Institute (SIWI), Stockholm

Lyndhurst B, WRAP (2012) Helping consumers reduce food waste-a retail survey 2011. Final Report, Waste and Resources Action Programme, Banbury

Mao X, Hu J, Liu X (2011) Epidemiological burden of bacterial foodborne diseases in China-preliminary study. Chin J Food Hyg 23:132–136. https://doi.org/10.13590/j.cjfh.2011.02.018

Matthews RH, Garrison YJ (1975) Food yields summarized by different stages of preparation. Agriculture handbook, Agricultural Research Service, (102 revised), US Department of Agriculture

Mattsson L, Williams H (2015) Waste reduction potential of fresh fruit and vegetables at retail stores: a case study of three supermarkets in Sweden. In: 2nd International Conference on Global Food Security, October 11–14, 2015, Ithaca, USA

Monier V, Mudgal S, Escalon V, O'Connor C, Anderson G, Montoux H, Reisinger H, Dolley P, Oglivie S, Morton G (2011) Preparatory study on food waste across EU 27. European Commission, Paris

Muth MK, Karns SA, Nielsen SJ, Buzby JC, Wells HF (2011) Consumer-Level food loss estimates and their use in the ERS loss-adjusted food availability data. Economic Research Service, Washington, DC, USA, Technical Bulletin No. 1927, US Department of agriculture, USA

Nature (2010) How to feed a hungry world, Editorial. Nature 266:531–532. https://doi.org/10.1038/466531a

Nellemann C, MacDevette M, Manders T, Eickhout B, Svihus B, Prins AG, Kaltenborn BP (2009) The environmental food crisis – the environment's role in averting future food crises. A UNEP rapid response assessment. United Nations Environment Programme, GRID-Arendal, Rome

NSC (1980) Postharvest food losses in Sri Lanka. Report of a Workshop. National Science Council (NSC), Colombo

Parfitt J, Barthel M, Macnaughton S (2010) Food waste within food supply chains: quantification and potential for change to 2050. Philos Trans R Soc B 365(1554):3065–3081. https://doi.org/10.1098/rstb.2010.0126

PBL (2009) Growing within limits: a report to the global assembly 2009 of the Club of Rome. Netherlands Environmental Assessment Agency, Bilthoven

Porter SD, Reay DS, Higgins P, Bomberg E (2016) A half-century of production-phase greenhouse gas emissions from food loss & waste in the global food supply chain. Sci Total Environ 571:721–729. https://doi.org/10.1016/j.scitotenv.2016.07.041

Putnam J, Kantor LS, Allshouse J (2000) Per capita food supply trends: progress toward dietary guidelines. Food Rev 23(3):2–14

Rembold F, Hodges R, Bernard M, Knipschild H, Léo O (2011) The African postharvest losses information system (APHLIS). Scientific and Technical Report, Institute for Environment and Sustainability, Joint Research Centre, European Commission, Luxembourg

SAG (2011) The "twelfth five-year" special construction plan of scientific grains storage on-farm, State Administration of Grain, China. Available at: http://www.chinagrain.gov.cn/

Schneider F (2013) Review of food waste prevention on an international level. In Proceedings of the Institution of Civil Engineers-Waste and Resource Management, ICE Publishing 166(4): 187–203

Song G, Li M, Semakula HM, Zhang S (2015a) Food consumption and waste and the embedded carbon, water and ecological footprints of households in China. Sci Total Environ 529:191–197. https://doi.org/10.1016/j.scitotenv.2015.05.068

Song H, Zhang H, Li J, Wu Z (2015b) Loss of harvest of china's grain-a case study of wheat in Henan Province. J Huazhong Agric Univ (social sciences edition) (04): 1–6. DOI: https://doi.org/10.13300/j.cnki.hnwkxb.2015.04.001

Stensgård A, Hanssen OJ (2015) Food waste in Norway 2014: status and trends. 2009-14, ForMat, Østfoldsforskning

Stewart CP (1946) Loss of nutrients in cooking. Proc Nutr Soc 4(02):164–171

Tai J, Zhang W, Che Y, Feng D (2011) Municipal solid waste source-separated collection in China: a comparative analysis. Waste Manag 31(8):1673–1682. https://doi.org/10.1016/j.wasman.2011.03.014

Tesco PLC (2014) Tesco and society report: using our scale for good. Annual Report and Financial Statements, TESCO available at: http://www.mrfarrugia.com/uploads/4/1/6/6/41664763/tesco_2014-_annual_report.pdf

Thylmann D, Druzhinina E, Deimling S (2013) The ecological footprint of cassava and maize post-harvest-losses in Nigeria: a life cycle assessment. On behalf of German Federal Ministry for Economic Cooperation and Development (BMZ), division rural development, agriculture and food security. Deutsche Gesellschaft für, Internationale Zusammenarbeit (GIZ) GmbH, Germany

Tilman D, Balzer C, Hill J, Befort BL (2011) Global food demand and the sustainable intensification of agriculture. Proc Natl Acad Sci 108:20260–20264. https://doi.org/10.1073/pnas.1116437108

UNEP (2012) Avoiding future famines: strengthening the ecological foundation of food security through sustainable food systems. A UNEP synthesis report, United Nations Environment Programme. Available at http://www.unep.org/publications/ebooks/avoidingfamines/

United Nations, Department of Economic and Social Affairs, Population Division (2014) World urbanization prospects: The 2014 revision, highlights (ST/ESA/SER.A/352)

Venkat K (2011) The climate change and economic impacts of food waste in the United States. Int J Food Syst Dyn 2(4):431–446

Vermeulen SJ, Campbell BM, Ingram JSI (2012) Climate change and food systems. Annu Rev Environ Resour 37(1):195–222. https://doi.org/10.1146/annurev-environ-020411-130608

Wang KL, Xu W (2012) Prediction of food waste production and management system research in Chengdu. Sichuan Environ 31(1):91–96(in Chinese). https://doi.org/10.14034/j.cnki.schj.2012.01.023

Wang W, Jaeger F, Li X, Wang X, Zhang J (2013) China's food production and cold chain logistics. In: 5th international workshop on cold chain management at University of Bonn, Germany, pp 10–11

Wechsler H, Devereaux RS, Davis M, Collins J (2000) Using the school environment to promote physical activity and healthy eating. Prev Med 31(2):S121–S137. https://doi.org/10.1006/pmed.2000.0649

Whitehair KJ, Shanklin CW, Brannon LA (2013) Written messages improve edible food waste behaviors in a university dining facility. J Acad Nutr Diet 113(1):63–69. https://doi.org/10.1016/j.jand.2012.09.015

WRAP (2009) Household food and drink waste in the UK. Final report, Waste & Resources Action Programme, UK

WRAP (2010) Waste arisings in the supply of food and drink to UK households. Final report, Waste & Resources Action Programme, UK

WRAP (2011a) New estimates for household food and drink waste in the UK. Final report (version 1.1), Waste & Resources Action Programme, UK

WRAP (2011b) Food waste in schools. Final report, Waste and Resources Action Programme, Banbury, UK

WRAP (2012) New estimates for household food and drink waste in the UK. Waste & Resources Action Program (WRAP), Banbury

WRAP (2013) Consumer attitudes to food waste and food packaging. Final report, Waste and Resources Action Programme, Banbury

Wu W, Xing R (2003) The study of municipal swill management. Munic Adm Technol 5(2):66–68. https://doi.org/10.16242/j.cnki.umst.2003.02.007

Wu K, Xu WF, Yu Z (2006) Perspective of catering waste resources management in Hefei city. Res Renewable Resour 4:31–33 (in Chinese)

Xu S (2005) Analysis on food consumption and waste of China. Food Nutr China 11:4–8 (in Chinese)

Xu S (2007) The severity of food waste in China. The countryside, agriculture and farmers (version B) (1): 31–32 (in Chinese)

Xu D, Shen DS, Feng HJ, Wang MZ, Deng YH (2011) Investigation and countermeasures analysis of catering waste in southern city in China. Environ Sci 32(7):2164–2168(in Chinese). https://doi.org/10.13227/j.hjkx.2011.07.048

Yu Z, Han S, Chu J, Xu Z, Zhu C, Guo X (2012) Trends in overweight and obesity among children and adolescents in China from 1981 to 2010: a meta-analysis. PLoS One 7(12):e51949. https://doi.org/10.1371/journal.pone.0051949

Zhan YR (1995) Sampling and investigation of food losses and waste in food supply chain in China. China Grain Econ 04:44–47 (in Chinese)

Zhang BX, Fu WX (2010) The investigation and analysis on per capita output of food waste in Beijing. Environ Sci Technol 33(12F):651–654. https://doi.org/10.3969/j.issn.1003-6504.2010.12F170

Zhang H, Wang Q, Mortimer SR (2012) Waste cooking oil as an energy resource: review of Chinese policies. Renew Sust Energ Rev 16(7):5225–5231. https://doi.org/10.1016/j.rser.2012.05.008

Zhang WF, Dou ZX, He P, Ju XT, Powlson D, Chadwick D, Norse D, Lu YL, Zhang Y, Wu L (2013) New technologies reduce greenhouse gas emissions from nitrogenous fertilizer in China. Proc Natl Acad Sci 110(21):8375–8380. https://doi.org/10.1073/pnas.1210447110

Zhang D, Cheng SK, Gao L, Liu X, Cao X, Liu Y, Bai J, Xu S, Yu W, Qin Q (2016) The carbon footprint of catering industry food waste: a Beijing case study. Acta Ecologica Sinica 36(18):5937–5948. https://doi.org/10.5846/stxb201504150769

Zhao XP (2008) Study on the development of foodstuff cold chain logistics in Dalian. Master thesis, Dalian Maritime University, Dalian (in Chinese)

Zhou J, Sun J (2015) Strategy research on perfecting the cold-chain logistics system of fresh food. International Conference on Education Technology, Management and Humanities Science (ETMHS), pp 738–742

Chapter 3
Beneficial Microorganisms for the Management of Soil Phosphorus

Ioannis Ipsilantis, Mina Karamesouti, and Dionisios Gasparatos

Abstract The dependence of all life on phosphorus (P) availability has revealed serious P challenges such as the P deficiency in weathered and eroded soils, the high cost of phosphate fertilizers, the scarcity and unequal global distribution of rock phosphates, the regional over-accumulation of P and the agricultural non-point source P pollution. In this context, microorganisms capable of mobilizing P in the soil system may be applied as a low-cost technology to enhance plant growth and crop yields. Here we review the beneficial role of microorganisms, namely arbuscular mycorrhizal fungi and P–mobilizing bacteria, to mediate P availability and transform legacy P (insoluble, bound) into soluble forms.

We found that the arbuscular mycorrhizal symbiosis improves plant P nutrition, however, high soil P concentration is also known to suppress it. The effectiveness of the symbiosis depends on the richness of arbuscular mycorrhizal fungal species. Most studies show that arbuscular mycorrhizal fungal diversity and effectiveness are modified by soil management practices. Fertilization with slow-release inorganic fertilizers, organic fertilizers and mycorrhizal symbiosis gives satisfactory crop yields, but long-term studies are few. Bacteria are the predominant microorganisms that mobilize native and applied P in soils, as compared to fungi or actinomycetes. Strains from the genera *Pseudomonas*, *Bacillus* and *Rhizobium* have so far been recognized as the most powerful phosphate solubilizers. The principal mechanism for mineral phosphate solubilization is the production of organic acids, e.g. oxalic, citric, gluconic, tartaric, lactic, fumaric; and enzymes: phosphatases, phytases, phosphonatases and C-P lyases.

I. Ipsilantis · D. Gasparatos (✉)
Soil Science Laboratory, School of Agriculture, Aristotle University of Thessaloniki, Thessaloniki, Greece
e-mail: gasparatos@agro.auth.gr

M. Karamesouti
Department of Natural Resources Management & Agricultural Engineering, Agricultural University of Athens, Athens, Greece

E. Lichtfouse (ed.), *Sustainable Agriculture Reviews 32*, Sustainable Agriculture Reviews 32, https://doi.org/10.1007/978-3-319-98914-3_3

Keywords Soil phosphorus · Availability · Plant nutrition · Arbuscular mycorrhizal fungi · P – Mobilizing bacteria · Sustainable management

3.1 Introduction

3.1.1 Phosphorus Issues

Phosphorus (P) is a mineral nutrient with a key role in supporting global alimentation requirements. It is listed among the most essential elements for flora and fauna growth and evolution (Elser 2012). It is naturally occurring mainly from apatite and it is released through weathering processes (Lajtha and Schlesinger 1988). The released P ions can be either absorbed by $CaCO_3$, Fe and Al oxides, or be incorporated into living organisms. Although P is usually abundant in soils, its availability to plants is limited, due to low solubility and soil fixation phenomena, resulting in plant growth restrictions (Gasparatos et al. 2006).

Although P fertilization is commonly used for achieving high agricultural productivity, in some areas, P fertilizers are used to prevent further land degradation and maintain even a basic level of productivity (Weikard and Seyhan 2009). The soil status, and mainly the type of parent material and the organic matter content, in combination with the climatic conditions, are important factors affecting the initial soil P concentration and availability. Indicatively, in areas with tropical climates, limited P concentration in soils suggests a significant issue for plant growth (Solomon and Lehmann 2000). In arid regions, the organic P compounds are typically in low percentages, due to the low soil organic matter (SOM) and water contents. Increases in temperature may also have a negative impact on the long-term soil P availability (Yu et al. 2016). Concerning the soil characteristics, soil aeration and saturated hydraulic conductivity appear as positively correlated to the available soil P, in contradiction to clay content and pH (Yu et al. 2016). In cases of compacted soils, limitations in plant root penetration reduce plants' capability to access the available soil P (Barzegar et al. 2016).

In an attempt to improve plant intake performance, P application from external sources was used as a common agricultural practice. Historically, soil fertilization with P was initially based on guano, human excreta and manure, while since the second half of the twentieth century, P fertilizers became the major source of this nutrient (Fig. 3.1).

In many developed countries, food production processes are highly dependent on phosphate fertilizers, with many reported cases of unsustainable over-fertilization (Yan et al. 2013). The irresponsible use of P, mainly due to its outwardly abundance and its low price, is lately attempted to be restricted (Sharpley et al. 2013). The cost of P-related environmental problems rehabilitation can no longer be considered insignificant. Governments are forced to increase investments, in an attempt to combat eutrophication caused by high quantities of phosphates which end up in aquatic

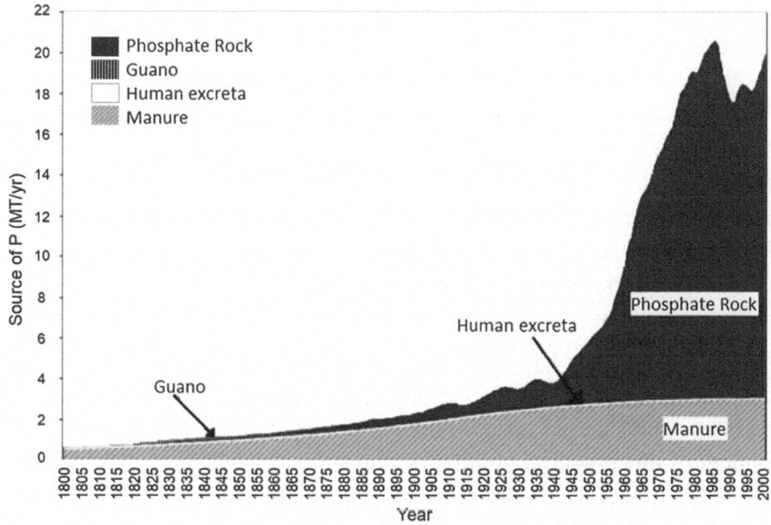

Fig. 3.1 Historical global sources of phosphorus (P) fertilizers (1800–2000). Increased food demand during nineteenth century, was met using manure as a P source. After mid-twentieth century, the Green Revolution was achieved through new agricultural practices including very high quantities of P fertilizers from mined phosphate rock. (Modified after Cordell et al. 2009)

systems (Dodds et al. 2008). However, the initial concerns for environmental problems, such as water pollution at the beginning and soil over-accumulation later, are lately evolved into Global awareness for potential socio-economic issues. Changes in alimentation preferences leading to increasing P demand, the significantly uneven spatial distribution of P on Earth and the uncertainty over the potential peak of global P production, suggest a new threat on food security (FAO 2015; ECSCU 2013; Cordell et al. 2012). Heckenmüller et al. (2014), discuss the unprecedented increase, in 2007–2008, of more than 900% of the prices in the global P trade. The export controls, such as those imposed by China in 2008, in order to protect its domestic supplies, can cause technical shortages and broader socio-economic turbulences. Under these circumstances countries with low or no P deposits are highly vulnerable to potential fluctuations in fertilizer and mineral P prices.

These concerns have opened an extensive debate about the quest of alternative sustainable use of P (Schröder et al. 2011; McLaughlin et al. 2011; Karamesouti and Gasparatos 2017). Some of these are summarized in P recycling procedures, such as P recovery from municipal and industrial wastewater, from other organic wastes, or from the soil itself, and in improved agricultural practices, targeting losses reduction from livestock rearing, crop cultivation or from any other loses in any level in the food chain (Rowe et al. 2016; Elser 2012).

Focusing on P recovery from soil, multiple approaches, providing economically feasible and environmentally friendly solutions, are already being developed (Table 3.1). The main common concept framing all these approaches is the reversibility of P fixation and the efficient use of its residual form.

Sattari et al. (2012) and Rowe et al. (2016) defined this residual P as the legacy P (P_{legacy}), described as

$$P_{legacy} = P_{inputs} - P_{outputs} - P_{losses}$$

Where

P_{legacy} = Legacy phosphorus
P_{inputs} = Phosphorus inputs to soils (fertilizers, manures)
$P_{outputs}$ = Phosphorus removed by crop production
P_{losses} = Phosphorus losses (runoff and leaching)

and represents the P that has been accumulated in soils after continuous inputs of fertilizers and manures. In many cases, legacy P was reported as pollutant and major source of eutrophication of water bodies (Sharpley et al. 2013). However, efficient management of this potential pollutant could support adequate agricultural production without additional fertilization for many years (Rowe et al. 2016).

3.1.2 Beneficial Microorganisms and Soil Phosphorus

Among the strategies listed in Table 3.1, the use of microorganisms, such as fungi, bacteria and endophytes, in mobilizing legacy soil P is considered particularly environmental-friendly and of lower cost, compared to fertilization practices (Adhya et al. 2015; Sharma et al. 2013). Their populations are highly variable among different soil types, with the bacteria significantly predominating in proportion, compared to fungi, while the former might incorporate approximately 10–15% of the soil organic P (Richardson 2007; Khan et al. 2007). A plethora of microorganisms can be usually identified in close distance from plant rooting system, with which reciprocity relations are being established. The fundamental role of microorganisms is

Table 3.1 Strategies for improving the utilization of legacy soil P. Adapted from Rowe et al. 2016; Bindraban et al. 2015; Whithers et al. 2014; Shen et al. 2011

Soil – crop management	Plant breeding	Microorganisms
Depletion of readily available P at critical levels	Root genetic traits such as root elongation, branching and development of root hairs, and enhancement of early root growth	*Bacillus*, *Pseudomonas* and *Penicillium* genera
Maintain soil quality	Release of exudates, such as organic acids, carbon substrates and enzymes	Arbuscular mycorrhizal fungi
Modification of soil pH	Physiological alterations, i.e. low metabolic P demand and low photosynthetic needs	Bio-inoculant products
Tillage practices	Selecting crop varieties for high P – use efficiency	
Fertilizer inputs		
Crop rotation		

not restricted in making P accessible to plants, but they can also be used to improve plant growth and agricultural productivity with no further need for additions of chemicals (Sharma et al. 2013; Puente et al. 2009). Organic matter decomposition, soil detoxification and critical nutrients' assurance for the plants are some other main functions conducted by microorganisms' communities.

Since early twentieth century, the role of microorganisms in P mobilization processes was already discussed (Khan et al. 2007), while in 1948, Gerretsen had mentioned the contribution of bacteria in P plant nutrition under controlled conditions. In 1988, McLaughlin et al. studied the incorporation of native soil P, accumulated due to fertilization and plant residues, into the microbial biomass. However, the wide-scale applicability in uncontrolled environment is still a particularly challenging task. The microorganisms can be highly affected by inherent soil characteristics, such as soil temperature, pH, soil moisture content etc., and also by interactions between different species (Sharma et al. 2013; Richardson and Simpson 2011). In this regard, Richardson (2007), suggested two main strategies focusing on (a) *the management of existing microbial populations in order to optimize their capacity to mobilize phosphorus*, and (b) *the use of specific microbial inoculants that can increase phosphorus mobilization*.

Arbuscular mycorrhizal fungi and P – mobilizing bacteria are two broad categories of microorganisms mobilizing inorganic P and converting it into readily available to plant forms (Fig. 3.2). In this review, the mechanisms and the effectiveness of these microbes in P acquisition will be discussed.

The recognition of microbials' role to the improvement of soil P availability, highlights the need for detailed studies on the characteristics of the microbial communities, the interactions within the various soil environments, the impact of land management practices, as well as development of new technologies, in order to reach for promising solutions for efficient soil P management.

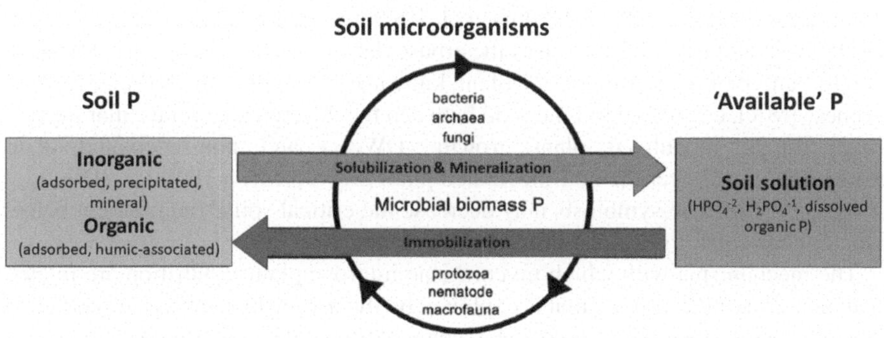

Fig. 3.2 The critical role of soil microorganisms in the cycling of phosphorus (P), which is the result of the biogeochemical processes of mobilization (solubilation/ mineralization) and immobilization (adsorption/precipitation). Modified after Richardson and Simpson 2011

3.2 Arbuscular Mycorrhizal Fungi

3.2.1 The Symbiosis and P Uptake

The arbuscular mycorrhizal fungi form symbiotic, mutually beneficial relations with plants in their roots. The plant provides photosynthetic carbon to the fungus and the fungus supplies nutrients with poor mobility in soil, particularly P (Smith and Read 2008). However, there is more to the symbiosis, as it is also known to suppresses losses to pathogens (Pozo and Azcón-Aguilar 2007; Karagiannidis et al. 2002), provides drought (Augé 2001) and salt resistance (Evelin et al. 2009), may increase soil aggregation (Rillig and Mummey 2006) and increase resistance to potentially toxic elements (Burghelea et al. 2015; Hildebrandt et al. 2007). There is a complex interaction between the partners and both the plant and the fungus, as well as the soil environment contribute to this complexity. As a result, although there are some well-known paradigms about the symbiosis, there are many cases that do not seem fit to the general dogma. For instance, a particular arbuscular mycorrhizal fungus-plant relationship was reported as parasitic (Modjo and Hendrix 1986).

It is well established that arbuscular mycorrhizal fungi improve plant P nutrition in soils with moderate or poor soluble P levels. At higher P levels the symbiosis is suppressed, manifested through lower root length colonization (Jensen and Jakobsen 1980). There is a threshold value of available P above which the carbon cost of the symbiosis is higher than the benefit of the arbuscular mycorrhizal pathway of P uptake and the plant response to arbuscular mycorrhizae is negative (Kahiluoto et al. 2000). This threshold may vary with the fungus, the plant and the environmental conditions, and may be determined by dose-response experiments showing how much P fertilizer can be saved by arbuscular mycorrhizal fungi, or how much P-fertilizer needs to be applied to achieve maximum yield with arbuscular mycorrhizal fungi (Ping et al. 2014; Medina et al. 1990; Elbon and Whalen 2015). At very low P levels addition of P increases the arbuscular mycorrhizal fungal colonization and the benefit of mycorrhiza to the plant, known as arbuscular mycorrhizae responsiveness (MR), calculated by: the ratio between the difference of arbuscular mycorrhizal fungal inoculated plant growth (DW_{AMF}) and non-inoculated plant growth ($DW_{control}$) over the non-inoculated plant growth, $MR = (DW_{AMF} - DW_{control}/DW_{control}) \times 100$. The symbiosis may decrease the critical soil P requirement below which plants cannot grow (Ryan et al. 2016; Janos 2007).

The mechanisms with which mycorrhizae improve plant P nutrition are related to an increase of the volume that is exploited by the mycorrhizosphere, compared to the soil volume of the rhizosphere (Marschner 1995; Li et al. 1991), or by synergism with P-solubilizing microorganism (Antunes et al. 2007). It seems that both the plant and fungus exploit the same sources of P (Blal et al. 1990). However, the fungus may have access to smaller soil pores, may produce more active phosphatases and have higher affinity for P (Bolan 1991). Accessing P from rock phosphate or other bedrock is more studied for ectomycorrhizae (van Schöll et al. 2008), while Koele et al. (2014), have shown that this may equally well happen by arbuscular mycorrhizae. Increased nutrient uptake in arbuscular mycorrhizal plants has been shown from basaltic and rhyolitic parent rocks, with P uptake improvement from the latter (Burghelea et al. 2015). Use of rock phosphate in combination with arbuscular mycorrhizal fungi may be more beneficial in Oxisols where soluble chemical fertilizer P is rapidly converted to insoluble forms (Lin and Fox 1992) and mycorrhizal plants may have access to the NaOH-P pools (Cardoso et al. 2006).

3.2.2 Conventional-High Input Agriculture and Arbuscular Mycorrhizal Fungi

Mycorrhizae and conventional, well fertilized agriculture may seem incompatible. However, there are reports where mycorrhizae were effective in spite of high P soil concentrations. It is not high P in soil solution that reduces arbuscular mycorrhizal fungal presence, but high P in the root tissues, as it has been shown by Menge et al. (1978). On the other hand, plants with root systems like onion may depend on arbuscular mycorrhizae for P even at high soil nutrient levels (Galván et al. 2009). Miller et al. (1995) summarizing the Guelf field experiments suggested that arbuscular mycorrhizal fungi may be important for well fertilized crops. They explain that the inhibition of the symbiosis by high field P was not found to be that great, also observed by others (Hayman et al. 1976; Gryndler et al. 1989), that mycorrhizae may improve P nutrition at the early plant stages when P in plant tissues is not high, but critical for high yields, and that localized fertilizer application may reduce arbuscular mycorrhizal fungal colonization only at the part of the root system that is exposed to it, and not in the whole root system. In strawberries produced in high P compost substrate, a particular cultivar- arbuscular mycorrhizal fungal combination produced ~50% more daughter plants than the non-mycorrhizal control, although colonization never exceeded 12% but other inoculated cultivars had a reduced number of daughter plants (Stewart et al. 2005). Douds et al. (2016) found in soils with available P at 214–258 mg/kg, that on the average of 7 years and different pre-inoculated tomato cultivars, there was a 6% increase in yield, but for individual years, arbuscular mycorrhizal plants had lower to much higher yield than the uninoculated control.

In high P soils arbuscular mycorrhizal fungi may be effective when plants are under some stress (Douds et al. 2016). For example, well fertilized field maize under water stress (reduced irrigation) was well colonized and benefited by arbuscular mycorrhizal fungi compared to the fumigated control and the fully irrigated treatment (Sylvia et al. 1993). Mycorrhizal citrus plants grown in high P conditions recovered faster from moisture stress and had higher leaf P concentration than non-inoculated controls (Fidelibus et al. 2001). In sweat potato, low temperature suppressed the plant P uptake leaving the fungal pathway as the sole source of P showing increased yield with arbuscular mycorrhizal fungi in spite of the high P soil (242–599 mg available P /kg soil) (Douds et al. 2015). The same may be the case for salinity/sodicity, a common problem of greenhouses, or other plant stresses that need to be examined. Furthermore, under low N, high P did not seem to inhibit arbuscular mycorrhizal fungi (Sylvia and Neal 1990) and at high P increasing intra-specific density increased the growth benefit to arbuscular mycorrhizal fungi (Hetrick et al. 1994), but this was not always the case (Schroeder and Janos 2005).

Under well fertilized conditions inoculation of plants with arbuscular mycorrhizal fungi may be effective if the indigenous population is ineffective (Medina et al. 1990). Combined application of NPK chemical fertilizers with arbuscular mycorrhizal fungal inoculum may improve plant P acquisition efficiency in P deficient soils than that with indigenous inoculum (Hu et al. 2010).

Soil management such as tillage and fertilization may impact arbuscular mycorrhizal fungal communities, changing species richness and evenness. In agricultural soils there is selection of arbuscular mycorrhizal fungal species with the genus Glomus being most prevalent, most likely based on its ability to sporulate relatively faster than the other genera and survive tillage disturbance and perhaps short plant cycles (Oehl et al. 2009; Voříšková et al. 2016). Furthermore, pesticides used in agriculture may affect the symbiosis (Trappe et al. 1984). There is evidence of reduced spore numbers, species diversity and selection for less effective arbuscular mycorrhizal fungi after long term fertilization using classic techniques based on spore morphologies and enumeration (Johnson 1993; Ortas and Coskan 2016; Oehl et al. 2004; Wang et al. 2011). These seem to be confirmed with molecular techniques (van Geel et al. 2016; Lin et al. 2012; Chen et al. 2014). On the contrary, species richness increased with fertilization in an alpine meadow (Liu et al. 2015) or was not different between organically and conventionally managed forage fields (Schneider et al. 2015). Cross inoculation experiments in soils with different histories of cumulative fertilization have shown that inorganic soluble P fertilization may decrease the infectivity and effectiveness of field arbuscular mycorrhizal fungal communities, with an indication of selection for arbuscular mycorrhizal fungi less sensitive to P (Kahiluoto et al. 2000). Changes in the arbuscular mycorrhizal fungal communities due to fertilization may take long time (Cheng et al. 2013), as the effects of intensive agricultural management may also last for more than 20–25 years (Voříšková et al. 2016; Schneider et al. 2015). Adoption of farm management practices that enhance the functioning of the arbuscular mycorrhizal fungal community indigenous to the soil and on farm inoculum production are the primary options available to row crop farmers for efficiently employing arbuscular mycorrhizal fungi (Douds et al. 2016).

Fertilization with slow release fertilizers, or less soluble forms of P such as organic amendments (manures, composts, biosolids, crop/legume residues) could supply sufficient P for plant growth and satisfactory yields and allow the benefits of mycorrhizae. Cavagnaro (2015) has recently reviewed the compatibility of composts with arbuscular mycorrhizal fungi. Composts seem to have a positive or neutral effect to arbuscular mycorrhizal fungal colonization and in most cases are compatible with arbuscular mycorrhizal fungal inoculum application. The impact of soil application of high rates of organic amendments may be negative and in any case unpredictable (Gosling et al. 2006) but this has not been studied extensively

(Cavagnaro 2015). A study from a long term (19 years) field inorganic fertilization experiment showed that application of organic amendments alone or in combination with inorganic fertilizers together with arbuscular mycorrhizal fungal inoculum had lower levels of arbuscular mycorrhizal colonization, but higher yield and total P acquisition than control and NK treatments and about the same levels as the NPK treatment (Hu et al. 2010). Jensen and Jakobsen (1980) found that farmyard manure and inorganic NPK fertilizer after 10 years of application both reduced colonization at the same low levels. Others have found after 74 years of experimentation that moderate quantities of farm yard manure may be less suppressive than equal amounts of NPK fertilizer (Joner 2000). Legume residues decreased colonization relative to the non-fertilized control, but some residues also decreased plant P uptake and results varied with residue quality (Hasbullah et al. 2011). High organic fertilizer rates may cause a temporary suppression of root colonization and an inhibition of plant growth when combined with arbuscular mycorrhizal fungal inoculum (Zhang et al. 2012). Use of slow instead of quick release fertilizers may increase arbuscular mycorrhizal fungal richness (van Geel et al. 2016), but organic, low P fertilizers seem to be more compatible with arbuscular mycorrhizal fungi than P rich, slow release inorganic fertilizers (Linderman and Davis 2004).

3.2.3 Legacy P and Arbuscular Mycorrhizal Fungi

Given the ability of arbuscular mycorrhizal fungi to mobilize soil P one would expect them to play a role in mobilizing legacy P, however, there is not a great volume of research targeting this. First we should identify the case scenario: A Legacy-P based agriculture in order to achieve satisfactory yields should cover at least part of the P needs by legacy P and for the rest it would either use chemical fertilizers, but in lower quantities than those used today, or no chemical fertilizers, but composts, manures or other organic P sources, or some combination of different forms of P. In any case, we would not expect a change of land management, as from agriculture to grassland.

The issue of legacy P may be examined in systems where soluble P fertilizer application has seized, such as in organic agriculture or in biodynamic systems. However, the potentially enhanced levels of arbuscular mycorrhizal fungi or other organisms may not compensate for decreased yields due to lack of fertilization (Ryan and Ash. 1999). The use of arbuscular mycorrhizal fungi in organic farming has been reviewed by Gosling et al. (2006). Research on previously well fertilized abandoned agricultural land is relevant, but not focused on the issue, since there is a change in land use after abandonment. Spohn et al. (2016), studied changes in soil P fractions and microbial community structure after abandonment of vineyeards in Tokaj, Hungary using a chronosequence of fields covering 200 years of abandonment. The concentration of labile P decreased during the first 50 years after abandonment and then stabilized at a low level and vegetation changed to grassland, shrubby grassland, shrubland to forest, while arbuscular mycorrhizal fungi decreased due to establishment of trees that form ectomycorrhizae. In a study of P fractions in arable or pasture soils not fertilized for at least 5 years, there was very low arbuscular mycorrhizal colonization in soils with the highest P, and this negatively affected the relationship between plant P and soil P (Fransson et al. 2003). In low P Danish soils, omitting P application for 10 years resulted in reduction of H_2SO_4 soluble P by 15–25% compared to the initial 8–12 ppm P levels and increased arbuscular mycorrhizal fungal infection that seemed to compensate for P nutrition (Jensen and Jakobsen 1980). In Bohemia, 25 years after abandonment of an intensively managed land, grassland plant species did not spread in and the arbuscular mycorrhizal fungal community was still different than that of the neighboring grasslands. In addition, the abandoned field arbuscular mycorrhizal fungi were fast colonizers, and had a lower number of propagules relative to that of grassland arbuscular mycorrhizal fungal communities, although operational taxonomic unit richness did not differ (Voříšková et al. 2016).

3.2.4 Plant Breeding for More Efficient Mycorrhizal Response

There is generally a large variation in the extent of plant root colonization among different plant species and within a plant species, among different landraces, lines, cultivars and accessions. This has led to exploring the possibility of breeding for better mycorrhizal response (van de Wiel et al. 2016). However, breeding plants for conventional, high input agriculture could have already made selections on the opposite direction, for plants dependent on fertilizers rather than on mycorrhizae. Such evidence was found for wheat (*Triticum aestivum*) (Hetrick et al. 1992; Zhu et al. 2001), with older landraces found to be more responsive to arbuscular mycorrhizal fungi than modern wheat cultivars. However, this was not the case for onion (Galván et al. 2009), durum wheat (Ellouze et al. 2016), sorghum (Leiser et al. 2016) and maize (Hao et al. 2008).

A strategy for breeding for more effective mycorrhizal symbiosis involves screening for plant genotypes that support the greatest extent of arbuscular mycorrhizal fungal root colonization. However, the level of colonization is not necessarily the best indicator of effectiveness of the symbiosis (Mensah et al. 2015) and very often it does not correlate with extractable soil P (Ryan et al. 2016), shoot dry weight (Baon et al. 1993), plant growth response (leaf tissue NPK) (Ellouse et al. 2016) and grain yield, while it may be highly variable and with low heritability (Leiser et al. 2016). Highly colonized plants may allocate a lot of photosynthetic carbon towards the fungus, and this cost may render the symbiosis ineffective, depending on soil nutrient levels and their availability via the plant uptake pathway. The N level in soil and the shoot N:P ratio should also be considered, since they may affect colonization (Liu et al. 2000). Selection of cultivars and arbuscular mycorrhizal fungi that form mycorrhizae rapidly, so that it would cover early plant P needs that are crucial for the final yield, has been proposed (Singh et al. 2012; Zhu et al. 2001), while high performance arbuscular mycorrhizal fungal isolates increase plant N content (Mensah et al. 2015). Another strategy involves selection of plants that change root traits associated with P uptake upon formation of the symbiosis. There was not much difference in clover (Ryan et al. 2016) and maize lines with different root architecture did not seem to respond differently to arbuscular mycorrhizal fungal inoculation, but there were differences with arbuscular mycorrhizal fungal species used as inoculum (Hao et al. 2008). Selection for better rooting system did not lead to lower plant benefit by arbuscular mycorrhizal fungi in onion (Galván et al. 2009).

Breeding evaluations may involve indigenous arbuscular mycorrhizal fungal communities which may be practical for experimentation and realistic for broad scale application. However, results may vary with different arbuscular mycorrhizal fungi and indigenous arbuscular mycorrhizal fungal communities may not be very effective or at sufficient quantity, compared to applying arbuscular mycorrhizal fungal inocula (Ryan et al. 2016). A rich indigenous arbuscular mycorrhizal fungal community at the experimental area (Leiser et al. 2016) may not be as effective as the best possible cultivar-arbuscular mycorrhizal fungus combination. However, involvement of arbuscular mycorrhizal fungi may require a different kind of soil management that maintains arbuscular mycorrhizal fungal soil inoculum and supports more effective arbuscular mycorrhizal fungal communities. Adoption of farm management practices that enhance the functioning of the arbuscular mycorrhizal fungal community indigenous to the soil and on farm inoculum production are the primary options available to row crop farmers for efficiently employing arbuscular mycorrhizal fungi (Douds et al. 2016). Such practices may be the answer to the problem of lack of yield stability in breeding for more efficient symbiosis, since breeders prefer cultivars that may still be highly productive with or without arbuscular mycorrhizal fungi (Galván et al. 2009; Leiser et al. 2016; Singh et al. 2012).

3.3 Phosphorus Mobilizing Bacteria

3.3.1 Soil Bacteria Mediating Phosphorus Mobilization

Bacteria play a critical role in biogeochemical cycles and are the predominant microorganisms that mobilize native and applied P in soils, as compared to fungi or actinomycetes. Many bacteria genera such as *Alcaligenes, Arthrobacter, Azotobacter, Bradyrhizobium, Bacillus, Burkholderia, Chromobacterium, Enterobacter, Erwinia, Escherichia, Flavobacterium, Micrococcus, Pantoea, Pseudomonas, Salmonella, Serratia, Streptomyces,* and *Thiobacillus* have been isolated and studied for their ability to solubilize and mineralize inorganic and organic P respectively.

Among the soil bacterial communities the most important phosphate mobilizing strains belong to genera *Bacillus* and *Pseudomonas* due to their superior capacity and stability. *Bacillus megaterium, B. firmus, B. circulans, B. coagulans, B. licheniformis, B. subtilis, B. polymyxa, B. sircalmous,* and *Pseudomonas aeruginosa*, *P. chlororaphis*, *P. fluorescens*, *P. liquifaciens*, *P. pickettii*, *P. putida*, *P. rathonis*, *P. savastanoi*, *P. striata*, and *P. stutzeri* could be referred to as the most important strains. In addition to *Bacillus* and *Pseudomonas* other P-solubilizing bacteria include *Nitrobacter*, *Nitrosomonas*, *Rhodococcus*, *Thiobacillus*, *Arthrobacter*, *Serratia*, *Synechococcus*, *Chryseobacterium*, *Gordonia*, *Phyllobacterium*, *Delftia*, *Micrococcus*, *Xanthomonas*, *Enterobacter*, *Pantoea*, and *Klebsiella*. Many studies have been published on plant growth promotion by inoculating P-solubilizing bacteria. Pereira and Castro (2014) screened five bacterial strains for solubilization of phosphate in order to enhance *Zea mays* growth in an agricultural P-deficient soil. The best P-solubilizing strains were *Pseudomonas* sp. EAV and *Arthrobacter nicotinovorans* EAPAA, since both highly increased P availability in soils and promoted maize growth. Recently, Panta et al. (2016) studied the native population of phosphate solubilizing bacteria in the rhizospheres of maize, rice, ginger and large cardamom grown in different regions of Sikkim (India). Among the 26 isolates, *Bacillus*, *Pseudomonas*, *Micrococcus*, *Staphylococcus*, *Microbacterium* and *Delftia* were the main bacteria found, with a phosphate solubilization capacity that varied between 30.2 and 203.7 mg/L.

Numerous reports have shown that $Ca_3(PO_4)_2$ is most easily solubilized followed by $FePO_4$, $AlPO_4$ and rock phosphate. According to Henri et al. (2008), *Pseudomonas fluorescens* can solubilize 100 mg P/L containing $Ca_3(PO_4)_2$ or 92 and 51 mg P/L containing $AlPO_4$ and $FePO_4$, respectively. Strains of *Pseudomonas* and of *Acetobacter diazotrophicus* were found to release 160.5–162.5 and 142–431 mg P/L from tri-calcium phosphate, respectively. Sindhu et al. (2014) reported that *Pseudomonas striata* is more efficient to solubilizing TCP (tri calcium phosphate) than *Bacillus* spp. and *Aspergillus awamorii*. Recently, Ahmed et al. (2016) evaluated the effects of six P mobilizing bacterial strains (*Pantoea ananatis*, *Pantoea agglomerans*, *Pantoea* sp. *Burkholderia* sp.) and three P sources of tricalcium phosphate on yield and P uptake of wheat. They found that all the selected P-mobilizing bacteria increased the grain yield of wheat significantly as these bacteria mobilized insoluble mineral P. Sharon et al. (2016) reported the highest level of phosphate solubilization from the insoluble tri-calcium complex by *Pantoea* sp. Pot1, at a rate of 956 mg/L. Figure 3.3 shows a schematic summary of phosphate solubilization capacity of different bacteria genera.

3.3.2 Mechanisms of P Mobilization by Soil Bacteria

Bacteria are responsible for mobilizing the soil P through multiple mechanisms that are expressed with direct and indirect effects. In direct processes, the solubilization of inorganic and mineralization of organic P are the main mechanisms of P release. The solubilization of inorganic P related to the production and the action of low molecular weight organic acids such as acetic, lactic, oxalic, tartaric, malic, succinic, citric, formic, gluconic, ketogluconic, and glycolic acid (Table 3.2). These acids help in lowering the pH through excretion of H^+ and at the same time, hydroxyl/

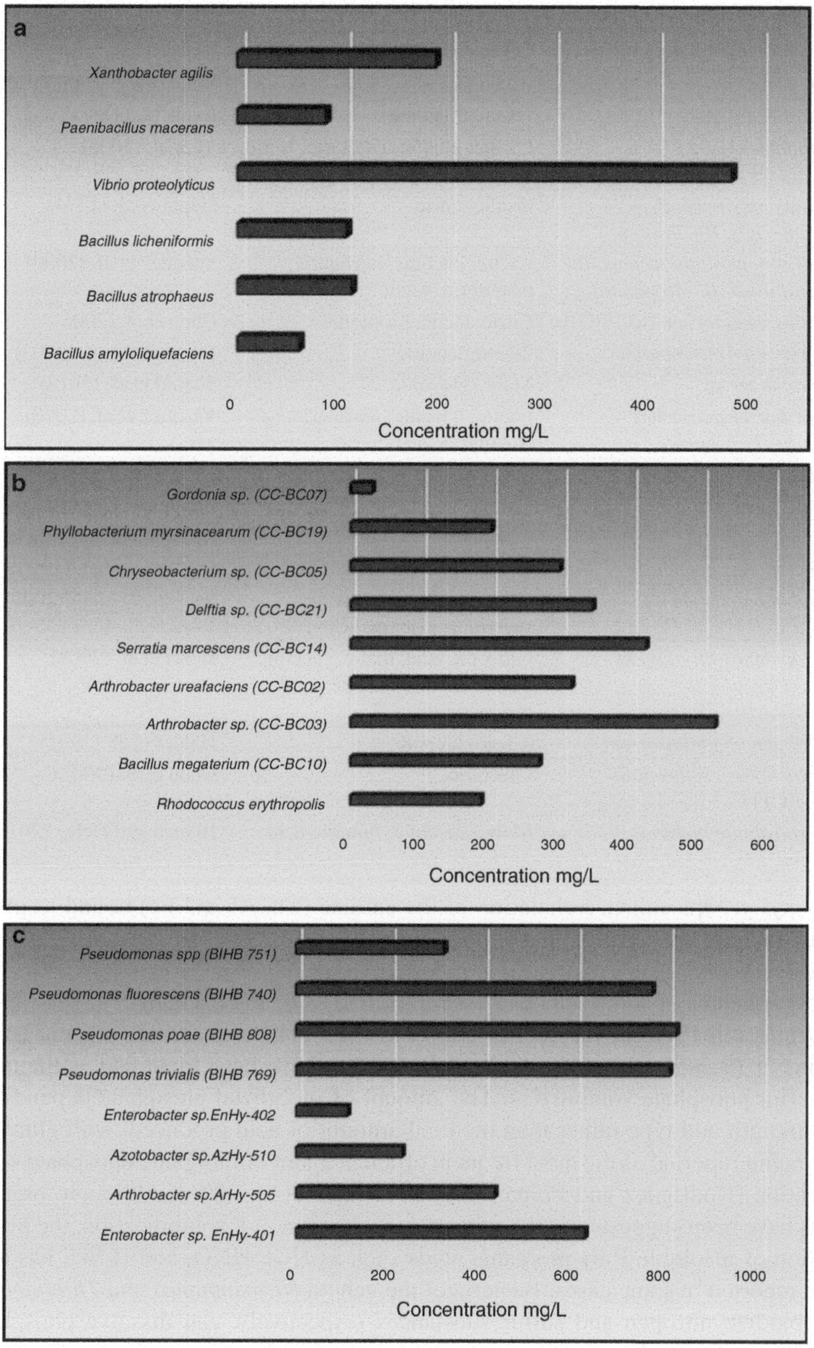

Fig. 3.3 Phosphorus solubilization capacity of bacterial cultures after (**a**) 24 h (**b**) 72 h and (**c**) 120 h of incubation. Data from Khan et al. 2013

Table 3.2 Production of organic acids by phosphate-solubilizing bacteria

Phosphate solubilizing bacteria	Organic acids	References
Acetobacter sp	Gluconic	Galar and Bolardi (1995)
Arthrobacter sp.	Oxalic, malonic	Banik and Dey (1982)
Azotobacter Hy-510	Oxalic, gluconic, tartaric, lactic, succinic, fumaric	Yi et al. (2008)
Bacillus polymyxa, B. licheniformis, Bacillus spp.	Oxalic, citric	Gupta et al. (1994)
Bacillus amyloliquefaciens, B. licheniformis, B. atrophaeus	Lactic, itaconic, isovaleric, isobutyric, acetic	Vazquez et al. (2000)
Bacillus megaterium (CC-BC10)	Citric, lactic, propionic	Chen et al. (2006)
Enterobacter intermedium	2-ketogluconic	Hwangbo et al. (2003)
Enterobacter sp.	Malic, gluconic	Shahid et al. (2012)
Enterobacter aerogenes, E. taylorae, E. asburiae	Lactic, itaconic, isovaleric, isobutyric, acetic	Vazquez et al. (2000)
Delftia (CC-BC21)	Succinic	Chen et al. (2006)
Micrococcus spp	Oxalic	Banik and Dey (1982)
Pantoea eucalypti	Cluconic	Castagno et al. (2011)
Pseudomonas cepacia	Gluconic, 2-ketogluconic	Bar-Yosef et al. (1999)
Pseudomonas poae	Gluconic, 2-ketogluconic, succinic, citric, malic	Vyas and Gulati (2009)
Pseudomonas fluorescens	Gluconic acid, malic, succinic, lactic, fumaric, tartaric, and transaconitic	Henri et al. (2008)
Rhizobium leguminosarum	2-ketogluconic	Halder et al. (1991)
Rhodococcus erythropolis (CC-BC11)	Gluconic	Chen et al. (2006)
Sinorhizobium meliloti	Malic, succinic, fumaric acid	Bianco and Defez (2010)

carboxyl groups enhance chelation of the cations (Ca, Al and Fe) bound to phosphate, thereby converting it to soluble forms.

Moreover, as Zhang et al. (2014) reported, these anions are competing with P for adsorption sites of soil by the process referred to as ligand exchange.

While soil bacteria vary considerably in their ability to secrete organic acids, strains of *Pseudomonas*, *Bacillus* and *Rhizobium* have been identified as the most powerful phosphate solubilizers. The amount of mobilized phosphate depends on the strength and type rather than the total amount of acid produced, with gluconic acid being reported as the most frequent efficient agent of inorganic phosphate solubilization (Rodríguez and Fraga 1999; Rodríguez et al. 2006). Although, organic acids have been suggested as the principal mechanism of P solubilization, the mobilization of insoluble P by inorganic acids such as HCl, HNO_3 and H_2SO_4 has also been reported in some cases. Bacteria of the genera *Nitrosomonas* and *Thiobacillus* that oxidize nitrogen and sulfur substances respectively, can dissolve phosphate compounds by producing nitric and sulphuric acids. However, as Kim et al. (1997) reported, inorganic acids (HCl) are less effective to solubilize P from hydroxyapatite compared to organic acids (citric or oxalic) at the same pH.

Organic P constitutes between 30 and 65% of the total P in soil and is an important source of P for plants. Organic P in soil is largely in the form of phosphate monoesters (inositol phosphate) accounting for up to 50% of the total organic P. To make these organic P compounds available for plant nutrition they must undergo mineralization (hydrolysis). The microbial mineralization of organic P is strongly influenced by pedoenvironmental parameters and conditions favoring nitrogen mineralization also support the mineralization of organic P. In this process, P can be mobilized by means of three groups of enzymes a) phosphatases (e.g. acid and alkaline phosphatases) b) phytases, which cause P release from insoluble phytates and c) phosphonatases and C-P lyases that are released by soil microorganisms (Sindhu et al. 2014). Soil bacteria strains from the genera *Pseudomonas, Bacillus, Rhizobium, Enterobacter, Serratia, Citrobacter, Proteus* and *Klebsiela* produce a range of acid phosphatases that catalyzes the release of available P for plant nutrition from organic P (Hayat et al. 2010). Nevertheless, bacteria with phytase activity have been also isolated from rhizosphere belonging to genera *Tetrathiobacter* and *Bacillus* which promoted the growth of Indian mustard and significantly increased the P content (Kumar et al. 2013). About 30–50% of bacterial isolates from soil synthesize phytase which causes release of P from phytic acid and species of *Arthrobacter*, *Streptomyces* and *Pseudomonas* have been found capable to form this enzyme.

Jorquera et al. (2008) have isolated bacteria with both activities, P solubilization with production of organic acids and mineralization of organic P with production of phytase, from the rhizospheres of wheat, oat, perennial ryegrass, yellow lupin and white clover. According to Tao et al. (2008) the solubilization activity of soil bacteria strains as *Bacillus megaterium*, *Burkholderia caryophylli*, *Pseudomonas cichorii*, and *Pseudomonas syringae* ranges between 25.4–41.7 mg P/L while the mineralization of organic P of bacteria strains as *Bacillus cereus* and *Bacillus megaterium* ranges between 8.2–17.8 mg P/L.

Besides the direct mobilization of inorganic (solubilization) and organic (mineralization) P, there are several ways through which indirect mobilization of P can also occur, (Owen et al. 2015; Rashid et al. 2016) due to

(i) The formation of carbonic acid through the release of CO_2 as result of microbial respiration.
(ii) The nitrogen assimilation, where bacteria release H^+ that cause decrease of soil pH and solubilization of P.
(iii) The reduction of metals bound to phosphate as the result of the redox activity of microorganisms and/or exudates.
(iv) The ability of P solubilizing bacteria to remove and assimilate P from the soil and thus stimulate the indirect dissolution of Ca-phosphates in order to re-establish the P-equilibrium (sink theory).

It is clear that P mobilizing bacteria have the capacity to convert insoluble forms of P (legacy P) into soluble and available P for plants through complex and dynamic mechanisms that must be well understood in order to predict how these bacteria may respond when applied to field conditions.

3.4 Conclusion

The strategy of employing microorganisms for mobilizing legacy soil P has a great volume of research behind it. Microorganisms with such a potential have been isolated and identified and the mechanisms with which they may mobilize legacy P have been studied. In addition, there is an even greater volume of research regarding microorganisms with other useful abilities (Plant Growth Promoting Microorganisms) that may lead to multifunctional inocula (Richardson et al. 2011). Nevertheless, the complexity of the subject is such that there is not much commercialization of biofertilizers yet (Vessey 2003), or for arbuscular mycorrhizal fungi the cost of production is still too high for broad scale agriculture, unless inoculum production technology is improved or there is on farm inoculum production (Douds et al. 2015, 2016). In addition, high variability with different host cultivars and field sites may lead to inconsistent results (Vessey 2003), or positive results may be attributed to random events (Karamanos et al. 2010). The further improvement of our knowledge on the interactions between plants, soil, inocula and indigenous microorganisms, the further exploration of the yet undiscovered wealth of the microbial world and introduction of the concept of soil management with the aim of enhancing effective microbial communities may eventually lead to success in employing microorganisms for mobilizing legacy soil P.

References

Adhya TK, Kumar N, Reddy G, Podile AR, Bee H, Samantaray B (2015) Microbial mobilization of soil phosphorus and sustainable P management in agricultural soils. Curr Sci 108:1280–1287

Ahmed MF, Kennedy IR, Choudhury (2016) Effects of bacterial inoculants and sources of phosphorus on yield and phosphorus uptake of wheat. J Plant Nutr 39:508–522. https://doi.org/10.1080/01904167.2015.1086794

Antunes PM, Schneider K, Hillis D, Klironomos JN (2007) Can the arbuscular mycorrhizal fungus *Glomus intraradices* actively mobilize P from rock phosphates? Pedobiologia 51:281–286. https://doi.org/10.1016/j.pedobi.2007.04.007

Augé RM (2001) Water relations, drought and vesicular-arbuscular mycorrhizal symbiosis. Mycorrhiza 11:3–42. https://doi.org/10.1007/s005720100097

Banik S, Dey BK (1982) Available phosphate content of an alluvial soil as influenced by inoculation of some isolated phosphate solubilizing microorganisms. Plant Soil 69:353–364. https://doi.org/10.1007/BF02372456

Baon JB, Smith SE, Alston AM (1993) Mycorrhizal responses of barley cultivars differing in P efficiency. Plant Soil 157:97–105. https://doi.org/10.1007/BF02390231

Bar-Yosef B, Rogers RD, Wolfam JH, Richman E (1999) *Pseudomonas cepacia* mediated rock phosphate solubilization in kaolinite and montmorillonite suspensions. Soil Sci Soc Am J 63:1703–1708. https://doi.org/10.2136/sssaj1999.6361703x

Barzegar A, Yousefi A, Zargoosh N (2016) Water stress and soil compaction impacts on clover growth and nutrient concentration. Eurasian J Soil Sci 5:139–145. https://doi.org/10.18393/ejss.2016.2.139-145

Bianco C, Defez R (2010) Improvement of phosphate solubilization and Medicago plant yield by an indole-3-acetic acid-overproducing strain of *Sinorhizobium meliloti*. Appl Environ Microbiol 76:4626–4632. https://doi.org/10.1128/AEM.02756-09

Bindraban PS, Dimkpa C, Nagarajan L, Roy A, Rabbinge R (2015) Revisiting fertilisers and fertilization strategies for improved nutrient uptake by plants. Biol Fertil Soils 51(8):897–911

Blal B, Morel C, Gianinazzi-Pearson V, Fardeau JC, Gianinazzi S (1990) Influence of vesicular-arbuscular mycorrhizae on phosphate fertilizer efficiency in two tropical acid soils planted with micropropagated oil palm *(Elaeis guineensis* jacq.). Biol Fertil Soils 9:43–48. https://doi.org/10.1007/BF00335860

Bolan N (1991) A critical review on the role of mycorrhizal fungi in the uptake of phosphorus by plants. Plant Soil 134:189–207. https://doi.org/10.1007/BF00012037

Burghelea C, Zaharescu DG, Dontsova K, Maier R, Huxman T, Chorover J (2015) Mineral nutrient mobilization by plants from rock: influence of rock type and arbuscular mycorrhiza. Biogeochemistry 124:187–203. https://doi.org/10.1007/s10533-015-0092-5

Cardoso IM, Boddington CL, Janssen BH, Oenema O, Kuyper TW (2006) Differential access to phosphorus pools of an oxisol by mycorrhizal and nonmycorrhizal maize. Commun Soil Sci Plant 37:1537–1551. https://doi.org/10.1080/00103620600710074

Castagno LN, Estrella MJ, Sannazzaro AL, Grassano AE, Ruiz OA (2011) Phosphate solubilization mechanism and in vitro plant growth promotion activity mediated by *Pantoea eucalypti* isolated from Lotus tenuis rhizosphere in the Salado river basin (Argentina). J Appl Microbiol 110:1151–1165. https://doi.org/10.1111/j.1365-2672.2011.04968.x

Cavagnaro TR (2015) Biologically regulated nutrient supply systems: compost and arbuscular mycorrhizas-a review. Adv Agron 129:293–321. https://doi.org/10.1016/bs.agron.2014.09.005

Chen YP, Rekha PD, Arun AB, Shen FT, Lai WA, Young CC (2006) Phosphate solubilizing bacteria from subtropical soil and their tricalcium phosphate solubilizing abilities. Appl Soil Ecol 34:33–41. https://doi.org/10.1016/j.apsoil.2005.12.002

Chen YL, Zhang X, Ye JS, Han HY, Wan SQ, Chen B (2014) Six year fertilization modifies the biodiversity of arbuscular mycorrhizal fungi in a temperate steppe in Inner Mongolia. Soil Biol Biochem 69:371–381. https://doi.org/10.1016/j.soilbio.2013.11.020

Cheng Y, Ishimoto K, Kuriyama Y, Osaki M, Ezawa T (2013) Ninety-year-, but not single, application of phosphorus fertilizer has a major impact on arbuscular mycorrhizal fungal communities. Plant Soil 365:397–407. https://doi.org/10.1007/s11104-012-1398-x

Cordell D, Drangert JO, White S (2009) The story of phosphorus: global food security and food for thought. Glob Environ Chang 19:292–305. https://doi.org/10.1016/j.gloenvcha.2008.10.009

Cordell D, Neset TSS, Prior T (2012) The phosphorus mass balance: identifying 'hotspots' in the food system as a roadmap to phosphorus security. Curr Opin Biotechnol 23:839–845. https://doi.org/10.1016/j.copbio.2012.03.010

Dodds WK, Bouska WW, Eitzmann JL, Pilger TJ, Pitts KL, Riley AJ, Schloesser JT, Thornbrugh DJ (2008) Eutrophication of US freshwaters: analysis of potential economic damages. Environ Sci Technol 43:12–19. https://doi.org/10.1021/es801217q

Douds DD Jr, Lee J, Shenk JE, Ganser S (2015) Inoculation of sweet potatoes with AM fungi produced on-farm increases yield in high P soil. J Appl Hortic 17:171–175

Douds DD Jr, Leea J, McKeevera L, Ziegler-Ulshb C, Ganserca S (2016) Utilization of inoculum of AM fungi produced on-farm increases the yield of *Solanum lycopersicum*: a summary of 7 years of field trials ona conventional vegetable farm with high soil phosphorus. Sci Hort Amsterdam 207:89–96. https://doi.org/10.1016/j.scienta.2016.05.013

ECSCU (2013) Science for environment policy indepth report: sustainable phosphorus use. Report produced for the European Commission DG Environment. University of the West of England, Bristol

Elbon A, Whalen JK (2015) Phosphorus supply to vegetable crops from arbuscular mycorrhizal fungi: a review. Biol Agric Hortic 31:73–90. https://doi.org/10.1080/01448765.2014.966147

Ellouze W, Hamel C, DePauw RM, Knox RE, Cuthbert RD, Singh AK (2016) Potential to breed for mycorrhizal association in durum wheat. Can J Microbiol 62:263–271. https://doi.org/10.1139/cjm-2014-0598

Elser JJ (2012) Phosphorus: a limiting nutrient for humanity? Curr Opin Biotechnol 23:833–838. https://doi.org/10.1016/j.copbio.2012.03.001

Evelin H, Kapoor R, Giri B (2009) Arbuscular mycorrhizal fungi in alleviation of salt stress: a review. Ann Bot 104:1263–1280. https://doi.org/10.1093/aob/mcp251

FAO (2015). World fertilizer trends and outlook to 2018. 66

Fidelibus MW, Martin CA, Stutz JC (2001) Geographic isolates of *Glomus* increase root growth and whole-plant transpiration of Citrus seedlings grown with high phosphorus. Mycorrhiza 10:231–236. https://doi.org/10.1007/s005720000084

Fransson AM, van Aarle IM, Olsson PA, Tyler G (2003) *Plantago lanceolata* L. and *Rumex acetosella* L. differ in their utilisation of soil phosphorus fractions. Plant Soil 248:285–295. https://doi.org/10.1023/A:1022331409517

Galar ML, Bolardi JL (1995) Evidence for a membrane bound pyrrolquinoline quinine-linked glucose dehydrogenase in *Acetobacter diazotrophicus*. Appl Environ Microbiol 43:713–716. https://doi.org/10.1007/BF00164778

Galván GA, Parádi I, Burger K, Baar J, Kuyper TW, Scholten OE, Kik C (2009) Molecular diversity of arbuscular mycorrhizal fungi in onion roots from organic and conventional farming systems in the Netherlands. Mycorrhiza 19:317–328. https://doi.org/10.1007/s00572-009-0237-2

Gasparatos D, Haidouti C, Haroulis A, Tsaousidou P (2006) Estimation of phosphorus status of soil Fe-enriched concretions with the acid ammonium oxalate method. Commun Soil Sci Plant 37:2375–2387. https://doi.org/10.1080/00103620600819891

Gerretsen FC (1948) The influence of microorganisms on the phosphate intake by the plant. Plant Soil 1:51–81. https://doi.org/10.1007/BF02080606

Gosling P, Hodge A, Goodlass G, Bending GD (2006) Arbuscular mychorrhizal fungi and organic farming. Agric Ecosyst Environ 113:17–35. https://doi.org/10.1016/j.agee.2005.09.009

Gryndler H, Leština J, Moravec V, Přikryl Z, Lipavsky J (1989) Colonization of maize roots by VAM-fungi under conditions of long-term fertilization of varying intensity. Agric Ecosyst Environ 29:183–186. https://doi.org/10.1016/0167-8809(90)90274-H

Gupta RR, Singal R, Shanker A, Kuhad RC, Saxena RK (1994) A modified plate assay for secreening phosphate solubilizing microorganisms. J Gen Appl Microbiol 40:255–260. https://doi.org/10.2323/jgam.40.255

Halder AK, Mishra AK, Chakrabarty PK (1991) Solubilizing of inorganic phosphates by *Bradyrhizobium*. Indian J Exp Biol 29:28–31. https://doi.org/10.1007/BF02898602

Hao L, Zhang J, Chen F, Christie P, Li X (2008) Response of two maize inbred lines with contrasting phosphorus efficiency and root morphology to mycorrhizal colonization at different soil phosphorus supply levels. J Plant Nutr 31:1059–1073. https://doi.org/10.1080/01904160802115227

Hasbullah, Marschner P, McNeill A (2011) Legume residue influence arbuscular mycorrhizal colonisation and P uptake by wheat. Biol Fertil Soils 47:701–707. https://doi.org/10.1007/s00374-011-0581-1

Hayat R, Ali S, Amara U, Khalid R, Ahmed I (2010) Soil beneficial bacteria and their role in plant growth promotion: a review. Ann Microbiol 60:579–598. https://doi.org/10.1007/s13213-010-0117-1

Hayman DS, Barea JM, Azcon R (1976) Vesicular-arbuscular mycorrhiza in southern Spain: its distribution in crops growing in soil of different fertility. Phytopathol Mediterr 15:1–6

Heckenmüller M, Narita D, Klepper G (2014) Global availability of phosphorus and its implications for global food supply: an economic overview. Kiel Working paper no. 1897, Kiel Institute for the World Economy, Kiel, 26 p

Henri F, Laurette NN, Annette D, John Q, Wolfgang M, Francois-Xavier E, Dieudonne N (2008) Solubilization of inorganic phosphates and plant growth promotion by strains of *Pseudomonas fluorescens* isolated from acidic soils of Cameroon. Afr J Microbiol Res 2:171–178

Hetrick BAD, Wilson GWT, Cox TS (1992) Mycorrhizal dependence of modern wheat varieties, landraces and ancestors. Can J Bot 70:2032–2040

Hetrick BAD, Hartnett DC, Wilson GWT, Gibson DJ (1994) Effects of mycorrhizae, phosphorus availability, and plant density on yield relationships among competing tallgrass prairie grasses. Can J Bot 72:168–176. https://doi.org/10.1139/b94-023

Hildebrandt U, Regvar M, Bothe H (2007) Arbuscular mycorrhiza and heavy metal tolerance. Phytochemistry 68:139–146. https://doi.org/10.1139/b92-253

Hu J, Lin X, Wang J, Cui X, Dai J, Chu H, Zhang J (2010) Arbuscular mycorrhizal fungus enhances P acquisition of wheat (*Triticum aestivum* L.) in a sandy loam soil with long-term inorganic fertilization regime. Appl Microbiol Biotechnol 88:781–787. https://doi.org/10.1007/s00253-010-2791-0

Hwangbo H, Park RD, Kim YW, Rim YS, Park KH, Kim TH, Such JS, Kim KY (2003) 2-ketogluconic acid production and phosphate solubilization by *Enterobacter intermedium*. Curr Microbiol 47:87–92. https://doi.org/10.1007/s00284-002-3951-y

Janos DP (2007) Plant responsiveness to mycorrhizas differs from dependence upon mycorrhizas. Mycorrhiza 17:75–91. https://doi.org/10.1007/s00572-006-0094-1

Jensen A, Jakobsen I (1980) The occurrence of vesicular-arbuscular mycorrhiza in barley and wheat grown in some Danish soils with different fertilizer treatments. Plant Soil 55:403–414. https://doi.org/10.1007/BF02182701

Johnson NC (1993) Can fertilization of soil select less mutualistic mycorrhizae? Ecol Appl 3:749–757. https://doi.org/10.2307/1942106

Joner EJ (2000) The effect of long-term fertilization with organic or inorganic fertilizers on mycorrhiza-mediated phosphorus uptake in subterranean clover. Biol Fertil Soils 32:435–440. https://doi.org/10.1007/s003740000279

Jorquera MA, Hernández MT, Rengel Z, Marschner P, Mora ML (2008) Isolation of culturable phosphobacteria with both phytate-mineralization and phosphate-solubilization activity from the rhizosphere of plants grown in a volcanic soil. Biol Fertil Soils 44:1025–1034. https://doi.org/10.1007/s00374-008-0288-0

Kahiluoto H, Ketoja E, Vestberg M (2000) Promotion of utilization of arbuscular mycorrhiza through reduced P fertilization 1. Bioassays in a growth chamber. Plant Soil 227:191–206. https://doi.org/10.1023/A:1026555717663

Karagiannidis N, Bletsos F, Stavropoulos N (2002) Effect of *Verticilliym* wilt (*Verticillium dahliae* Kleb.) and mycorrhiza (*Glomus mosseae*) on root colonization, growth and nutrient uptake in tomato and eggplant seedlings. Sci Hortic-Amsterdam 94:145–156. https://doi.org/10.1016/S0304-4238(01)00336-3

Karamanos RE, Flore NA, Harapiak JT (2010) Re-visiting use of *Penicillium bilaii* with phosphorus fertilization of hard red spring wheat. Can J Plant Sci 90:265–277. https://doi.org/10.4141/CJPS09123

Karamesouti M, Gasparatos D (2017) Sustainable management of soil phosphorus in a changing world. In: Rakshit A, Abhilash P, Singh H, Ghosh S (eds) Adaptive soil management : from theory to practices. Springer, Singapore, pp 189–214 doi.org/10.1007/978-981-10-3638-5_9

Khan MS, Zaidi A, Wani PA (2007) Role of phosphate-solubilizing microorganisms in sustainable agriculture - a review. Agron Sustain Dev 27:29–43. https://doi.org/10.1051/agro:2006011

Khan MS, Ahmad E, Zaidi A, Oves M (2013) Functional aspect of phosphate-solubilizing bacteria: importance in crop production. In: Maheshwari DK et al (eds) Bacteria in agrobiology: crop productivity. Springer, Berlin, pp 237–265. https://doi.org/10.1007/978-3-642-37241-4_10

Kim KY, McDonald GA, Jordan D (1997) Solubilization of hydroxyapatite by *Enterobacter agglomerans* and cloned *Escherichia coli* in culture medium. Biol Fertil Soils 24:347–352. https://doi.org/10.1007/s003740050256

Koele N, Dickie IA, Blum JD, Gleason JD, de Graaf L (2014) Ecological significance of mineral weathering in ectomycorrhizal andarbuscular mycorrhizal ecosystems from a field-based comparison. Soil Biol Biochem 69:63–70. https://doi.org/10.1016/j.soilbio.2013.10.041

Kumar V, Singh P, Jorquera MA (2013) Isolation of phytase producing bacteria from Himalayan soils and their effect on growth and phosphorus uptake of Indian mustard (*Brassica juncea*). World J Microbiol Biotechnol 29:1361–1369. https://doi.org/10.1007/s11274-013-1299-z

Lajtha K, Schlesinger WH (1988) The biogeochemistry of phosphorus cycling and phosphorus availability along a desert soil chronosequence. Ecology 69:24–39. https://doi.org/10.2307/1943157

Leiser WL, Olatoye MO, Rattunde FWR, Neumann G, Weltzien E, Haussmann BIG (2016) No need to breed for enhanced colonization by arbuscular mycorrhizal fungi to improve low-P adaptation of West African sorghums. Plant Soil 401:51–64. https://doi.org/10.1007/s11104-015-2437-1

Li XL, George E, Marschner H (1991) Extension of the phosphorus depletion zone in VA mycorrhizal white clover in a calcareous soil. Plant Soil 136:41–48

Lin ML, Fox RL (1992) The comparative agronomic effectiveness of rock phosphate and superphosphate for banana. Fert Res 31:131–135. https://doi.org/10.1007/BF01063287

Lin XG, Feng YZ, Zhang HY, Chen RR, Wang JH, Zhang JB, Chu HY (2012) Long-term balanced fertilization decreases arbuscular mycorrhizal fungal diversity in an arable soil in North China revealed by 454 pyrosequencing. Environ Sci Technol 46:5764–5771. https://doi.org/10.1021/es3001695

Linderman RG, Davis EA (2004) Evaluation of commercial inorganic and organic fertilizer effects on arbuscular mycorrhizae formed by *Glomus intraradices*. Hortechnology 14:196–202

Liu A, Hamel C, Hamilton RI, Smith DL (2000) Mycorrhizae formation and nutrient uptake of new corn (*Zea mays* L.) hybrids with extreme canopy and leaf architecture as influenced by soil N and P levels. Plant Soil 221:157–166. https://doi.org/10.1023/A:1004777821422

Liu Y, Johnson NC, Mao L, Shi G, Jiang S, Ma X, Du G, An L, Feng H (2015) Phylogenetic structure of arbuscular mycorrhizal community shifts in response to increasing soil fertility. Soil Biol Biochem 89:196–205. https://doi.org/10.1016/j.soilbio.2015.07.007

Marschner H (1995) Mineral nutrition of higher plants, 2nd edn. Academic, London

Mclaughlin MJ, Alston AM, Martin JK (1988) Phosphorus cycling in wheat pasture rotations. II. The role of the microbial biomass in phosphorus cycling. Aust J Soil Res 26:333–342. https://doi.org/10.1071/SR9880333

McLaughlin MJ, McBeath TM, Smernik R, Stacey SP, Ajiboye B, Guppy C (2011) The chemical nature of P accumulation in agricultural soils—implications for fertiliser management and design: an Australian perspective. Plant Soil 349:69–87. https://doi.org/10.1007/s11104-011-0907-7

Medina OA, Kretschmer AE Jr, Sylvia DM (1990) Growth response of field-grown Siratro (*Macroptilium atropurpureum* Urb.) and *Aeschynomene americana* L. to inoculation with selected vesicular-arbuscular mycorrhizal fungi. Biol Fertil Soils 9:54–60. https://doi.org/10.1007/BF00335862

Menge J, Steirle D, Bagyaraj D, Johnson E, Leonard R (1978) Phosphorus concentrations in plants responsible for inhibition of mycorrhizal infection. New Phytol 80:575–578. https://doi.org/10.1111/j.1469-8137.1978.tb01589.x

Mensah JA, Koch AM, Antunes PM, Kiers ET, Hart M, Bücking H (2015) High functional diversity within species of arbuscular mycorrhizal fungi is associated with differences in phosphate and nitrogen uptake and fungal phosphate metabolism. Mycorrhiza 25:533–546. https://doi.org/10.1007/s00572-015-0631-x

Miller RM, Reinhardt DR, Jastrow JD (1995) External hyphal production of vesicular-arbuscular mycorrhizal fungi in pasture and tallgrass prairie communities. Oecologia 103:17–23. https://doi.org/10.1007/BF00328420

Modjo HS, Hendrix JW (1986) The mycorrhizal fungus *Glomus macrocarpus* as a cause of tobacco stunt disease. Phytopathology 76:688–691

Oehl F, Sieverding E, Mäder P, Dubois D, Ineichen K, Boller T, Wiemken A (2004) Impact of long-term conventional and organic farming on the diversity of arbuscular mycorrhizal fungi. Oecologia 138:574–593. https://doi.org/10.1007/s00442-003-1458-2

Oehl F, Sieverding E, Ineichen K, Mäder P, Wiemken A, Boller T (2009) Distinct sporulation dynamics of arbuscular mycorrhizal fungal communities from different agroecosystems in long-term microcosms. Agric Ecosyst Environ 134:257–268. https://doi.org/10.1016/j.agee.2009.07.008

Ortas I, Coskan A (2016) Precipitation as the most affecting factor on soil–plant environment conditions affects the mycorrhizal spore numbers in three different ecological zones in Turkey. Acta Agr Scand B–S P 66:369–378. https://doi.org/10.1080/09064710.2015.1132005

Owen D, Williams AP, Griffith GW, Withers PJA (2015) Use of commercial bio-inoculants to increase agricultural production through improved phosphorus acquisition. Appl Soil Ecol 86:41–54. https://doi.org/10.1016/j.apsoil.2014.09.012

Panta B, Rahman H, Panda J (2016) Phosphate solubilizing bacteria from the acidic soils of Eastern Himalayan region and their antagonistic effect on fungal pathogens. Rhizosphere 2:62–71. https://doi.org/10.1016/j.rhisph.2016.08.001

Pereira SI, Castro PM (2014) Phosphate-solubilizing rhizobacteria enhance *Zea mays* growth in agricultural P-deficient soils. Ecol Eng 73:526–535. https://doi.org/10.1016/j.ecoleng.2014.09.060

Ping X, Lin-Zhou L, Xiao-Ying D, Jing X, Pei-Kun J, Ren-Fang S (2014) Response of soil phosphorus required for maximum growth of *Asparagus officinalis* L. to inoculation of arbuscular mycorrhizal fungi. Pedosphere 24:776–782

Pozo MJ, Azcón-Aguilar C (2007) Unraveling mycorrhiza-induced resistance. Curr Opin Plant Biol 10:393–398. https://doi.org/10.1016/S1002-0160(14)60064-3

Puente ME, Li CY, Bashan Y (2009) Endophytic bacteria in cacti seeds can improve the development of cactus seedlings. Environ Exp Bot 66:402–408. https://doi.org/10.1016/j.envexpbot.2009.04.007

Rashid MI, Mujawar LH, Shahzad T, Almeelbi T, Ismail IM, Oves M (2016) Bacteria and fungi can contribute to nutrients bioavailability and aggregate formation in degraded soils. Microbiol Res 183:26–41. https://doi.org/10.1016/j.micres.2015.11.007

Richardson AE (2007). "Making microorganisms mobilize soil phosphorus". First international meeting on microbial phosphate solubilization. Springer Netherlands, 2007. doi: https://doi.org/10.1007/978-1-4020-5765-6_10

Richardson AE, Simpson RJ (2011) Soil microorganisms mediating phosphorus availability update on microbial phosphorus. Plant Physiol 156:989–996. https://doi.org/10.1104/pp.111.175448

Richardson AE, Lynch JP, Ryan PR, Delhaize E, Smith FA, Smith SE, Harvey PR, Ryan MH, Veneklaas EJ, Lambers H, Oberson A, Culvenor RA, Simpson RJ (2011) Plant and microbial strategies to improve the phosphorus efficiency of agriculture. Plant Soil 349:121–156. https://doi.org/10.1007/s11104-011-0950-4

Rillig M, Mummey D (2006) Mycorrhizas and soil structure. New Phytol 171:41–53. https://doi.org/10.1111/j.1469-8137.2006.01750.x

Rodríguez H, Fraga R (1999) Phosphate solubilizing bacteria and their role in plant growth promotion. Biotechnol Adv 17:319–339. https://doi.org/10.1016/S0734-9750(99)00014-2

Rodríguez H, Fraga R, Gonzalez T, Bashan Y (2006) Genetics of phosphate solubilization and its potential applications for improving plant growth-promoting bacteria. Plant Soil 287:15–21. https://doi.org/10.1007/978-1-4020-5765-6_2

Rowe H, Withers P, Baas P, Chan N, Doody D, Holiman J, Jacobs B, Li H, MacDonald G, McDowell R, Sharpley A, Shen J, Taheri W, Wallenstein M, Weintraub MN (2016) Integrating legacy soil phosphorus into sustainable nutrient management strategies for future food, bioenergy and water security. Nutr Cycl Agroecosyst 104:393–412. https://doi.org/10.1007/s10705-015-9726-1

Ryan MH, Ash J (1999) Effects of phosphorus and nitrogen on growth of pasture plants and VAM fungi in SE Australian soils with contrasting fertilizer histories (conventional and biodynamic). Agric Ecosyst Environ 73:51–62. https://doi.org/10.1016/S0167-8809(99)00014-6

Ryan MH, Kidd DR, Sandral GA, Yang Z, Lambers H, Culvenor RA, Stefanski A, Nichols PGH, Haling RE, Simpson RJ (2016) High variation in the percentage of root length colonised by arbuscular mycorrhizal fungi among 139 lines representing the species subterranean clover (*Trifolium subterraneum*). Appl Soil Ecol 98:221–232. https://doi.org/10.1016/j.apsoil.2015.10.019

Sattari SZ, Bouwman AF, Giller KE, van Ittersum MK (2012) Residual soil phosphorus as the missing piece in the global phosphorus crisis puzzle. P Natl Acad Sci USA 109:6348–6353

Schneider KD, Lynch DH, Dunfield K, Khosla K, Jansa J, Voroney RP (2015) Farm system management affects community structure of arbuscular mycorrhizal fungi. Appl Soil Ecol 96:192–200. https://doi.org/10.1016/j.apsoil.2015.07.015

Schröder JJ, Smit AL, Cordell D, Rosemarin A (2011) Improved phosphorus use efficiency in agriculture: a key requirement for its sustainable use. Chemosphere 84:822–831. https://doi.org/10.1016/j.chemosphere.2011.01.065

Schroeder MS, Janos DP (2005) Plant growth, phosphorus nutrition, and root morphological responses to arbuscular mycorrhizas, phosphorus fertilization, and intraspecific density. Mycorrhiza 15:203–216. https://doi.org/10.1007/s00572-004-0324-3

Shahid M, Hameed S, Imran A, Ali S, van Elsas JD (2012) Root colonization and growth promotion of sunflower (*Helianthus annuus* L.) by phosphate solubilizing *Enterobacter* sp. Fs-11. World J Microbiol Biotechnol 28:2749–2758. https://doi.org/10.1007/s11274-012-1086-2

Sharma SB, Sayyed RZ, Trivedi MH, Gobi TA (2013) Phosphate solubilizing microbes: sustainable approach for managing phosphorus deficiency in agricultural soils. Springerplus 2:587. https://doi.org/10.1186/2193-1801-2-587

Sharon JA, Hathwaik LT, Glenn GM, Imam SH, Lee CC (2016) Isolation of efficient phosphate solubilizing bacteria capable of enhancing tomato plant growth. J Soil Sci Plant Nutr 16:525–536. https://doi.org/10.4067/S0718-95162016005000043

Sharpley A, Jarvie HP, Buda A, May L, Spears B, Kleinman P (2013) Phosphorus legacy: overcoming the effects of past management practices to mitigate future water quality impairment. J Environ Qual 42:1308–1326. https://doi.org/10.2134/jeq2013.03.0098

Shen J, Yuan L, Zhang J, Li H, Bai Z, Chen X, Zhang W, Zhang F (2011) Phosphorus dynamics: from soil to plant. Plant Physiol 156:997–1005

Sindhu SS, Phour M, Choudhary SR, Chaudhary D (2014) Phosphorus cycling: prospects of using rhizosphere microorganisms for improving phosphorus nutrition of plants. Geomicrobiol Biogeochem 39:199–237. https://doi.org/10.1007/978-3-642-41837-2_11

Singh AK, Hamel C, DePauw RM, Knox RE (2012) Genetic variability in arbuscular mycorrhizal fungi compatibility supports the selection of durum wheat genotypes for enhancing soil ecological services and cropping systems in Canada. Can J Microbiol 58:293–302. https://doi.org/10.1139/w11-140

Smith SE, Read DJ (2008) Mycorrhizal Symbiosis, 3rd edn. Academic, Amsterdam

Solomon D, Lehmann J (2000) Loss of phosphorus from soil in semiarid northern Tanzania as a result of cropping: evidence from sequential extraction and 31P NMR spectroscopy. Eur J Soil Sci 51:699–708. https://doi.org/10.1111/j.1365-2389.2000.00326.x

Spohn M, Novák TJ, Incze J, Giani L (2016) Dynamics of soil carbon, nitrogen, and phosphorus in calcareous soils after land-use abandonment – A chronosequence study. Plant Soil 401:185–196. https://doi.org/10.1007/s11104-015-2513-6

Stewart L, Hamel C, Hogue R, Moutoglis P (2005) Response of strawberry to inoculation with arbuscular mycorrhizal fungi under very high soil phosphorus conditions. Mycorrhiza 15:612–619. https://doi.org/10.1007/s00572-005-0003-z

Sylvia DM, Neal LH (1990) Nitrogen affects the phosphorus response of VA mycorrhiza. New Phytol 115:303–310. https://doi.org/10.1111/j.1469-8137.1990.tb00456.x

Sylvia DM, Hammond LC, Bennett JM, Haas JH, Linda SB (1993) Field response of maize to a VAM fungus and water management. Agron J 85:193–198. https://doi.org/10.2134/agronj1993.00021962008500020006x

Tao GC, Tian SJ, Cai MY, Xie GH (2008) Phosphate-solubilizing and - mineralizing abilities of bacteria isolated from soils. Pedosphere 18:515–523. https://doi.org/10.1016/S1002-0160(08)60042-9

Trappe JM, Molina R, Castellano M (1984) Reactions of mycorrhizal fungi and mycorrhizal formation to pesticides. Annu Rev Phytopathol 22:331–359

van de Wiel CCM, van der Linden CG, Scholten OE (2016) Improving phosphorus use efficiency in agriculture: opportunities for breeding. Euphytica 207:1–22. https://doi.org/10.1007/s10681-015-1572-3

van Geel M, de Beenhouwer M, Ceulemans T, Caes K, Ceustermans A, Bylemans D, Gomand A, Lievens B, Honnay O (2016) Application of slow-release phosphorus fertilizers increases arbuscular mycorrhizal fungal diversity in the roots of apple trees. Plant Soil 402:291–301. https://doi.org/10.1007/s11104-015-2777-x

van Schöll L, Kuyper TW, Smits MM, Landeweert R, Hoffland E, Breemen N (2008) Rock-eating mycorrhizas: their role in plant nutrition and biogeochemical cycles. Plant Soil 303:35–47. https://doi.org/10.1007/s11104-007-9513-0

Vazquez P, Holguin G, Puente ME, Lopez-Cortes A, Bashan Y (2000) Phosphate-solubilizing microorganisms associated with the rhizosphere of mangroves in a semiarid coastal lagoon. Biol Fertil Soils 30:460–468. https://doi.org/10.1007/s003740050024

Vessey JK (2003) Plant growth promoting rhizobacteria as biofertilizers. Plant Soil 255:571–586. https://doi.org/10.1023/A:1026037216893

Voříšková A, Janoušková M, Slavíková R, Pánková H, Daniel O, Vazačová K, Rydlová J, Vosátka M, Münzbergová Z (2016) Effect of past agricultural use on the infectivity and composition of a community of arbuscular mycorrhizal fungi. Agric Ecosyst Environ 221:28–39. https://doi.org/10.1016/j.agee.2016.01.012

Vyas P, Gulati A (2009) Organic acid production in vitro and plant growth promotion in maize under controlled environment by phosphate-solubilizing fluorescent pseudomonas. BMC Microbiol 9:1–15. https://doi.org/10.1186/1471-2180-9-174

Wang FY, Hu JL, Lin XG, Qin SW, Wang JH (2011) Arbuscular mycorrhizal fungal community structure and diversity in response to long-term fertilization: a field case from China. World J Microbiol Biotechnol 27:67–74. https://doi.org/10.1007/s11274-010-0427-2

Weikard HP, Seyhan D (2009) Distribution of phosphorus resources between rich and poor countries: the effect of recycling. Ecol Econ 68:1749–1755. https://doi.org/10.1016/j.ecolecon.2008.11.006

Withers PJ, Sylvester-Bradley R, Jones DL, Healey JR, Talboys PJ (2014) Feed the crop not the soil: rethinking phosphorus management in the food chain. Environ Sci Technol 48(12):6523–6530

Yan Z, Liu P, Li Y, Ma L, Alva A, Dou Z, Chen Q, Zhang F (2013) Phosphorus in China's intensive vegetable production systems: Overfertilization, soil enrichment, and environmental implications. J Environ Qual 42:982–989. https://doi.org/10.2134/jeq2012.0463

Yi Y, Huang W, Ge Y (2008) Exo-polysaccharide: a novel important factor in the microbial dissolution of tricalcium phosphate. World J Microbiol Biotechnol 24:1059–1065. https://doi.org/10.1007/s11274-007-9575-4

Yu K, Chen X, Pan G, Zhang X, Chen C (2016) Dynamics of soil available phosphorus and its impact factors under simulated climate change in typical farmland of Taihu Lake region, China. Environ Monit Assess 188:1–8. https://doi.org/10.1007/s10661-015-5087-0

Zhang GY, Ran W, Zhang LP, Huang QW, Wie MF, Fan QL, Liu Z, Shen QR, Xu GH (2012) Effect of *Gomus mosseae* on maize growth at different organic fertilizer application rates. J Plant Nutr 35:165–175. https://doi.org/10.1080/01904167.2012.636120

Zhang L, Fan J, Ding X, He X, Zhang F, Feng G (2014) Hyphosphere interactions between an arbuscular mycorrhizal fungus and a phosphate solubilizing bacterium promote phytate mineralization in soil. Soil Biol Biochem 74:177–183. https://doi.org/10.1016/j.soilbio.2014.03.004

Zhu YG, Smith SE, Barritt AR, Smith FA (2001) Phosphorus (P) efficiencies and mycorrhizal responsiveness of old and modern wheat cultivars. Plant Soil 237:249–255. https://doi.org/10.1023/A:1013343811110

Chapter 4
New Insights into the Yields of Underexploited Grain Legume Species

C. Cernay, D. Makowski, and E. Pelzer

Abstract Protein-rich grain legumes are grown for both human food and animal feed, and their multiple benefits to the environment. Pea (*Pisum sativum*) is the most widely cultivated grain legume in Europe. In the world, several field experiments have compared the yields of a broad range of grain legumes in contrasting environments, but these experiments have never been synthesized. We address two questions: 1) What is the yield levels of pea compared to other grain legume species in Europe? 2) Which grain legume species with good yield performances in North America and Oceania are candidates for future European experiments? We conducted a statistical analysis of five variables – grain yield, total aerial biomass, grain crude protein, grain gross energy, grain nitrogen content – comparing 22 grain legume species with pea, based on experimental data extracted from 61 peer-reviewed articles and nutritional data. We identify soybean (*Glycine max*), narrowleaf lupin (*Lupinus angustifolius*), and faba bean (*Vicia faba*), as alternative grain legumes to pea in Europe. Grain legume species grown in North America do not outperform pea for most of the criteria. In Oceania, faba bean has significantly higher yield than pea, whereas several species do not differ significantly from pea. Based on data collected in North America and Oceania, we suggest assessing the relative productivities of several vetches and lupins (*Lathyrus*, *Lupinus*, and *Vicia* species) in European experiments. Our findings reveal new insights into the yields of as yet underexploited grain legume species for potential future use in Europe.

Keywords Grain legume species · Comparison · Yield · Meta-analysis

C. Cernay · D. Makowski · E. Pelzer (✉)
UMR Agronomie, INRA, AgroParisTech, Université Paris-Saclay, Thiverval-Grignon, France
e-mail: Elise.Pelzer@inra.fr

E. Lichtfouse (ed.), *Sustainable Agriculture Reviews 32*, Sustainable Agriculture Reviews 32, https://doi.org/10.1007/978-3-319-98914-3_4

4.1 Introduction

Legumes are often cultivated for the grains as a source of protein and energy for human and animal consumption, and may indeed represent a nutritious complement to cereal grains for human diet and animal feed (Duranti 2006; Vaz Patto et al. 2015; Temba et al. 2016). It is assumed that legumes provide environmental benefits and potential economic returns (Graham and Vance 2003; Sinclair and Vadez 2012; Jensen et al. 2012; Reckling et al. 2016). Legume crops are able to fix atmospheric nitrogen and generally require no nitrogen input (Graham and Vance 2003; Sinclair and Vadez 2012). Grain legume cultivation may also enhance soil nitrogen supply through mineralization (Chalk 1998; Evans et al. 2001; Peoples et al. 2009), and increase the yields of subsequent cereals and oilseeds in the crop sequence (Jensen et al. 2004; Williams et al. 2014; Preissel et al. 2015).

From 2004 to 2014, the area under grain legumes accounted for less than 2% of the cultivated area in Europe (FAOSTAT 2016). Between 1961 and 2014, pea (*Pisum sativum*; Fig. 4.1) was the most widely cultivated grain legume species in Europe (FAOSTAT 2016; Zander et al. 2016). More recently, there has been a substantial increase in the cultivated area under soybean (*Glycine max*) in several European countries (FAOSTAT 2016; Zander et al. 2016). In 2014, pea and soybean together accounted for 76% of the area under grain legumes in Europe, whereas only 15% of the area under grain legumes was covered by chickpea (*Cicer arietinum*), faba bean (*Vicia faba*), garden vetch (*Vicia sativa*), kidney bean (*Phaseolus vulgaris*), lentil (*Lens culinaris*), and lupins (*Lupinus* species) (FAOSTAT 2016).

This large imbalance in the areas under the different grain legume species is the consequence of two concomitant phenomena in Europe. First, only a small number of grain legume species, including pea, have been investigated as model species in

Fig. 4.1 Spring pea field in Burgundy, France

research programs (*e.g.*, Gepts et al. 2005; Murphy-Bokern et al. 2014; Foyer et al. 2016; Magrini et al. 2016). Second, the legume breeding sector is fragmented into a small number of private companies, working on only a few species (Murphy-Bokern et al. 2014; LMC International 2009; Wiggering et al. 2012), despite reports of potential environmental and nutritional benefits associated with the cultivation of alternative legumes species (Jensen et al. 2012; Lucas et al. 2015; Multari et al. 2015).

Europe could learn valuable lessons for the diversification of grain legume crops from the expansion of legume production in North America and Oceania. Both these regions have ranges of climatic conditions similar to those found in Europe (Peel et al. 2007). However, North America and Oceania together exported 11 times more legumes than Europe over a period extending from 2003 to 2013 (considering that exports are linked to harvested areas) (FAOSTAT 2016). Soybean is the predominant grain legume in North American agriculture, but alternative grain legume species have become more popular since the 1980s, for example in the northern Great Plains as a means of diversifying cereal-fallow rotations (Miller et al. 2001; Zentner et al. 2002; Angadi et al. 2008; Gan et al. 2015). In Australia, crop breeders, agronomists, and industry, have been working closely together since the 1970s and 1980s, to develop the commercial production of a broad range of grain legume species (Siddique et al. 2013). This also relied on the establishment of experiments specifically carried out to compare the productivities of a large number of legume species but for a few criteria at a time from a single data source (*e.g.*, Laurence 1979; Silsbury 1975; Gregory 1998; Siddique et al. 1999; Malik et al. 2015). Yet, to the best of our knowledge, there is no multi-criteria comparison of a broad range of grain legume species based on a quantitative synthesis of published experimental data (Fig. 4.2).

In this study, we mainly focus on Europe, but we rely on data collected in both North America and Oceania in order to identify grain legume species that could be cultivated in Europe in the future. We address two key questions: (1) Which grain legume species have productivities lower than, similar to, and higher than, that of

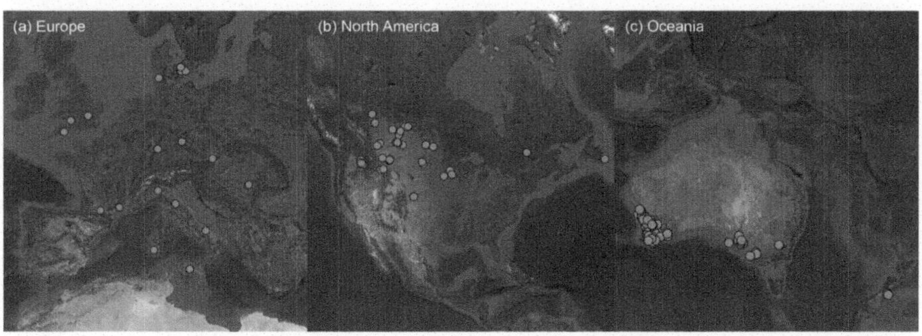

Fig. 4.2 Experimental field sites included in the statistical analysis. The study regions are: Europe (**a**), North America (**b**), and Oceania (**c**). Each red bullet point indicates the location of an experimental field site. In Europe, field sites are located in Austria, Denmark, France, Germany, Italy, the United Kingdom, and Romania. In North America, field sites are located in Canada, and the United States of America. In Oceania, field sites are located in Australia, and New Zealand. Basemap from Google Maps 2016

pea in Europe? (2) Which grain legume species that have been directly compared with pea in North America and Oceania but not in Europe may be considered as potential candidates for inclusion in new field European experiments? We perform a statistical analysis comparing 22 grain legume species with pea as a reference species. All grain legumes considered in this study are compared on the basis of five criteria related to different aspects of their productivity: (1) grain yield (hereafter referred to as grain biomass), (2) total aerial biomass, (3) grain crude protein, (4) grain gross energy, (5) grain nitrogen content.

4.2 Materials and Methods

4.2.1 Data

We used a subset of a global experimental dataset including 173 peer-reviewed articles selected from a systematic literature review on grain legumes, described in Cernay et al. (2016). The global dataset can be downloaded from http://datadryad.org/resource/doi:10.5061/dryad.mf42f. The subset used for this study includes data extracted from 61 peer-reviewed articles. Our dataset (Tables 4.1, 4.2 and 4.3) includes results from field experiments carried out in Europe, North America or Oceania (Fig. 4.2). Each experiment compares pea with at least one other grain legume species and reports data for the following productivity variables; grain biomass, total aerial biomass, grain nitrogen content, percentage of nitrogen in grains. Over all experiments, 22 grain legume species are compared to pea. We choose pea – the grain legume species most widely cultivated, on average, in Europe, over the 1961–2014 period – as a reference species for all regions. Both grain crude protein and grain gross energy are calculated for each legume species from the nutritional tables of the Animal Feed Resources Information System database (hereafter referred to as FEEDIPEDIA database 2016) (Table 4.4).

4.2.2 Statistical Analysis for Estimating Mean Ratios

All grain legume species are directly compared with pea (*i.e.*, the reference species) by estimating mean log-transformed ratios for five productivity criteria: (1) grain biomass, (2) total aerial biomass, (3) grain crude protein, (4) grain gross energy, (5) grain nitrogen content. For each criterion *XX*, the log-transformed ratio, as in (1): $L_{ijk} = \log\left(\frac{\overline{X}_{ijk}}{\overline{X}_{rjk}}\right)$, is calculated for each species *i* grown in each field site*growing season *k* in each article *j*. $\overline{X}_{ijk}$ is the value of XX averaged over crop management techniques (i.e., tillage, fertilization, pest control, and irrigation) for species *i* grown in field site*growing season *k* ikn article *j*, and $\overline{X}_{rjk}$ is the value of *XX* averaged over

Table 4.1 Scientific and common names for species by region, adapted from the United States Department of Agriculture Plants database

Scientific name	Common name	Region
Cicer arietinum	Chickpea	Europe, North America, Oceania
Glycine max	Soybean	Europe, North America
Lathyrus aphaca	Yellow pea	Oceania
Lathyrus cicera	Red pea	Oceania
Lathyrus clymenum	Cicercha purpurina	Oceania
Lathyrus ochrus	Cyprus vetch	Oceania
Lathyrus sativus	White pea	North America, Oceania
Lens culinaris	Lentil	Europe, North America, Oceania
Lupinus albus	White lupin	Europe, Oceania
Lupinus angustifolius	Narrowleaf lupin	Europe, North America, Oceania
Lupinus atlanticus	NA[a]	Oceania
Lupinus luteus	Yellow lupin	Europe, Oceania
Lupinus pilosus	Blue lupin	Oceania
Phaseolus vulgaris	Kidney bean	North America
Pisum sativum	Pea	Europe, North America, Oceania
Vicia articulata	Oneflower vetch	Oceania
Vicia benghalensis	Purple vetch	Oceania
Vicia ervilia	Blister vetch	Oceania
Vicia faba	Faba bean	Europe, North America, Oceania
Vicia hybrida	Hairy yellow vetch	Oceania
Vicia narbonensis	Purple broad vetch	Oceania
Vicia sativa	Garden vetch	Oceania
Vicia villosa	Winter vetch	Oceania

Available from: http://plants.usda.gov/java/. Accessed 29 Mar 2016. All species are ranked alphabetically according to scientific names
[a]Not Available

Table 4.2 Number of available experimental data for productivity variables by region

Region	Grain biomass	Total aerial biomass	Harvest index	Grain nitrogen content	Percentage of nitrogen in grains	Water use	Grain water use efficiency
Europe	265	82	21	41	0	0	0
North America	917	176	7	22	46	89	113
Oceania	1014	215	121	–	0	0	0
Total	2196	473	149	63	46	89	113

the same crop management techniques for pea (the subscript *r* refers to the reference species, i.e., pea) grown in the same field site*growing season in the same article.

The effect size *L* is equivalent to the log response ratio defined by Hedges et al. (1999), and compares control groups (corresponding here to pea crops) and experimental groups (corresponding here to non-pea legume species) over a set of studies.

Table 4.3 Name and definition for variables included in the data set

Name	Definition
Article_Index	Index of each article.
Article_Author_First	Name of the first author.
Article_Title	Article title.
Article_Year_Publication	Publication year or "NA"[a].
Article_Journal	Journal name or "NA".
Site_Index	Index of each site from each article.
Site_Name	Site name.
Site_Country	Site country.
Crop_Index	Index of each crop.
Crop_Site_Growing_Season_Index	Index for each crop grown at the same field site during the same growing seasons.
Crop_Growing_Season_Year_First	First calendar year at which the crop is seeded and/or the growing season starts or "NA". When values are averaged over more than one growing season, only the calendar year of the first growing season is reported.
Crop_Growing_Season_Year_Last	Last calendar year at which the crop is harvested and/or the growing season ends or "NA". When values are averaged over more than one growing season, only the calendar year of the last growing season is reported.
Crop_Growing_Season_Number	Number of growing seasons. When values are averaged over more than one growing season, the number of growing seasons is reported.
Crop_Species_Scientific_Name	Species scientific name.
Crop_Replicate_Number	Number of replicates or "NA".
Crop_Biomass_Grain	Grain biomass or "NA".
Crop_Biomass_Grain_Unit	Unit of grain biomass or "NA".
Crop_Biomass_Grain_DM_Percentage	Dry matter (DM) percentage of grain biomass or "NA".
Crop_Biomass_Aerial	Aerial biomass or "NA".
Crop_Biomass_Aerial_Unit	Unit of aerial biomass or "NA".
Crop_Biomass_Aerial_DM_Percentage	Dry matter (DM) percentage of aerial biomass or "NA".
Crop_Biomass_Aerial_Definition	Definition of components included in aerial biomass or "NA".
Crop_Biomass_Aerial_Stage_Detailed	Detailed phenology stage (i.e., originally reported in the selected article) at which aerial biomass is measured or "NA".
Crop_Biomass_Aerial_Stage_Simplified	Simplified phenology stage (i.e., either "Before physiological maturity" or "Physiological maturity") at which aerial biomass is measured or "NA".
Crop_Harvest_Index	Harvest index or "NA".
Crop_N_Content_Grain	Grain nitrogen (N) content or "NA".
Crop_N_Content_Grain_Unit	Unit of grain nitrogen (N) content or "NA".
Crop_N_Percentage_Grain	Percentage of nitrogen (N) in grains or "NA".

(continued)

Table 4.3 (continued)

Name	Definition
Crop_Water_Use_Balance	Water use or water balance or "NA".
Crop_Water_Use_Balance_Unit	Unit of water use or water balance or "NA".
Crop_Water_Use_Balance_Equation	Equation of water use or water balance or "NA".
Crop_Water_Use_Balance_Efficiency_Grain	Grain water use efficiency or grain water balance efficiency or "NA".
Crop_Water_Use_Balance_Efficiency_Grain_Unit	Unit of grain water use efficiency or grain water balance efficiency or "NA".

The global dataset can be downloaded from http://datadryad.org/resource/doi:10.5061/dryad.mf42f

[a]Not Available

Table 4.4 Mean grain crude protein (unit) and mean grain gross energy (unit) according to the FEEDIPEDIA database

Scientific name	Mean grain crude protein (%)[a]	Mean grain gross energy (10^9 J t^{-1})[a]	FEEDIPEDIA database item
Cicer arietinum[b]	22.20	19.60	Chickpea seeds, desi type
			Chickpea seeds, kabuli type
Glycine max	39.60	23.60	Soybean seeds
Lathyrus aphaca	NA[c]	NA	NA
Lathyrus cicera	23.80	18.80	Chick vetch (*Lathyrus cicera*), seeds
Lathyrus clymenum	NA	NA	NA
Lathyrus ochrus	26.10	18.90	Ochre vetch (*Lathyrus ochrus*), seeds
Lathyrus sativus	30.00	19.10	Grass pea (*Lathyrus sativus*), seeds
Lens culinaris	26.90	18.50	Lentil seeds
Lupinus albus	NA	NA	NA
Lupinus angustifolius	33.80	20.30	Lupin (*Lupinus angustifolius*), blue, seeds
Lupinus atlanticus	NA	NA	NA
Lupinus luteus	43.00	20.90	Lupin (*Lupinus luteus*), yellow, seeds
Lupinus pilosus	NA	NA	NA
Phaseolus vulgaris	24.80	18.60	Common bean seeds
Pisum sativum	23.90	18.30	Pea seeds
Vicia articulata	NA	NA	NA
Vicia benghalensis	NA	NA	NA
Vicia ervilia	NA	NA	NA
Vicia faba[d]	29.00	18.70	Faba bean (*Vicia faba*), all cultivars
Vicia hybrida	NA	NA	NA
Vicia narbonensis	27.20	19.00	NA

(continued)

Table 4.4 (continued)

Scientific name	Mean grain crude protein (%)[a]	Mean grain gross energy (10^9 J t^{-1})[a]	FEEDIPEDIA database item
Vicia sativa	28.40	18.90	Common vetch (*Vicia sativa*), seeds
Vicia villosa	29.30	NA	Hairy vetch (*Vicia villosa*), seeds

Available from: http://www.feedipedia.org/. Accessed 15 Mar 2016. All species are ranked alphabetically according to scientific names. The scientific names of the species correspond to the common names between parentheses: *Cicer arietinum* (chickpea), *Glycine max* (soybean), *Lathyrus aphaca* (yellow pea), *Lathyrus cicera* (red pea), *Lathyrus clymenum* (cicercha purpurina), *Lathyrus ochrus* (cyprus vetch), *Lathyrus sativus* (white pea), *Lens culinaris* (lentil), *Lupinus albus* (white lupin), *Lupinus angustifolius* (narrowleaf lupin), *Lupinus atlanticus* (Not Available), *Lupinus luteus* (yellow lupin), *Lupinus pilosus* (blue lupin), *Phaseolus vulgaris* (kidney bean), *Pisum sativum* (pea), *Vicia articulata* (oneflower vetch), *Vicia benghalensis* (purple vetch), *Vicia ervilia* (blister vetch), *Vicia faba* (faba bean), *Vicia hybrida* (hairy yellow vetch), *Vicia narbonensis* (purple broad vetch), *Vicia sativa* (garden vetch), and *Vicia villosa* (winter vetch)

[a]Mean grain crude protein and mean grain gross energy are expressed on a dry matter basis

[b]Mean grain crude protein and mean grain gross energy are determined from the FEEDIPEDIA database items [b]"Chickpea seeds, [b]desi type" and "Chickpea seeds, kabuli type"

[c]Not Available

[d]Crude protein and gross energy are for the grains

This effect size is frequently used in meta-analysis (Hedges et al. 1999). The mean log-transformed ratio is estimated here for each criterion *X* and each species *i* across all field site*growing season combinations and selected articles, with linear random-effects models (Mengersen et al. 2013). Random-effect models are commonly used for meta-analysis in agronomy (Philibert et al. 2012) and in ecology (Mengersen et al. 2013) to encompass both the variance between studies (due to differences of the true effect size between studies) as well as the variance of estimates within studies (due to sampling and estimation error) (Mengersen et al. 2013). Here, we use random-effects models to account for the variability of the mean ratio between articles, sites, and growing seasons. Model variants including one random "field site*growing season" effect or two random "article" and "field site*growing season" effects are fitted in turn.

Random-effects models are fitted using the "lme" (linear mixed-effect) function of the "nlme" package (version 3.1.111) (Pinheiro et al. 2013) of the R software (version 3.0.2, 2013). The "lme" function fits random-effects models and allows for nested random effects and data weighting. The best model variant is selected on the basis of the Akaike Information Criterion (hereafter referred to as AIC) (Akaike 1974). In order to justify the use of a random-effects model, the AIC criterion is also computed for a fixed-effect model that assumes that all studies share the same true mean log ratio. Mean ratios R_i Riand 95% confidence intervals are then calculated from the exponential of the estimated mean log-transformed ratios for each criterion *X* and each species *i*, as in (2)*:* Ri = eXiXr. Estimated mean ratios are considered to be significantly different from one (*i.e.*, the species show significant higher or lower performance levels compared with that of pea for a given criterion) if their 95% confidence intervals do not include one. This analysis is then repeated, weighting the log-transformed ratios L_{ijk} Lijkby their precisions. As the variances of $\overline{X}_{ijk}$ and $\overline{X}_{rjk}$ Xrjkare available for only a small number of the original measurements, the log-transformed ratios L_{ijk} Lijkare weighted on the basis of the number of replicates n_{ijk} and n_{rjk} nrjkused to calculate $\overline{X}_{ijk}$ and $\overline{X}_{rjk}$, respectively. In situations in which a given criterion X X*is* averaged over more than one growing season, the number of replicates is considered to be the sum of the number of replicates over all growing seasons. The weight ω_{ijk} ωijkof each log-transformed ratio L_{ijk} Lijkis calculated, as in (3): ωijk = nijknrjkXijk2Xrjk2nijkXijk2 + nrkXrjk2. This weight is a generalization of the sample size weight described by Wiebe et al. (2006). Species comparison obtained with and without data weighting are compared.

A lower AIC indicates a better model but, when the AIC differences are small, it is useful to assess the sensitivity of the results to the model assumptions. We use weighted values to assess sensitivity to model assumptions for the two model variants, separately: (1) inclusion of one random "field site*growing season" effect (Table 4.6), and (2) inclusion of two random "article" and "field site*growing season" effects (Table 4.7). We also use weighted values to analyse whether raw residuals by species, standardized residuals by species, and standardized residuals versus fitted values across all field site*growing season combinations and selected articles are symmetrically distributed in shape for each criterion in each region. Mean ratios are estimated independently for Europe, North America and Oceania.

4.3 Results

We focus below on the results obtained weighting log-transformed ratios. Very similar results are obtained unweighting log-transformed ratios (Fig. 4.3a–e, Fig. 4.4a–e, and Fig. 4.5a–d).

4.3.1 Good Performances of Soybean, Narrowleaf Lupin, and Faba Bean in Europe

The mean grain biomass ratios estimated for soybean (R = 1.37), faba bean (R = 0.90) and yellow lupin (R = 0.67) in Europe are not significantly different from one another (Fig. 4.3a and Table 4.5); their 95% confidence intervals include one. The grain biomasses of these three species are, therefore, on average, not significantly different from that of pea for the same field sites during the same growing seasons with the same crop management techniques. The grain biomass data are extracted from 12 articles for faba bean, but from only three articles for soybean, and one article for yellow lupin. These differences in sample size result in soybean having a larger 95% confidence interval, and yellow lupin having a much larger

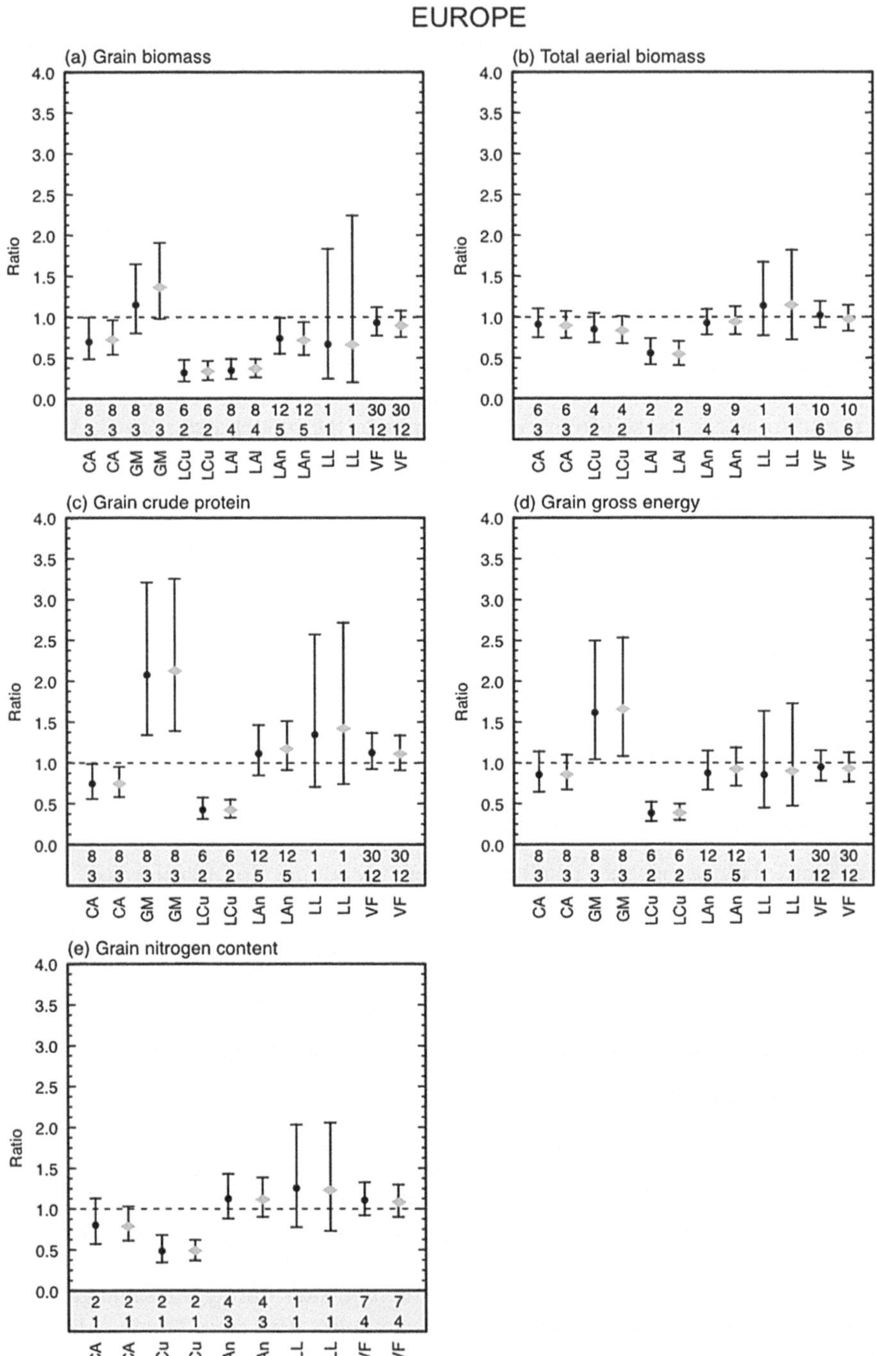
EUROPE
(a) Grain biomass
(b) Total aerial biomass
(c) Grain crude protein
(d) Grain gross energy
(e) Grain nitrogen content
Ratio
CA CA GM GM LCu LCu LAl LAl LAn LAn LL LL VF VF

confidence interval than that of faba bean. Thus, the results obtained for faba bean are more robust than those obtained for the other two species. The estimated mean grain biomasses of chickpea, narrowleaf lupin, and white lupin are significantly lower than that of pea, by −28%, −29%, and − 65%, respectively. Lentil also has a significant lower grain biomass than that of pea ($R = 0.32$), but only two articles are available for this species in Europe. The mean total aerial biomass ratios estimated for chickpea, narrowleaf lupin, and faba bean, are not significantly different from one (Fig. 4.3b and Table 4.5).

Results for the grain crude protein and grain gross energy are presented in Fig. 4.3c, d, and Table 4.5. The mean grain crude protein and mean grain gross energy ratios estimated for soybean are 2.13 and 1.66, respectively. The mean grain crude protein and mean grain gross energy ratios estimated for narrowleaf lupin and faba bean are not significantly different from one. The mean grain crude protein ratio estimated for chickpea is significantly lower than one, indicating a mean grain crude protein level − 26% lower than that of pea. The differences between species obtained for grain nitrogen content are similar to those obtained for grain crude protein (Fig. 4.3c, e, and Table 4.5), except chickpea.

4.3.2 Most of Grain Legume Species Considered Do Not Outperform Pea in North America

The other grain legume species grown in North America perform less well than pea for most of the criteria, with few exceptions. Estimated mean grain biomass is significantly lower than that of pea (*Pisum sativum*) for all species, except white pea (*Lathyrus sativus*) (Fig. 4.4a and Table 4.5). The uncertainty on the estimated mean grain biomass ratio for this species is very high, due to the small number of articles on white pea. All the other species have significantly lower mean grain biomasses than that of pea. The mean grain biomasses estimated for faba bean and lentil are −31% and − 40% lower than that of pea, respectively. Mean grain biomass is −45% lower for chickpea, −53% lower for soybean, −64% lower for narrowleaf lupin, and − 70% lower for kidney bean. These lower mean grain biomasses probably reflect

Fig. 4.3 Ratios of grain biomass (**a**), total aerial biomass (**b**), grain crude protein (**c**), grain gross energy (**d**), grain nitrogen content (**e**) estimated for seven grain legume species, relative to those estimated for pea (*Pisum sativum*) in Europe. The scientific names of the species concerned are abbreviated here as follows: *Cicer arietinum* (chickpea, CA), *Glycine max* (soybean, GM), *Lens culinaris* (lentil, LCu), *Lupinus albus* (white lupin, LAl), *Lupinus angustifolius* (narrowleaf lupin, LAn), *Lupinus luteus* (yellow lupin, LL), *Lupinus* species (lupins, LU), and *Vicia faba* (faba bean, VF). All species are presented alphabetically according to scientific names. Bullet points indicate mean ratios estimated unweighted (black rounds) or weighted (gray diamonds) values for each criterion. Error bars indicate 95% confidence intervals. Within the bottom box (gray), the lower and upper rows indicate the number of selected articles and the number of log-transformed ratios, respectively

Table 4.5 Productivity for all grain legume species relative to pea (*Pisum sativum*) in Europe, North America and Oceania

	Europe					North America					Oceania			
	Grain biomass ratio[a]	Total aerial biomass ratio[a]	Grain crude protein ratio[a,b]	Grain gross energy ratio[a,b]	Grain nitrogen content ratio[a]	Grain biomass ratio	Total aerial biomass ratio	Grain crude protein ratio	Grain gross energy ratio	Grain nitrogen content ratio	Grain biomass ratio	Total aerial biomass ratio	Grain crude protein ratio	Grain gross energy ratio
Cicer arietinum	–	0	–	0	(0)	–	–	–	–	(–)	–	–	–	–
Glycine max	0		+	+		–	(0)	0	–					
Lathyrus aphaca											(–)			
Lathyrus cicera											0	0	0	0
Lathyrus clymenum											(0)			
Lathyrus ochrus											0	(0)	0	0
Lathyrus sativus						(0)	(0)	(0)	(0)	(0)	0	0	0	0
Lens culinaris	(–)	(0)	(–)	(–)	(–)	–	–	–	–	–	–	–	–	–
Lupinus albus	–	(–)									–	–		
Lupinus angustifolius	–	0	0	0	0	–		–	–	(–)	0	0	0	0
Lupinus atlanticus											(0)	(0)		
Lupinus luteus	(0)	(0)	(0)	(0)	(0)						(–)	(0)	(–)	(0)
Lupinus pilosus											(0)	(0)		

Phaseolus vulgaris						–		–	–	(–)				
Vicia articulata											(0)			
Vicia benghalensis											(–)	(0)		
Vicia ervilia											(–)	(0)		
Vicia faba	0	0	0	0	0	–	(0)	0	–	(0)	+	0	+	+
Vicia hybrida											(–)			
Vicia narbonensis											0	(0)	0	0
Vicia sativa											0	0	0	0
Vicia villosa											(–)		(–)	

All species are presented alphabetically according to scientific names. The scientific names of the species correspond to the common names between parentheses: *Cicer arietinum* (chickpea), *Glycine max* (soybean), *Lathyrus aphaca* (yellow pea), *Lathyrus cicera* (red pea), *Lathyrus clymenum* (cicercha purpurina), *Lathyrus ochrus* (cyprus vetch), *Lathyrus sativus* (white pea), *Lens culinaris* (lentil), *Lupinus albus* (white lupin), *Lupinus angustifolius* (narrowleaf lupin), *Lupinus atlanticus* (no common name), *Lupinus luteus* (yellow lupin), *Lupinus pilosus* (blue lupin), *Phaseolus vulgaris* (kidney bean), *Pisum sativum* (pea), *Vicia articulata* (oneflower vetch), *Vicia benghalensis* (purple vetch), *Vicia ervilia* (blister vetch), *Vicia faba* (faba bean), *Vicia hybrida* (hairy yellow vetch), *Vicia narbonensis* (purple broad vetch), *Vicia sativa* (garden vetch), and *Vicia villosa* (winter vetch). Mean ratios are estimated from weighted experimental data. The best model variant including either one random "field site*growing season" effect or two random "article" and "field site*growing season" effects is used. The symbol "–" indicates that the value of the criterion for the given species is significantly lower than that of pea, and that more than two articles are available for the species concerned. The symbol "(–)" indicates that the value of the criterion for the given species is significantly lower than that of pea, and that there are no more than two articles for the species concerned. The symbol "0" indicates that the value of the criterion for the given species is not significantly different from that of pea, and that there are more than two articles for the species concerned. The symbol "(0)" indicates that the value of the criterion for the given species is not significantly different from that of pea, and that there are no more than two articles available for the species concerned. The symbol "+" indicates that the value of the criterion for the given species is significantly higher than that of pea, and that there are more than two articles available for the species concerned. The symbol "(+)" indicates that the value of the criterion for the given species is significantly higher than that of pea, and that there are no more than two articles available for the species concerned

[a]Experimental data from the selected articles are used to estimate the given criterion

[b]Nutritional data from the Feedipedia database are used to estimate the given criterion

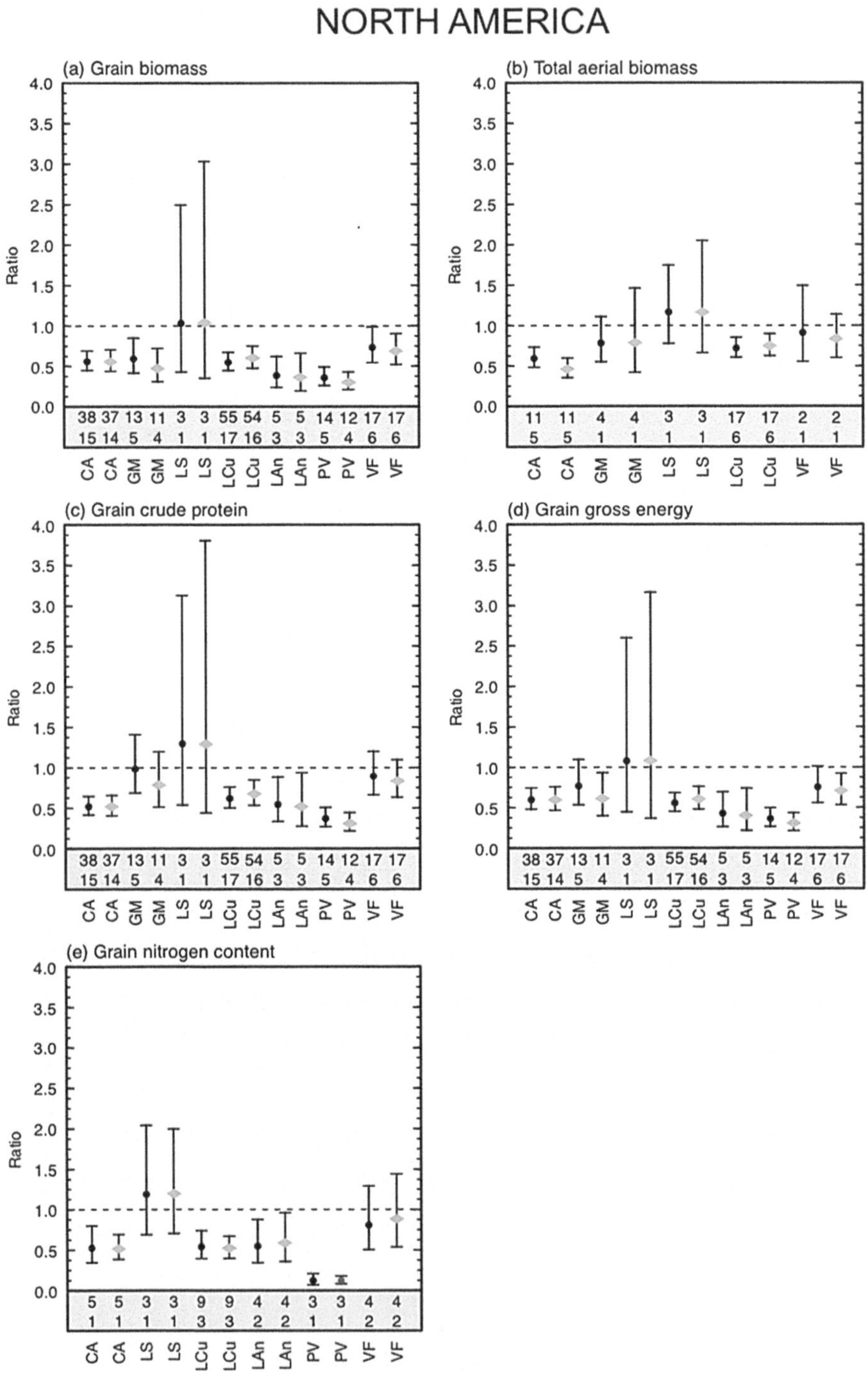
NORTH AMERICA
(a) Grain biomass
(b) Total aerial biomass
(c) Grain crude protein
(d) Grain gross energy
(e) Grain nitrogen content
Ratio
CA CA GM GM LS LS LCu LCu LAn LAn PV PV VF VF

the location of most of the experiments in the northern Great Plains, where the grain biomass of pea tends to be higher than those of other grain legume species (Miller et al. 2002). The mean total aerial biomass ratios estimated for lentil and chickpea are significantly lower than one (Fig. 4.4b and Table 4.5). The differences in this ratio are not significant for soybean, white pea, and faba bean, but few data are available for these three species.

The mean grain crude protein ratios estimated for soybean and faba bean are 0.78 and 0.83 (Fig. 4.4c and Table 4.5), respectively. Both these species have performances similar to that of pea for this criterion. For the other species, the estimated mean grain crude protein ratios are either significantly lower than one (*i.e.*, chickpea, lentil, narrowleaf lupin, and kidney bean) or very uncertain (*i.e.*, white pea). The mean grain gross energy estimated for all species is significantly lower than that of a reference species (Fig. 4.4d and Table 4.5), except white pea, but with more uncertainty. The estimated mean grain nitrogen content ratios (Fig. 4.4e and Table 4.5) are significantly lower than one for all species, with two exceptions: white pea and faba bean.

4.3.3 *Several Vetches and Lupins Tend to Outperform Pea in Oceania*

Faba bean has significantly higher mean grain biomass than that of pea, with estimated ratio of $R = 1.71$ (Fig. 4.5a and Table 4.5). Several species do not differ significantly from pea in terms of grain biomass: red pea (*Lathyrus cicera*), cicercha purpurina (*Lathyrus clymenum*), cyprus vetch (*Lathyrus ochrus*), white pea, narrowleaf lupin, *Lupinus atlanticus* (no common name), blue lupin (*Lupinus pilosus*), oneflower vetch (*Vicia articulata*), purple broad vetch (*Vicia narbonensis*), and garden vetch. The remaining nine grain legume species have significantly lower grain biomasses than that of pea. The mean total aerial biomasses of chickpea, lentil, and white lupin are significantly lower than that of pea.

The mean grain crude protein estimated for faba bean is significantly higher than for pea, by +104% (Fig. 4.5c and Table 4.5). Estimated mean grain crude protein ratio is not significantly different from one in red pea, cyprus vetch, white pea,

Fig. 4.4 Ratios of grain biomass (**a**), total aerial biomass (**b**), grain crude protein (**c**), grain gross energy (**d**), grain nitrogen content (**e**) estimated for seven grain legume species, relative to those estimated for pea (*Pisum sativum*) in North America. The scientific names of the species concerned are abbreviated here as follows: *Cicer arietinum* (chickpea, CA), *Glycine max* (soybean, GM), *Lathyrus sativus* (white pea, LS), *Lens culinaris* (lentil, LCu), *Lupinus angustifolius* (narrowleaf lupin, LAn), *Phaseolus vulgaris* (kidney bean, PV), *Phaseolus* and *Vigna* species (beans, PVi), and *Vicia faba* (faba bean, VF). All species are presented alphabetically according to scientific names. Bullet points indicate mean ratios estimated unweighted (blackrounds) or weighted (gray diamonds) values for each criterion. Error bars indicate 95% confidence intervals. Within the bottom box (gray), the lower and upper rows indicate the number of selected articles and the number of log-transformed ratios, respectively

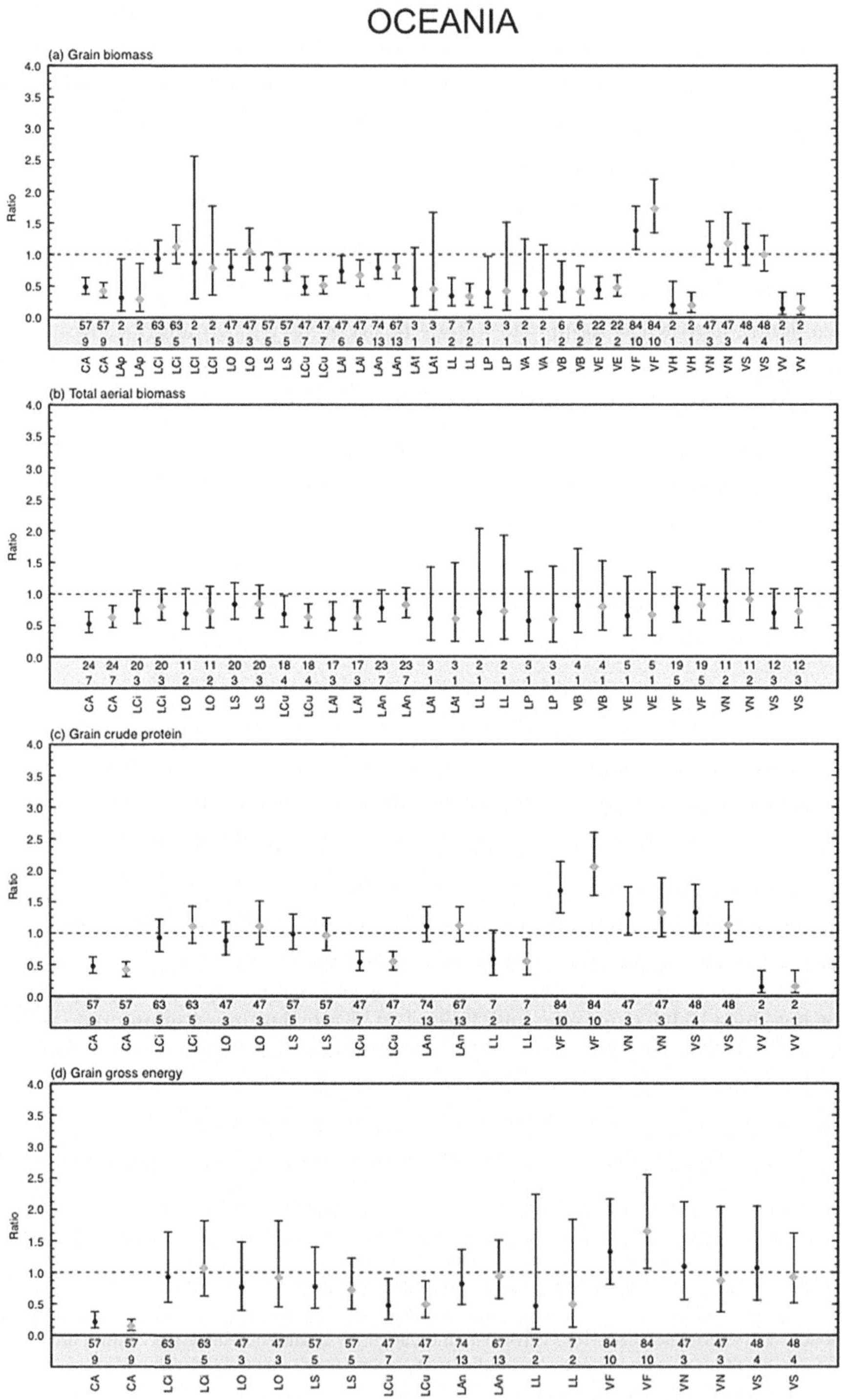
OCEANIA
(a) Grain biomass
(b) Total aerial biomass
(c) Grain crude protein
(d) Grain gross energy
Ratio

narrowleaf lupin, purple broad vetch, and garden vetch (*i.e.*, their 95% confidence intervals include one). Estimated mean grain crude protein for chickpea, lentil, yellow lupin, and winter vetch, is significantly lower than for pea. Only faba bean performs significantly better than pea for grain gross energy (Fig. 4.5d and Table 4.5).

4.4 Discussion

Our analysis provides European farmers, plant breeders, agricultural companies and policy makers with new comprehensive insights into grain legume productivity. In particular, our results identify the grain legumes with the greatest productivity potential for expansion in Europe. We have mainly focused on the yield potential and protein production of different crops compared to pea. To the best of our knowledge, our study presents the first multi-criteria comparison of a broad range of grain legume species based on a quantitative synthesis of published experimental data. However, the choice of a species by growers is not only guided by the productivity. The length of the growing season, the effect of a given crop species in the succeeding crops (diseases, weeds, N availability, etc.), the pedoclimatic conditions (for instance, soybean require a close to neutral or a basic soil pH, whereas lupins require slightly acidic, non-calcareous soils), the agricultural practices (*e.g.*, irrigation of soybean), the market conditions, the availability of technical advice and outlet are other elements accounted for. Extension services and regional experts could provide farmers with useful information to adjust their choice of species to local characteristics (*e.g.*, soil and weather conditions). Many grain legumes are well valued and recognized in many European countries for human food, most of them related to Mediterranean diets. In this case, the "production" point of view should not be used as a unique indicator of holistic performance of a crop within a cropping system.

Fig. 4.5 Ratios of grain biomass (**a**), total aerial biomass (**b**), grain crude protein (**c**), grain gross energy (**d**), estimated for twenty grain legume species, relative to those estimated for pea (*Pisum sativum*) in Oceania. The scientific names of the species concerned are abbreviated here as follows: *Cicer arietinum* (chickpea, CA), *Lathyrus aphaca* (yellow pea, LAp), *Lathyrus cicera* (red pea, LCi), *Lathyrus clymenum* (cicercha purpurina, LCl), *Lathyrus ochrus* (cyprus vetch, LO), *Lathyrus sativus* (white pea, LS), *Lens culinaris* (lentil, LCu), *Lupinus albus* (white lupin, LAl), *Lupinus angustifolius* (narrowleaf lupin, LAn), *Lupinus atlanticus* (no common name, LAt), *Lupinus luteus* (yellow lupin, LL), *Lupinus pilosus* (blue lupin, LP), *Vicia articulata* (oneflower vetch, VA), *Vicia benghalensis* (purple vetch, VB), *Vicia ervilia* (blister vetch, VE), *Vicia faba* (faba bean, VF), *Vicia hybrida* (hairy yellow vetch, VH), *Vicia narbonensis* (purple broad vetch, VN), *Vicia sativa* (garden vetch, VS), and *Vicia villosa* (winter vetch, VV). All species are presented alphabetically according to scientific names. Bullet points indicate mean ratios estimated unweighted (black rounds) or weighted (gray diamonds) values for each criterion. Error bars indicate 95% confidence intervals. Within the bottom box (gray), the lower and upper rows indicate the number of selected articles and the number of log-transformed ratios, respectively

Some of our results are based on a small sample size. Here, only studies including at least two grain legume species were selected in order to limit the risk of confounding effect between species and site-year characteristics. Thus, our dataset does not include all published available data on productivity of grain legume species. Moreover, grey literature may be a source of additional data but is not easy to get. The main conclusions of our statistical analysis are not particularly sensitive to model assumptions (Tables 4.6 and 4.7). Model variants based on weighted values tend to produce narrower confidence intervals, and more accurate results. Distributions of residuals are symmetric in shape for all species and criteria in all regions and do not show any evidence of heterogeneous residual variance.

We use experimental data collected in agronomic experiments where several grain legume species are directly compared at the same field sites during the same growing seasons with the same crop management techniques (*i.e.*, tillage, fertilization, pest control, and irrigation). Based on this data source, we are able to compare the yield levels of grain legume species cultivated in the same agronomic and environmental conditions. Nutritional data extracted from feed and food composition tables enables us to compare grain legume species according to their grain crude protein and gross grain energy. Grain composition depends on the grain legume species (FEEDIPEDIA database), but also on cultivars, and on soil and climatic conditions (Duc et al. 2011; Vadez et al. 2012). The grain quality of grain legume species was collected and included in our dataset when available (Cernay et al. 2016), but quality criteria were not reported in many of the selected articles. Finally, based on these data sources, we are able to perform a multi-criteria analysis of the yield performances of a broad range of grain legume species.

Faba bean has performances similar to those of pea for all criteria in Europe. Faba bean may, therefore, be considered an interesting alternative to pea in this world region. This species is well adapted to the diverse climate conditions and soil types of Europe (Flores et al. 2013). It is recognized as a valuable source of protein for the human diet and animal feed (Crépon et al. 2010; Multari et al. 2015; Koivunen et al. 2016). Soybean and narrowleaf lupin also perform well relative to pea, especially for grain crude protein and grain gross energy. However, only a small number of field experiments have directly compared these species in Europe. New experiments are thus warranted to identify the geographical areas where both these species perform best under European conditions. Measurements of total aerial biomass (or harvest index), grain nitrogen content (or percentage of nitrogen into grains), as well as grain quality would also be useful to further refine our estimates, and decrease the uncertainty levels associated with the yield performances of soybean and lupins in Europe.

Narrowleaf and yellow lupins display good relative performances for most criteria in Europe. Lucas et al. (2015) stressed the importance of advanced breeding and processing operations to turn European-grown lupins into attractive marketable ingredients for human food and animal feed. Chickpea and lentil have lower overall performances than pea for most criteria. Both chickpea and lentil are suitable for incorporation into animal feed, but both are widely used for human consumption (*e.g.*, Boye et al. 2010; Jukanti et al. 2012). Hence, these grain legume species

Table 4.6 Productivity rankings for all grain legume species relative to pea in Europe, North America and Oceania (inclusion of one random "field site*growing season" effect)

	Europe					North America					Oceania			
	Grain biomass ratio[a]	Total aerial biomass ratio[a]	Grain crude protein ratio[a,b]	Grain gross energy ratio[a,b]	Grain nitrogen content ratio	Grain biomass ratio	Total aerial biomass ratio	Grain crude protein ratio	Grain gross energy ratio	Grain nitrogen content ratio	Grain biomass ratio	Total aerial biomass ratio	Grain crude protein ratio	Grain gross energy ratio
Cicer arietinum	–	0	–	0	(0)	–	–	–	–	(–)	–	–	–	–
Glycine max	0		+	+		–	(0)	0	–					
Lathyrus aphaca											(–)			
Lathyrus cicera											0	0	0	0
Lathyrus clymenum											(0)			
Lathyrus ochrus											0	(0)	0	0
Lathyrus sativus						(0)	(0)	(0)	(0)	(0)	–	0	–	0
Lens culinaris	(–)	(0)	(–)	(–)	(–)	–	–	–	–	–	–	–	–	–
Lupinus albus	–	(–)									–	–		
Lupinus angustifolius	–	0	0	0	0	–		–	–	(–)	–	0	0	0
Lupinus atlanticus											(0)	(0)		
Lupinus luteus	(0)	(0)	(0)	(0)	(0)						(–)	(0)	(–)	(0)
Lupinus pilosus											(0)	(0)		
Phaseolus vulgaris						–		–	–	(–)				
Vicia articulata											(–)			
Vicia benghalensis											(–)	(0)		
Vicia ervilia											(–)	(0)		

(continued)

Table 4.6 (continued)

	Europe					North America					Oceania			
	Grain biomass ratio[a]	Total aerial biomass ratio[a]	Grain crude protein ratio[a,b]	Grain gross energy ratio[a,b]	Grain nitrogen content ratio	Grain biomass ratio	Total aerial biomass ratio	Grain crude protein ratio	Grain gross energy ratio	Grain nitrogen content ratio	Grain biomass ratio	Total aerial biomass ratio	Grain crude protein ratio	Grain gross energy ratio
Vicia faba	0	0	0	0	0	–	(0)	–	–	(0)	+	0	+	+
Vicia hybrida											(–)			
Vicia narbonensis											0	(0)	0	0
Vicia sativa											–	0	0	0
Vicia villosa											(–)		(–)	

All species are presented alphabetically according to scientific names. The scientific names of the species correspond to the common names between parentheses: *Cicer arietinum* (chickpea), *Glycine max* (soybean), *Lathyrus aphaca* (yellow pea), *Lathyrus cicera* (red pea), *Lathyrus clymenum* (cicercha purpurina), *Lathyrus ochrus* (cyprus vetch), *Lathyrus sativus* (white pea), *Lens culinaris* (lentil), *Lupinus albus* (white lupin), *Lupinus angustifolius* (narrowleaf lupin), *Lupinus atlanticus* (Not Available), *Lupinus luteus* (yellow lupin), *Lupinus pilosus* (blue lupin), *Phaseolus vulgaris* (kidney bean), *Pisum sativum* (pea), *Vicia articulata* (oneflower vetch), *Vicia benghalensis* (purple vetch), *Vicia ervilia* (blister vetch), *Vicia faba* (faba bean), *Vicia hybrida* (hairy yellow vetch), *Vicia narbonensis* (purple broad vetch), *Vicia sativa* (garden vetch), and *Vicia villosa* (winter vetch). Mean ratios are estimated from weighted experimental data. Only the model variant including one random "field site*growing season" effect is used. The symbol "–" indicates that the value of the criterion for the given species is significantly lower than that for pea, and that more than two articles are available for the species concerned. The symbol "(–)" indicates that the value of the criterion for the given species is significantly lower than that of pea, and that there are no more than two articles for the species concerned. The symbol "0" indicates that the value of the criterion for the given species is not significantly different from that of pea, and that there are more than two articles for the species concerned. The symbol "(0)" indicates that the value of the criterion for the given species is not significantly different from that of pea, and that there are no more than two articles available for the species concerned. The symbol "+" indicates that the value of the criterion for the given species is significantly higher than that of pea, and that there are more than two articles available for the species concerned. The symbol "(+)" indicates that the value of the criterion for the given species is significantly higher than that of pea, and that there are no more than two articles available for the species concerned

[a]Experimental data from the selected articles are used to estimate the given criterion

[b]Nutritional data from the FEEDIPEDIA database are used to estimate the given criterion

Table 4.7 Productivity rankings for all grain legume species relative to pea in Europe, North America and Oceania (inclusion of two random "article" and "field site*growing season" effects)

	Europe					North America					Oceania			
	Grain biomass ratio[a]	Total aerial biomass ratio[a]	Grain crude protein ratio[a,b]	Grain gross energy ratio[a,b]	Grain nitrogen content ratio[a]	Grain biomass ratio	Total aerial biomass ratio	Grain crude protein ratio	Grain gross energy ratio	Grain nitrogen content ratio	Grain biomass ratio	Total aerial biomass ratio	Grain crude protein ratio	Grain gross energy ratio
Cicer arietinum	–	0	–	0	(0)	–	–	–	–	(–)	–	–	–	–
Glycine max	0		+	+		–	(0)	0	–					
Lathyrus aphaca											(–)			
Lathyrus cicera											0	0	0	0
Lathyrus clymenum											(0)			
Lathyrus ochrus											0	(0)	0	0
Lathyrus sativus						(0)	(0)	(0)	(0)	(0)	0	0	0	0
Lens culinaris	(–)	(0)	(–)	(–)	(–)	–	0	–	–	–	–	–	–	–
Lupinus albus	–	(–)									–	–		
Lupinus angustifolius	–	0	0	0	0	–		–	–	(–)	0	0	0	0
Lupinus atlanticus											(0)	(0)		
Lupinus luteus	(0)	(0)	(0)	(0)	(0)						(–)	(0)	(–)	(0)
Lupinus pilosus											(0)	(0)		
Phaseolus vulgaris						–		–	–	(–)				
Vicia articulata											(0)			
Vicia benghalensis											(–)	(0)		

(continued)

Table 4.7 (continued)

	Europe					North America					Oceania			
	Grain biomass ratio[a]	Total aerial biomass ratio[a]	Grain crude protein ratio[a,b]	Grain gross energy ratio[a,b]	Grain nitrogen content ratio[a]	Grain biomass ratio	Total aerial biomass ratio	Grain crude protein ratio	Grain gross energy ratio	Grain nitrogen content ratio	Grain biomass ratio	Total aerial biomass ratio	Grain crude protein ratio	Grain gross energy ratio
Vicia ervilia											(–)	(0)		
Vicia faba	0	0	0	0	0	–	(0)	0	–	(0)	+	0	+	+
Vicia hybrida											(–)			
Vicia narbonensis											0	(0)	0	0
Vicia sativa											0	0	0	0
Vicia villosa											(–)		(–)	

All species are presentd alphabetically according to scientific names. The scientific names of the species correspond to the common names between parentheses: *Cicer arietinum* (chickpea), *Glycine max* (soybean), *Lathyrus aphaca* (yellow pea), *Lathyrus cicera* (red pea), *Lathyrus clymenum* (cicercha purpurina), *Lathyrus ochrus* (cyprus vetch), *Lathyrus sativus* (white pea), *Lens culinaris* (lentil), *Lupinus albus* (white lupin), *Lupinus angustifolius* (narrowleaf lupin), *Lupinus atlanticus* (Not Available), *Lupinus luteus* (yellow lupin), *Lupinus pilosus* (blue lupin), *Phaseolus vulgaris* (kidney bean), *Pisum sativum* (pea), *Vicia articulata* (oneflower vetch), *Vicia benghalensis* (purple vetch), *Vicia ervilia* (blister vetch), *Vicia faba* (faba bean), *Vicia hybrida* (hairy yellow vetch), *Vicia narbonensis* (purple broad vetch), *Vicia sativa* (garden vetch), and *Vicia villosa* (winter vetch). Mean ratios are estimated from weighted experimental data. Only the model variant including two random "article" and "field site*growing season" effects is used. The symbol "–" indicates that the value of the criterion for the given species is significantly lower than that for pea, and that more than two articles are available for the species concerned. The symbol "(–)" indicates that the value of the criterion for the given species is significantly lower than that of pea, and that there are no more than two articles for the species concerned. The symbol "0" indicates that the value of the criterion for the given species is not significantly different from that of pea, and that there are more than two articles for the species concerned. The symbol "(0)" indicates that the value of the criterion for the given species is not significantly different from that of pea, and that there are no more than two articles available for the species concerned. The symbol "+" indicates that the value of the criterion for the given species is significantly higher than that of pea, and that there are more than two articles available for the species concerned. The symbol "(+)" indicates that the value of the criterion for the given species is significantly higher than that of pea, and that there are no more than two articles available for the species concerned

[a]Experimental data from the selected articles are used to estimate the given criterion

[b]Nutritional data from the FEEDIPEDIA database are used to estimate the given criterion

would provide farmers with high added value market opportunities, which might offset their low productivities (Voisin et al. 2014; Magrini et al. 2016). The replacement of pea with other legume species, displaying higher levels of grain crude protein, could substantially increase overall crude protein production in European countries. An alternative scenario is to increase the proportion of cultivated land cropped with grain legumes species compared to cereal and oilseed crops, even if this scenario could lead to a reduction of cereal and oilseed productions.

The overall species comparisons obtained in North America and Oceania confirm the key findings for Europe. In North America, Miller et al. (2002) reported that the mean grain biomass of soybean was −47% lower than that of pea. This lower grain biomass is consistent with our results. In North America, soybean has a mean grain biomass −41% (unweighted values) or − 53% (weighted values) lower than that of pea (Fig. 4.4a and Table 4.5). However, soybean performs well for most of the other criteria, and, in addition, presents lower interannual variability in grain biomass than that of pea (Cernay et al. 2015).

The results obtained in North America and Oceania also confirm relatively good yield levels of faba bean in North America, and its interesting performances in Oceania. Siddique et al. (2013) reported that faba bean could be grown across a broad spectrum of contrasting environments in Australia. White pea has yield levels similar to those of pea for most of the criteria considered in both North America and Oceania, but with high levels of uncertainty due to the small number of experiments carried out. Calderón et al. (2012), and Vaz Patto and Rubiales (2014) argued that further research efforts are required to assess the agronomic potential of white pea. The grains of this legume species harbor several undesirable nutritional and neurotoxic factors, decreasing its likely suitability as a food and feed substitute for pea (Hanbury et al. 2000; Yan et al. 2006; Vaz Patto and Rubiales 2014).

The number of species directly compared with pea is about three times higher in Oceania (20 species) than in North America (7 species) and Europe (7 species). Several of the species that have been tested in Oceania but not in Europe perform well with respect to pea in terms of productivity, especially for red pea, *Cicercha purpurina*, cyprus vetch, white pea, *Lupinus atlanticus* (no common name), blue lupin, oneflower vetch, purple broad vetch, and garden vetch. It will be useful to include these grain legume species in future field experiments with the prospect of assessing their relative productive performances across European conditions.

Acknowledgments We thank Gaetano Amato, Chengci Chen, Adria L Fernandez, Yantai T Gan, Hans-Peter Kaul, Mark A Liebig, Perry Miller, Reinhard Neugschwandtner, Kadambot H M Siddique *and* Den Spaner for providing us with additional information and raw data. We thank Alain Bône for assistance with the literature search. We thank Tamara Ben-Ari, Guénaëlle Corre-Hellou, Fabien Ferchaud, Valérie Heuzé, Marie-Hélène Jeuffroy, Philippe Lescoat, Jean-Marc Meynard, Olivier Réchauchère, Anne Schneider, Gilles Tran, and Anne-Sophie Voisin, for comments. This work was supported by the French National Research Agency (ANR) under the "Investments for the Future" program (ANR-10-IDEX-0003-02), as part of the STIMUL (Scenarios towards integrating multi-scale land use tools) flagship project funded by the LabEx BASC (ANR–11–LABX–0034).

References

Akaike H (1974) A new look at the statistical model identification. IEEE Trans Autom Control 19:716–723. https://doi.org/10.1109/TAC.1974.1100705

Angadi SV, Brandt SA, McConkey BG, Cutforth HW, Miller PR, Ulrich D, Selles F, Volkmar KM, Entz MH, Brandt SA (2008) Adaptation of alternative pulse and oilseed crops to the semiarid Canadian Prairie: seed yield and water use efficiency. Can J Plant Sci 88:425–438. https://doi.org/10.4141/CJPS07078

Boye J, Zare F, Pletch A (2010) Pulse proteins: processing, characterization, functional properties and applications in food and feed. Food Res Int 43:414–431. https://doi.org/10.1016/j.foodres.2009.09.003

Calderón FJ, Vigil MF, Nielsen DC, Benjamin JG, Poss DJ (2012) Water use and yields of no-till managed dryland grasspea and yellow pea under different planting configurations. Field Crop Res 125:179–185. https://doi.org/10.1016/j.fcr.2011.08.006

Cernay C, Ben-Ari T, Pelzer E, Meynard JM, Makowski D (2015) Estimating variability in grain legume yields across Europe and the Americas. Sci Rep 5:11171. https://doi.org/10.1038/srep11171

Cernay C, Pelzer E, Makowski D (2016) A global experimental dataset for assessing grain legume production. Sci Data 3:160084. https://doi.org/10.1038/sdata.2016.84

Chalk PM (1998) Dynamics of biologically fixed N in legume-cereal rotations: a review. Aust J Agric Res 49:303–316. https://doi.org/10.1071/A97013

Crépon K, Marget P, Peyronnet C, Carrouée B, Arese P, Duc G (2010) Nutritional value of faba bean (*Vicia faba* L.) seeds for feed and food. Field Crop Res 115:329–339. https://doi.org/10.1016/j.fcr.2009.09.016

Duranti M (2006) Grain legume proteins and nutraceutical properties. Fitoterapia 77:67–82. https://doi.org/10.1016/j.fitote.2005.11.008

Duc G, Jeuffroy MH, Tivoli B (2011) Grain legumes to improve environmental services in grain crops production systems: main INRA achievements that contributed to the implementation of the supply chain and prospects. Innovations Agronomiques 12:157–180

Evans J, McNeill AM, Unkovich MJ, Fettell NA, Heenan DP (2001) Net nitrogen balances for cool-season grain legume crops and contributions to wheat nitrogen uptake: a review. Aust J Exp Agr 41:347–459. https://doi.org/10.1071/EA00036

FAOSTAT. Statistics Division of Food and Agriculture Organization of the United Nations (FAOSTAT database) (2016) Available from: http://faostat3.fao.org/home/E. Accessed 05 Apr 2016

FEEDIPEDIA. Animal Feed Resources Information System (FEEDIPEDIA database) (2016) Available from: http://www.feedipedia.org/. Accessed 15 Mar 2016

Flores F, Rubiales D, Hybl M, Knudsen JC, Marget P, Muel F, Nadal S, Narits L, Raffiot B, Sass O, Solis I, Winkler J, Stoddard FL, Rubiales D (2013) Adaptation of spring faba bean types across European climates. Field Crop Res 145:1–9. https://doi.org/10.1016/j.fcr.2013.01.022

Foyer CH, Considine MJ, Lam HM, Nguyen HT, Siddique KHM, Varshney RK, Colmer TD, Cowling W, Bramley H, Mori TA, Hodgson JM, Cooper JW, Miller AJ, Kunert K, Vorster J, Cullis C, Ozga JA, Wahlqvist ML, Liang Y, Shou H, Shi K, Yu JQ, Fodor N, Kaiser BN, Wong FL, Valliyodan B, Considine MJ (2016) Neglecting legumes has compromised human health and sustainable food production. Nat Plants 2:16112. https://doi.org/10.1038/nplants.2016.112

Gan Y, Poppy L, Hamel C, O'Donovan JT, Cutforth H, Zentner RP, Campbell CA, Niu Y, Poppy L (2015) Diversifying crop rotations with pulses enhances system productivity. Sci Rep 5:14625. https://doi.org/10.1038/srep14625

Gepts P, Young ND, Beavis WD, Brummer EC, Shoemaker RC, Stalker HT, Weeden NF, Young ND (2005) Legumes as a model plant family genomics for food and feed. Report of the cross-legume advances through genomics conference. Plant Physiol 137:1228–1235. https://doi.org/10.1104/pp.105.060871

Graham PH, Vance CP (2003) Legumes: importance and constraints to greater use. Plant Physiol 131:872–877. https://doi.org/10.1104/pp.017004

Gregory PJ (1998) Alternative crops for Duplex soils: growth and water use of some cereal, legume, and oilseed crops, and pastures. Aust J Agric Res 49:21–32. https://doi.org/10.1071/A97053

Hanbury CD, White CL, Mullan BP, Siddique KHM (2000) A review of the potential of *Lathyrus sativus* L. and *L. cicera* L. grain for use as animal feed. Anim Feed Sci Tech 87:1–27. https://doi.org/10.1016/S0377-8401(00)00186-3

Hedges LV, Gurevitch J, Curtis PS (1999) The meta-analysis of response ratios in experimental ecology. Ecology 80:1150–1156

Jensen CR, Petersen CT, Joernsgaard B, Andersen MN, Christiansen JL, Mogensen VO, Friis P, Petersen CT (2004) The effect of lupins as compared with peas and oats on the yield of the subsequent winter barley crop. Eur J Agron 20:405–418. https://doi.org/10.1016/S1161-0301(03)00057-1

Jensen ES, Morrison MJ, Peoples MB, Boddey RM, Gresshoff PM, Hauggaard-Nielsen H, Alves BJR, Morrison MJ (2012) Legumes for mitigation of climate change and the provision of feedstock for biofuels and biorefineries: a review. Agron Sustain Dev 32:329–364. https://doi.org/10.1007/s13593-011-0056-7

Jukanti AK, Gaur PM, Gowda CLL, Chibbar RN (2012) Nutritional quality and health benefits of chickpea (*Cicer arietinum* L.): a review. Br J Nutr 108:11–26. https://doi.org/10.1017/S0007114512000797

Koivunen E, Partanen K, Perttilä S, Palander S, Tuunainen P, Valaja J (2016) Digestibility and energy value of pea (*Pisum sativum* L.), faba bean (*Vicia faba* L.) and blue lupin (narrow-leaf) (*Lupinus angustifolius*) seeds in broilers. Anim Feed Sci Tech 218:120–127. https://doi.org/10.1016/j.anifeedsci.2016.05.007

Laurence R (1979) A comparison of the grain and protein yield potential of some annual legume species in South Australia. Aust J Exp Agr Anim Husb 19:495–503. https://doi.org/10.1071/EA9790495

LMC International (2009) Evaluation of measures applied under the Common Agricultural Policy to the protein crop sector. European Commission, Agriculture and Rural Development, Evaluation, Reports on Market and Income Policies. Available from: http://ec.europa.eu/agriculture/eval/reports/protein_crops/index_en.htm. Accessed 12 Apr 2016

Lucas MM, Stoddard FL, Annicchiarico P, Frías J, Martínez-Villaluenga C, Sussmann D, Duranti M, Seger A, Zander PM, Pueyo JJ (2015) The future of lupin as a protein crop in Europe. Front Plant Sci 6:705. https://doi.org/10.3389/fpls.2015.00705

Magrini MB, Walrand S, Anton M, Cholez C, Corre-Hellou G, Duc G, Jeuffroy MH, Meynard JM, Pelzer E, Voisin AS, Walrand S (2016) Why are grain-legumes rarely present in cropping systems despite their environmental and nutritional benefits? Analyzing lock-in in the French agrifood system. Ecol Econ 126:152–162. https://doi.org/10.1016/j.ecolecon.2016.03.024

Malik RS, Seymour M, French RJ, Kirkegaard JA, Lawes RA, Liebig MA (2015) Dynamic crop sequencing in western Australian cropping systems. Crop Pasture Sci 66:594–609. https://doi.org/10.1071/CP14097

Mengersen K, Schmid CH, Jennions MD, Gurevitch J (2013) Statistical models and approaches to inference. In: Koricheva J, Gurevitch J, Mengersen K (eds) Hanbook of meta-analysis in ecology and evolution. Princeton University Press, Princeton/Oxford, pp 89–107

Miller PR, McDonald CL, Derksen DA, Waddington J (2001) The adaptation of seven broadleaf crops to the dry semiarid prairie. Can J Plant Sci 81:29–43. https://doi.org/10.4141/P00-028

Miller PR, Neill KE, McConkey BG, Clayton GW, Brandt SA, Staricka JA, Johnston AM, Lafond GP, Schatz BG, Baltensperger DD, Neill KE (2002) Pulse crop adaptation in the northern Great Plains. Agron J 94:261–272. https://doi.org/10.2134/agronj2002.2610

Multari S, Stewart D, Russell WR (2015) Potential of fava bean as future protein supply to partially replace meat intake in the human diet. Compr Rev Food Sci Food Saf 14:511–522. https://doi.org/10.1111/1541-4337.12146

Murphy-Bokern D, Torres A, Watson CA, Stoddard FL, Lindström K, Zander P, Reckling M, Preissel S, Bues A, Torres A (2014) Outlook for knowledge and technology for legume-supported cropping systems. Legume Futures Report 5.3. European Commission, Seventh

Framework Programme for Research and Technological Development, Legume-Supported Cropping Systems for Europe (Legume Futures). Available from: http://www.legumefutures.de/images/Legume_Futures_Report_5.3.pdf. Accessed 19 Apr 2016

Peel MC, Finlayson BL, McMahon TA (2007) Updated world map of the Köppen-Geiger climate classification. Hydrol Earth Syst Sci 11:1633–1644. https://doi.org/10.5194/hess-11-1633-2007

Peoples MB, Jensen ES, Brockwell J, Herridge DF, Rochester IJ, Alves BJR, Urquiaga S, Boddey RM, Dakora FD, Bhattarai S, Maskey SL, Sampet C, Rerkasem B, Khan DF, Hauggaard-Nielsen H, Jensen ES (2009) The contributions of nitrogen-fixing crop legumes to the productivity of agricultural systems. Symbiosis 48:1–17. https://doi.org/10.1007/BF03179980

Philibert A, Loyce C, Makowski D (2012) Assessment of the quality of the meta-analysis in agronomy. Agriculture, Ecosystem and Environment 148:72–82

Pinheiro J, Bates D, DebRoy S, Sarkar D, R Development Core Team (2013) R package "nlme": linear and nonlinear mixed effects models. Version 3.1.111

Preissel S, Reckling M, Schläfke N, Zander P (2015) Magnitude and farm-economic value of grain legume pre-crop benefits in Europe: a review. Field Crop Res 175:64–79. https://doi.org/10.1016/j.fcr.2015.01.012

R Development Core Team (2013) R: a language and environment for statistical computing. R Foundation for Statistical Computing. Version 3.0.2. Available from: http://www.r-project.org/. Accessed 15 Oct 2013

Reckling M, Bachinger J, Bergkvist G, Watson CA, Stoddard FL, Zander PM, Walker R, Pristeri A, Toncea I, Bachinger J (2016) Trade-offs between economic and environmental impacts of introducing legumes into cropping systems. Front Plant Sci 7:669. https://doi.org/10.3389/fpls.2016.00669

Siddique KHM, Loss SP, Regan KL, Jettner RL (1999) Adaptation and seed yield of cool season grain legumes in Mediterranean environments of south-western Australia. Aust J Agric Res 50:375–388. https://doi.org/10.1071/A98096

Siddique KHM, Materne M, Erskine W, Hobson K, Knights EJ, Leonforte A, Khan TN, Paull JG, Redden R, Materne M (2013) Cool-season grain legume improvement in Australia – use of genetic resources. Crop Pasture Sci 64:347–360. https://doi.org/10.1071/CP13071

Silsbury JH (1975) Grain yield and flowering of some introduced grain legumes in south Australia. Aust J Exp Agr Anim Husb 15:556–560. https://doi.org/10.1071/EA9750556

Sinclair TR, Vadez V (2012) The future of grain legumes in cropping systems. Crop Pasture Sci 63:501–512. https://doi.org/10.1071/CP12128

Temba MC, Njobeh PB, Adebo OA, Olugbile AO, Kayitesi E (2016) The role of compositing cereals with legumes to alleviate protein energy malnutrition in Africa. Int J Food Sci Tech 51:543–554. https://doi.org/10.1111/ijfs.13035

Vadez V, Berger JD, Warkentin T, Asseng S, Ratnakumar P, Chandra Rao KP, Gaur PM, Munier-Jolain N, Larmure A, Voisin AS, Sharma HC, Pande S, Sharma M, Krishnamurthy L, Zaman MA (2012) Adaptation of grain legumes to climate change: a review Agron. Sustain Dev 32:31–44. https://doi.org/10.1007/s13593-011-0020-6

Vaz Patto MC, Rubiales D (2014) *Lathyrus* diversity: available resources with relevance to crop improvement – *L. sativus* and *L. cicera* as case studies. Ann Bot–London 113:895–908. https://doi.org/10.1093/aob/mcu024

Vaz Patto MC, Domoney C, Amarowicz R, Aryee ANA, Boye JI, Chung HJ, Martin-Cabrejas MA, Domoney C (2015) Achievements and challenges in improving the nutritional quality of food legumes. Crit Rev Plant Sci 34:105–143. https://doi.org/10.1080/07352689.2014.897907

Voisin AS, Guéguen J, Huyghe C, Jeuffroy MH, Magrini MB, Meynard JM, Mougel C, Pellerin S, Pelzer E (2014) Legumes for feed, food, biomaterials and bioenergy in Europe: a review. Agron Sustain Dev 34:361–380. https://doi.org/10.1007/s13593-013-0189-y

Wiebe N, Vandermeer B, Platt RW, Klassen TP, Moher D, Barrowman NJ (2006) A systematic review identifies a lack of standardization in methods for handling missing variance data. J Clin Epidemiol 59:342–353. https://doi.org/10.1016/j.jclinepi.2005.08.017

Wiggering H, Finckh M, Heß J, Wehling P, Michaelis T (2012) The Legumes Expert Forum. Science, economy and society – making ecosystem services from legumes competitive. A research strategy of the German Agricultural Research Alliance. German Agricultural Research Alliance (DAFA). Available from: http://www.dafa.de/fileadmin/dam_uploads/images/Fachforen/ff_leguminosen-en_2012.pdf. Accessed 12 Apr 2016

Williams CM, King JR, Ross SM, Olson MA, Hoy CF, Lopetinsky KJ (2014) Effects of three pulse crops on subsequent barley, canola, and wheat. Agron J 106:343–350. https://doi.org/10.2134/agronj2013.0274

Yan ZY, Li FM, Spencer PS, Li ZX, Liang YM, Wang YF, Li FM (2006) Lathyrus sativus (grass pea) and its neurotoxin ODAP. Phytochemistry 67:107–121. https://doi.org/10.1016/j.phytochem.2005.10.022

Zander P, Watson C, Amjath-Babu TS, Preissel S, Reckling M, Bues A, Schlaefke N, Kuhlman T, Bachinger J, Uthes S, Stoddard F, Murphy-Bokern D, Watson C (2016) Grain legume decline and potential recovery in European agriculture: a review. Agron Sustain Dev 36:1–20. https://doi.org/10.1007/s13593-016-0365-y

Zentner RP, Derksen DA, Wall DD, Nagy CN, Smith EG, Young DL, Miller PR, Campbell CA, McConkey BG, Brandt SA, Lafond GP, Johnston AM, Derksen DA (2002) Economics of crop diversification and soil tillage opportunities in the Canadian Prairies. Agron J 94:216–230. https://doi.org/10.2134/agronj2002.2160

Chapter 5
Grain Legumes for the Sustainability of European Farming Systems

Faisal Mahmood, Tanvir Shahzad, Sabir Hussain, Muhammad Shahid, Muhammad Azeem, and Jacques Wery

Abstract Grain legumes offer many agronomic, environmental and socio-economic benefits when grown in succession with cereals. They can increase the yields of following crops in the rotation. They fix indirectly atmospheric nitrogen, which makes them economical and environmentally friendly. Globally grain legumes are cultivated on an area of 201,728 thousand ha with a total production of 383,728 thousand tones. In Europe, grain legumes are cultivated on an area of 5726 thousand ha, which represents only 1.8% of total arable lands in Europe. Cultivated area of grain legumes is very low as compared to other words countries and, consequently, Europe imports yearly 20 million tons of soybean meals and 12 million tons of soybean grain. Farmers show lack of interest in cultivating grain legumes due to many climatic, soils, technical, agronomic and economic constraints. These constraints can be removed by technological innovations, provision of more premiums, increasing the sale price and grain yield, and reduction in yield variability of grain legumes.

Keywords Grain legumes · Biological N fixation · Alternative crops · Sustainable agriculture · Crop rotations

F. Mahmood (✉) · T. Shahzad · S. Hussain
Department of Environmental Sciences & Engineering, Government College University Faisalabad, Faisalabad, Pakistan

M. Shahid
Department of Bioinformatics & Biotechnology, Government College University Faisalabad, Faisalabad, Pakistan

M. Azeem
Department of Botany, Government College University Faisalabad, Faisalabad, Pakistan

J. Wery
SupAgro Montpellier UMR-System, Montpellier, France

E. Lichtfouse (ed.), *Sustainable Agriculture Reviews 32*, Sustainable Agriculture Reviews 32, https://doi.org/10.1007/978-3-319-98914-3_5

5.1 Introduction

The reconciliation of economy and environment is a key factor in achieving sustainability. The European Union (EU) wishes to achieve the sustainability of its agriculture in order to produce high quality food materials and to manage energy crisis and the risks related to climate and market fluctuations. These risks can be reduced by enforcing a reduction in the possible negative impacts of agricultural activities on the environment such as water quality, biodiversity, greenhouse gas emissions and public health (MP3-Grain Legumes 2010). Grain legumes are generally considered as key crops for sustainable agriculture (AEP 2004; Wani et al. 2003) due to their unique characteristic of nitrogen-fixing plants. This makes them economical and environmentally friendly crops compared to other arable crops (Reckling et al. 2016).

Synthetic nitrogen (N) fertilizers are considered as one of the most expensive input in modern agriculture, which account for 40–65% of on-farm commercial energy use, respectively for developed and less developed countries (Mudahar and Hignett 1987). Cereals are considered as the major N fertilizer user crops and it is estimated that approximately 50% of the world N fertilizers have been used only by cereal crops every year (Roberts 2009). Smil (2001) reported that worldwide about 1.3% of all the energy produced is used by various types of fertilizers and the cost of fertilizers is expected to increase due to increasing use of non-renewable energy resources for other purposes.

In this context, the crops which used very small or no N fertilizer should be promoted to limit mineral nitrogen and energy used in agriculture (Magrini et al. 2016). Cultivation of grain legumes in the EU could be one of the best alternative choices. Currently grain legumes are grown on only 1.8% of arable lands in the European Union (EU), which was 4.7% in 1961. Moreover, there is a substantial deficiency of vegetable proteins in Europe and every year this deficiency is compensated by importing 20 million tons of soybean meals and 12 million tons of soybean grain from America, which cause a heavy load on import budget (Roman et al. 2016; Magrini et al. 2016). Deficiency of grain legumes production in Europe was due to the General Agreement on Tarif and Trade in 1947 followed by Blair House Agreement in 1992. According to these agreements, the EU was allowed to import protein from America without any tax on import, which discouraged the farmers to grow grain legumes on their farms because these crops were less competitive to the cereals (Cernay et al. 2015). Moreover, different institutional, agronomic (Preissel et al. 2015), technical, climatic and economic constraints also discouraged the farmers to grow grain legumes on their farms (Von Richthofen et al. 2006).

The most frequent problems cited for legumes are: provision of less subsidies compared to other cereal crops, higher susceptibility to pest and diseases (Gueguen et al. 2008; Wery and Ahlawat 2007), need of greater technicality for their production (Carrouee et al. 2003) and low or fluctuating prices and crop yield (Jeuffroy 2006), inducing an overall low competitiveness with cereal crops in farming systems. Due to these constraints, the EU grain legumes sector has strongly declined over the last decade. In France, their area has now reached its lowest level

(165,000 ha) since the 80s with 63% decrease observed only between 2004 and 2008 (Magrini et al. 2016). It is challenging to propose and evaluate strategies that would allow in increasing grain legumes area in Europe by considering all above mentioned constraints. This should be addressed in a variety of contents by keeping in mind the institutional, socioeconomic and environmental factors (Reckling et al. 2016). In this chapter, the author has tried to compile all that data and analysis that can help for promoting grain legumes area in Europe.

5.1.1 Growing of Grain Legumes in World and in Europe

Globally grain legumes are cultivated on an area of 201,728 thousand hectares (thou ha) with a total production of 383,728 thousand tones (thou tones). Soybean is the most dominant crop cultivated on an area of 111,272 thou tons (72.1% of global area under grain legumes) with a production of 276,405 thou tones. It is mainly cultivated in South and Central America with an area of 52,106 thou ha and yield of 146,149 thou tons in Brazil and Argentina and North America (32,523 thou ha and 94,681 thou tons) mainly in USA and Asia(20,629 thou ha and 27,294 thou tons) mainly in China (Roman et al. 2016).

Except oceanic countries, area under grain legumes is much lower in Europe as compared to other regions of the world. In Europe, grain legumes are cultivated only on an area of 5726 thou ha which is only 2.8% of the total global area, with a total production of 10,575 thou tones in 2015. Soybean is the most dominant crop cultivated in the EU, with total area of 3176 thou ha and total production 5943 thou tons, followed by field bean and broad beans (Tables 5.1 and 5.2). The area dedicated to grain legumes in the EU represents between 1% and 7% of the arable crop area sown with grain legumes in European countries (Von Richthofen and GL-Pro partners 2006). This is very low compared to Brazil (44%), USA (32%), India

Table 5.1 Grain legumes growing area in the world (thousand ha)

Species	North America	South and Central America	Europe	Africa	Asia	Oceania	Total
Pea	1634	152	1723	812	1875	181	6377
Field beans	616	5957	260	5695	14,237	62	26,827
Soybean	32,523	52,106	3176	1797	20,629	41	111,272
Lentils	1095	20	84	178	2820	146	4343
Chick pea	158	161	74	483	12,079	574	13,529
Broad bean	0	163	238	570	964	112	2047
Lupine	0	34	153	14	0	450	651
Peanuts	421	686	11	12,405	11,871	14	25,408
Cow pea	16	16	7	11,075	160	0	11,274
TOTAL	36,463	60,295	5726	33,029	64,635	1580	201,728

Roman et al. (2016)

Table 5.2 Grain legumes total production in the world (thousand tons)

Species	North America	South and Central America	Europe	Africa	Asia	Oceania	Total
Pea	4558	188	3021	720	2229	263	10,979
Field beans	1317	5590	500	4860	10,635	53	22,855
Soybean	94,681	146,149	5943	2246	27,294	92	276,405
Lentils	2108	12	71	186	2246	327	4950
Chick pea	327	270	94	531	11,068	813	13,103
Broad bean	0	192	663	738	1494	297	3381
Lupine	0	55	251	21	1	459	787
Peanuts	1893	1759	8	11,547	29,951	28	45,160
Cow pea	29	19	24	5422	193	0	5687
TOTAL	104,913	154,134	10,575	26,271	85,108	2306	383,307

Roman et al. (2016)

(18%), Canada (13%) and Australia (9%) (Schneider 2008). Whereas the potential of these crops is estimated to be 15–25% like other countries of the world (GL-Pro partners 2007).

5.2 Benefits of Growing Grain Legumes

5.2.1 *Nutritional Value*

Primarily, grain legumes are grown for their grains, which are used either for human consumption as a food or for animal as a feed (Singh et al. 2007). They are the cheapest sources of supplementary proteins for humans compared to meat. For example, the cost of a unit of legume protein is 50% lower than a unit of meat protein in Brazil, 70% in Egypt, 75% in Rwanda and 60% in India (Graham and Vance 2003; Joshi et al. 2002; Byerlee and White 2000). They occupy an important place in human nutrition, especially in low-income groups of people in developing countries, which is why they are often called poor man‘s meat. They are generally good sources of slow-release carbohydrates and are rich in proteins i.e. ~18–25% by weight, which is twice the protein contents of wheat and three times that of rice. Soybean is unique in this family, containing about 35–43% protein in addition to oil (Tharanathan and Mahadevamma 2003). They also contain high levels of macro- and micro-nutrients (Ca, P, K, Fe, and Zn), vitamins, fiber and complex carbohydrates all of which contribute to balanced nutrition. Moreover, they complement the consumption of cereals since they provide an amino acid balance and better protein utilization. An optimum nutritional balance diet is composed of cereals and legumes in an approximate ratio of 2 to 1 (MP3-Grain Legumes 2010). Legume consumption has also been shown to lower cholesterol levels and to reduce the risk of diabetes, breast and colon cancer and heart attacks. Kabagambe et al. (2005) reported that legumes may protect against myocardial in fraction by 38% with the use of one third cup of cooked beans on a daily basis. Soybean and lupins are useful to reduce blood cholesterol and thus can protect from hypercholesterolemia and atherosclerosis (Harland and Haffner 2008; Marchesi et al. 2008; Sirtori et al. 2012). Lupins are also considered to have potential for antidiabetic effect (Bertoglio et al. 2011). Phytoestrogens which is obtained from legumes is considered to have positive effects on reducing the risk of cancer and harmful effect on the uterus, thyroid gland and mammary gland (Gierus et al. 2012).

5.2.2 *Biological N Fixation*

The key strength of grain legumes is their specific characteristic as nitrogen-fixing plants, which fulfil their N requirement from the fixation process (Graham and Vance 2003). From an agro-economic point of view, it is considered as an additional output from the grain legume which is cost-saving in terms of mineral or organic fertilizer purchases and application. This ability of grain legumes makes them special in reducing the need for synthetic N fertilizers to almost zero in the legume crop (Zander et al. 2016).

Biological N fixation by grain legumes is carried out by a symbiotic association with the *Rhizobium* or *Bradyrhizobium* bacteria within the root nodules of legumes. This process is made possible by an enzyme complex, the nitrogenase, which supports the organic N production process from gaseous N_2 (Crew and Peoples 2004; Salisbury and Ross 1978). It is estimated that during the growing season, legumes fix N at the rate of 1–2 kg N ha^{-1} day^{-1} (Giller 2001; Unkovich and Pate 2000; Van Kessel and Hartley 2000). Smil (1999) reported that legumes annually fix about 40 to 60 million metric tons of N in agricultural contexts and 3 to 5 million metric tons N in natural ecosystems. It is estimated that legumes grown for grain, hay, pasture and other agricultural purposes account for almost half of the annual quantity of the N fixed by biological system (Anonymous 1984). Burris and Roberts (1993) reported that biological processes contribute to 65% of the N used in agriculture. Through biological N fixation process, grain legumes can accumulate N in their above as well as belowground biomass. Fababean and field pea are the most widely grown grain legumes in Europe which can accumulate on average 130 and 153 kg N ha^{-1} in their aboveground biomass and large quantities in their belowground biomass accounting 30–60% of the total N accumulated by legumes (Peoples et al. 2009). Figure 5.1 shows the total amount of N fixed in the EU27 by grain legume crops in 2009. Pea, fababean and soybean were the main grain legumes that fixed the maximum amount of N_2. A large amount was also fixed by pulses. Remaining legumes fixed the small amount of N_2.

In combination with plant photosynthesis and potential growth (Wery 1987), the availability of Phosphorus (P) is considered as a major driving force behind N_2 fixation for signal transduction and membrane biosynthesis and also for ATP requirements for nodule development and function (Ribet and Drevon 1996).

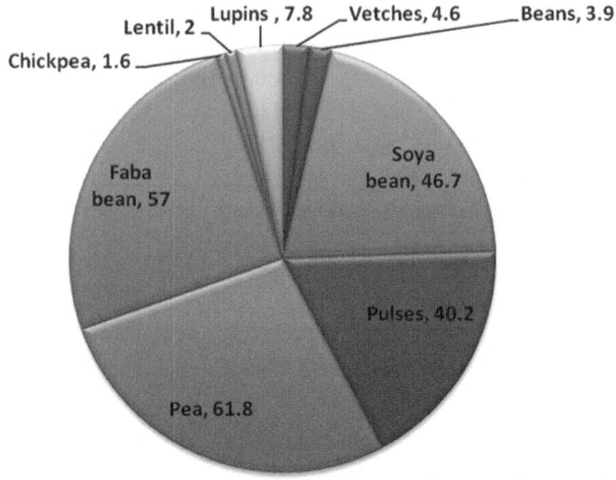

Fig. 5.1 Calculated quantities of total N fixed in Europe (EU27) by grain legume crops in 2009 (Gg) as reported by Baddeley et al. (2014)

About 33% of the world 's arable land is deficient in P (Sanchez and Euhara 1980), while maximum benefits from N_2 fixation depend on the availability of P in the soil (Kennedy and Cocking 1997). The other limitations to N_2 fixation are drought, salinity, N fertilization, and nutrient limitations (Graham and Vance 2003) through their direct effects on nodules or indirect effect on potential growth and N requirements (Wery 1987). There is also a genetic variability in N_2 fixation (Sinclair et al. 1987).

Biological nitrogen fixation by grain legumes is considered a very good alternative of synthetic fertilizer application and its efficiency can be enhanced by integrated nutrient management by adding manure fertilizer, biosoilds and recycling of crop residues (Lal 2004). Direct benefits of grain legumes in term of nitrogen fixation are well documented as compared to their effect on following crops, for example cereals. Peoples et al. (2015) reported that inclusion of grain legumes in rotations of cereals can provide an additional 40 to 90 kg N ha^{-1} in the first year and 20–35 kg of N ha^{-1} in second year of crop rotations.

5.3 Advantages of Growing Grain Legumes Within Cereal-Based Rotations

Grain legumes offer many agronomic, environmental and socio-economic benefits when grown in succession with cereals. Most of the work on grain legumes is done at field scale by comparing their strengths and weaknesses with those of other crops (mainly cereals) (Wery and Ahlawat 2007). But to quantify the benefits of grain legumes and to improve their production and their contribution to sustainable farming systems, the entire crop rotation must be considered. Only the analysis of whole rotation allows a correct and adequate evaluation of grain legume cropping systems (Von Richthofen et al. 2006). As compared to cereals, grain legumes are considered as substitute of N fertilizers and enhancers of soil organic matter content due to the N_2 fixation process. Due to this unique characteristic of grain legumes, crop rotation with grain legumes improves soil health, diversifies cropping systems and maintains soil fertility resulting in many economic, agronomic and environmental advantages (Preissel et al. 2015).

These advantages can be classified into specific and non-specific advantages (Preissel et al. 2015). The production of grains without any fertilization within a rotation is a specific advantage of grain legumes provided by the symbiotic fixation process. The other advantages are non-specific because they are shared with some non legume crops i.e. reduction in amount of N fertilizer for the following crop in the rotation, increase in soil organic matter and hence soil fertility, suppression of weeds, insects and diseases due to break cycle (Stevenson and van Kessel 1997), which results in decreasing the negative environmental impacts of insecticides and pesticides applications (Reckling et al. 2014). Some of these specific and non-specific agronomic, environmental and socio-economic benefits are discussed in the following sections.

5.3.1 Agronomic Benefits

Grain legumes are known to increase the yields for the following crops in the rotation (Rochester et al. 1998). Legumes often increase the yield of the subsequent crops in the rotations as compared to cereals grown after a non-legume crop (Rao et al. 1996; Peoples and Crasswell 1992). It is common in Europe to find cereals' rotation with grain legumes cultivation within the rotations which could lead to increased crop yield of subsequent cereal crops in the rotation. European experiments show that in temperate and in Mediterranean conditions, grain yield of cereals following grain legumes were increased by 0.5–1.6 t $ha^{-1}$and 0.2–1.0 t ha^{-1} respectively (Preissel et al. 2015). The yield of the second or even the third crop in the rotation after legume may also increase. Moreover, the increase in yield following legumes in a rotation is not limited in European cropping systems (Evans et al. 2003). An increase of 25–40% in yield has been observed in maize cultivated after soybeans and common beans in the eastern region of Central Africa (MP3-Grain Legumes 2010). Dakora et al. (1987) reported that in African savanna, the rotations cowpea-maize and groundnut-maize have increased the maize yields by 95% and 89% respectively. Introduction of cowpea as cropping preceded to sorghum in sorghum rotations increased the sorghum yield by 65% (Salez and Martin 1992). A survey in Europe showed that when farmers were asked about the impact of grain legumes in crop rotations, they referred to them as good crops for improving soil fertility and leading to high additional grain yields for the following crops (Von Richthofen et al. (2006). On average they found that wheat after grain legumes produces 0.6–0.9 t ha^{-1} more yield as compared to wheat after cereals. Haque et al. (1995) explained this yield increase by the positive effect of legume on soil's chemical, physical and biological properties. However, positive effect on yield of subsequent crop in the cereal based grain legumes rotations will depend on the species, amounts of N fixed and environmental conditions (Zander et al. 2016). It is reported that on different soil types in Denmark, N uptake in subsequent crops increased by 23–59% after field pea and narrow-leafed lupin (Jensen et al. 2004). Howevr, the increase in yield of durum wheat following vetch in a semi-arid Mediterranean environment was only 14–15% (Giambalvo et al. 2004). Low fertility situation is also one of the important factors in N related yield effects on the subsequent crop (Preissel et al. 2015). Fertile soils with adequate N supply can lead to increased yield of the crop following a legume in a rotation (Bachinger and Zander 2007; Kirkegaard et al. 2008). Yield increased by 0.5 and 1 t ha^{-1} was found in wheat and barley grown after grain legumes in a farmer survey in Germany (Alpmann et al. 2013). However, in semi-arid conditions of central Spain yield increases was only by 0.2 t ha^{-1} for barley following vetch (López-Fando and Almendros 1995).

Application of plant nutrient loaded (Paustian et al. 1997c; Glendining and Powlson 1995) organic amendments and inclusion of legume in continuous cereals rotations help in improving the soil quality and building up the soil carbon pool that consequently increases the crop yields and amount of crop residues returned to the soil (Wani et al. 1994, 2003; Paustian et al. 1997b). Mvondo et al. (2007) and Peoples et al. (1995) stated that legumes rotations with other crops also increase the biological activity of soil by enhancing the presence of fine roots, millipedes, earthworms and ants. This ultimately may result in improving the soil fertility and crop yield of the following crops in the rotations. It is difficult to evaluate the role of grain legumes in changing the total soil N pool, because total soil N pool is very large and annual changes are small (Van Kessel and Hartley 2000). Therefore, long term rotational studies are necessary to quantify such changes. Although such studies are limited, Campbell et al. (2000) evaluated the impact of legume-based cropping systems on total soil N, C and net mineralization over a period of 14 years. They showed an increase of total soil N from 3.26 to 3.58 t ha^{-1} for wheat-lentil rotation as compared to wheat monocrop. Similarly, total soil C was increased from 34.6 to 36.6 t ha^{-1} for the same rotations but with fertilized wheat. They also observed that net mineralization was higher for wheat lentil rotation as compared to wheat monocrop. Despite a reduction of 13 kg N ha^{-1} per year in wheat-lentil compared to wheat monocrop, the total soil N pool increased at a higher rate of 23 kg N ha^{-1} per year compared to 8 kg N ha^{-1} per year for fertilized wheat. Another study with cowpea, pigeon pea and chickpea rotated with sorghum and sunflower showed that total soil N contents were increased after 10 years (Rego and Seeling 1996; Wani et al. 1996). Belowground plant residues are also very crucial for total soil N pool and grain legumes are also very important for that belowground total N pool (Rego and Seeling 1996; Wani et al. 1996) which depends on N_2 fixation by grain legumes (Van Kessel and Hartley 2000). From the above discussion, it can be concluded that grain legumes can increase the total soil N pool. However, this effect is more obvious on poor soils due to increased N_2 fixation (Van Kessel and Hartley 2000).

5.3.2 Socio-economic Benefits

The maximum economic benefits from grain legumes are obtained with long-term rotations because their beneficial effects become apparent only over long periods (Chalk 1998). The reasoning of the rotation is too often based on "the most profitable crops" without considering the entire rotation of which they form a part. The profitability of a crop is considered independently of the succession of different crops that make up the rotation. The isolated comparison of a crop's gross margin does not reveal the monetary value of grain legumes for the following crop (Von Richthofen et al. 2006). The calculation of rotation gross margin demonstrates that inclusion of grain legumes in intensive cereal rotations does not cause a drop in farmers 'income. On the contrary, in most cases grain legumes rotations offers slightly higher gross margins than intensive rotations with 75% or more cereals, as shown in Fig. 5.2 for different rotations in Europe (Von Richthofen et al. 2006). In Saxony-Anhalt (Germany), inclusion of peas in five-year rotations with 80% cereals increased the gross margin by 29 €/ha (11%). Similarly, for four-year rotations this advantage was still 11 €/ha (4%) higher (Von Richthofen et al. 2006). Rao and Mathuva (1999) and Von Richthofen et al. (2006) also found that crop rotations including grain legumes (cowpeas and pigeon pea) have gross margins equal to/or greater than cereals rotations without grain legumes. Carrouée et al. (2002)

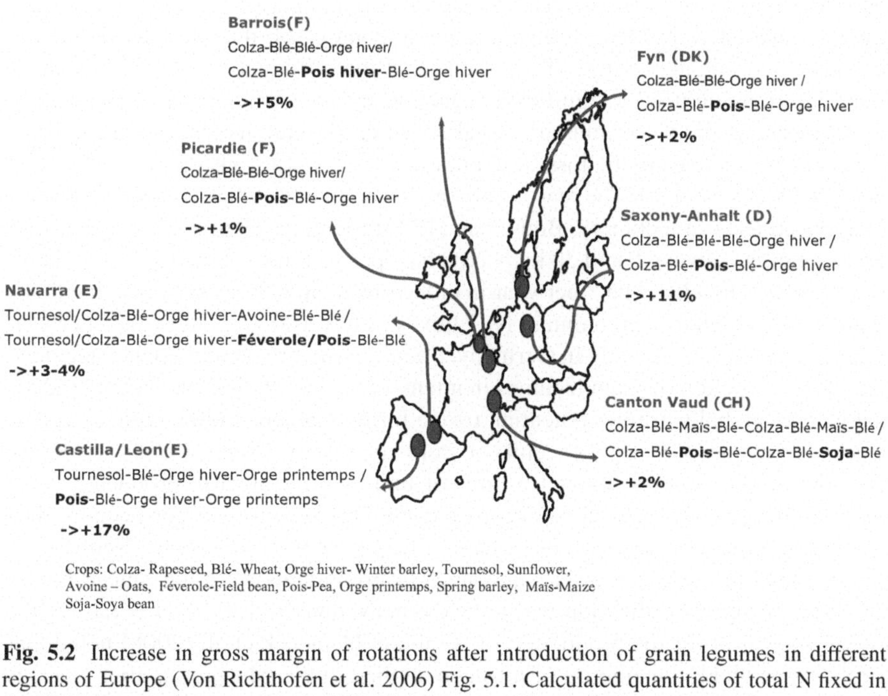

Fig. 5.2 Increase in gross margin of rotations after introduction of grain legumes in different regions of Europe (Von Richthofen et al. 2006) Fig. 5.1. Calculated quantities of total N fixed in the EU27 by grain legume crops in 2009 (Gg) as reported by Baddeley et al. (2014)

compiled different available sources, and discussed the benefits and impacts of introducing grain legumes into crop rotations. They came to a generally positive assessment. Rao and Mathuva (1999) found that cropping systems based on annual grain legumes were 32–49% more profitable than continuous maize cropping. Von Richthofen and GL-Pro partners (2006) found that pesticide and soil tillage costs can be reduced by 20–30% and 25–30% respectively by including the legumes as preceding crop in cereals rotations. They also found that total cost can be reduced by 50 €/ha for pea-cereal rotations as compared to 5 year cereals rotations. Another study conducted by UNIP (2008) in Indre-and-Loire region of France showed that overall peas-wheat rotation can save 60–150 €/ha as compared to continuous cereal rotatio.

5.3.3 Environmental Benefits

Legumes can play a critical role in natural ecosystems, agriculture, and agro-forestry due to their ability to fix atmospheric N_2, which makes them economical and environmentally friendly crops (Graham and Vance 2003). The ability of grain legumes to fix atmospheric nitrogen saves non-renewable energy resources used for synthesis of N fertilizers, as manufacturing nitrogen fertilizer is a high energy-consuming process (Nemecek and Erzinger 2005). Nemecek et al. (2008) stated that introducing grain legumes into European crop rotations offers interesting options for reducing environmental burdens, especially in a context of depleted fossil energy resources and climate change. They found that the introduction of peas in cereal-based rotations induced a significant reduction in; (i) consumption of fossil fuels (14%) as compared to continuous cereal-based crop rotations and (ii) nitrogenous emissions by decreasing the losses of ammonia (−26%), nitrous oxide (−10%) and nitrogen oxides (−11%). The reasons are the lower quantity of N-fertilizers and also the reduced use of machinery. Bouwman (1996) found on 87 plots, N_2O emissions fluxes ranging between 0 and 30 kg N N_2O ha^{-1} per year for fertilized plots, in comparison with 0 to 4 kg N ha^{-1} per year in unfertilized plots. It is estimated that fields planted with legumes can maintain N_2O fluxes as low as 0–0.07 kg N ha^{-1} per year (Conrad et al. 1983). A study in Germany, France, Switzerland and Spain concluded that the introduction of grain legumes in intensive cereal rotations is likely to reduce energy use, global warming potential, ozone formation and acidification as well as eco- and human toxicity per unit of cultivated area (Nemecek et al. 2008). Considering that it takes about 1.5 litres of fuel oil equivalent to produce 1 kg of mineral nitrogen, and that cereal crops receive 180 kg nitrogen per hectare, thus growing legumes can save 270 litres/ha of oil equivalent (UNIP 2008). Ncube et al. (2009) found that when cowpea, pigeonpea or groundnuts were introduced before sorghum, nitrogen fertilization was reduced on average by 130 kg of N ha^{-1} in the following season for the production of sorghum. Nemecek et al. (2008) noted that for the same yield, the amount of nitrogen applied to the wheat crop after pea was 14% lower than the single wheat rotation. He also found that the amount of nitrogen applied to wheat following pea was reduced from 180 kg N ha^{-1} to 157 kg N ha^{-1}. This is confirmed in a study by UNIP (2008) which showed that pea rotated with wheat can save between 20–50 Kg N ha^{-1} as compared to wheat-wheat rotation. Wery and Ahlawat (2007) stated that grain legumes can save 20–60 kg/ha of N for the following cereal with a supplemental yield of 1 t ha^{-1}. Jensen (1997) also found an average N benefit of about 20 kg N ha^{-1} from peas in a crop rotation. He also found that after a pea harvest, greater quantities of mineral N are found in the soil than after a cereal harvest, which can be used by the following crop. Food legumes such as cowpea, mung bean, moth bean, pigeon pea, groundnut and fodder legumes such as berseem were found to increase yields of subsequent cereal crops in semi-arid India by an equivalent effect of 30–40 kgN ha^{-1} (Lal et al. 1978; Rao et al. 1983). It is assumed that in intensive cropping systems the introduction of grain legumes could help in reducing the weeds, insects and diseases, due to breaks in the cycle of these agents (Mwanamwenge et al. 1998; Peoples et al. 1995). Bulson et al. (1997) and Liebman and Dyck (1993) also stated that crop rotations with legumes could provide successful strategies for weed, insects and diseases suppression due to disruption of conditions suitable for their development and may lead to reduce the applications of pesticides and fungicides as compared to continuous cereal rotations (MP3-Grain legumes 2010). Nemecek et al. (2008) showed that inclusion of peas in cereal-based rotations of wheat canola-wheat-wheat-winter barley in Saxony-Anhalt (Germany) has reduced the use of pesticides by 10%. This reduced use of pesticides resulted in significant environmental benefits because it reduced terrestrial eco-toxicity by 7%. The introduction of legumes in continuous cereal-based cropping systems can also improve biodiversity, although as stated by Munier-Jolain and Collard (2006) this effect is not specific to grain legumes. In regions where crop rotations are fairly diverse, as in Switzerland, no additional break-crop effect can be found after the introduction of grain legumes. But in regions where crop rotations are not very diverse, legumes can help in introducing biological diversity. Nemecek et al. (2008) stated that legumes can contribute to the conservation of biological diversity by promoting diversity of crops. The biodiversity points given by the SALCA assessment method (Jeanneret et al. 2006) were higher (7.3) for rotations with grain legumes as compared to rotation without grain legumes (7.1). Grain legumes are also considered valuable crops in reducing the soil eosion by improving soil structure, improved water infiltration, and water holding capacity (Bruce et al. 1987; Jensen and Hauggaard-Nielsen 2003; Peoples et al. 2009; Jensen et al. 2011).

5.4 Disadvantages of Grain Legumes

Although legumes have many advantages, they also have some disadvantages when sown within cereals rotations. Some of these disadvantages are detailed in following sections.

5.4.1 Nitrate Leaching

It is generally considered that the reduction in number of N fertilizer applications and total amount of N fertilizer over the legume-based rotation reduces the risk of nitrate leaching (Drinkwater et al. 1998). But this is not always true. N leaching occurs on both legume and cereal-based cropping systems (Dinnes et al. 2002; Fillery 2001; Poss and Saragoni 1992; White 1988). However, this can differ with soil type, climate and growing season. Crew and Peoples (2004) found that N leaching was higher for soils with high hydraulic conductivity, drained soil exposed to flood irrigation or high rainfall. Fillery (2001) stated that there is a higher chance of N leaching during summer or winter fallow in legume-based systems. Moreover, N leaching risks are higher in first growing phase of subsequent crop after the harvest of legume crop due to lower demand of N for the subsequent crop (Fillery 2001; Peoples et al. 2009). Nemecek et al. (2008) showed that crop rotations with peas caused a 4% higher nitrate leaching. They gave several reasons for this behaviour: longer period of bare soil, higher amount of mineral nitrogen in soil after the pea crop, shallow root system of pea crop, more N content of pea straw than wheat straw that leads to higher N mineralization. Von Richthofen et al. (2006) also found that the risk of nitrate leaching is often increased by the inclusion of a grain legume crop in cereal rotations. However, where possible it can be reduced by efficient catch crop or cover crop management, cereal legumes intercropping (Pappa et al. 2012; Jensen and Hauggaard-Nielsen 2003) or early sowing of winter crops (Rapeseed) just after the harvest of grain legumes. Drinkwater et al. (1998) found the reverse results, with cereal-based systems giving an average N leached 7% higher that of legume-based systems. The situation is different with perennial forage legumes, which are growing for a longer period during the year and therefore extract nitrate from soil. For example, Owens et al. (1994) showed a 48–76% reduction of nitrate leaching by including alfalfa in the rotation of cereal crops. One should not draw definite conclusions from such studies because of the use of the best management practices in most of such studies and the use of different rates of N fertilizer (Sinclair and Cassman 1999). Some researchers argue that N derived from legumes has the same negative effects as N derived from chemical fertilizers, and the increased production obtained from N fixed by legumes seems to be insufficient to match the requirement of increasing population (Cassman et al. 2002; Smil 2001; Sinclair and Cassman 1999). However, Crew and Peoples (2004) compared the sustainability of both sources of N in terms of ecological integrity, energy balance and food security and found that N derived from legumes is potentially more sustainable than chemical sources of N.

5.4.2 Labour Requirements

Rao and Mathuva (1999) reported that maize rotated with cowpea required similar labour as a maize monocrop rotation. He also found that maize rotated with different legumes as intercrop resulted in change in labour use. For example, maize crops rotated with cowpea and pigeon pea required respectively 15% and 32% less labour as compared to continuous maize rotation. Wery and Ahlawat (2007), on the other hand, arrived at the opposite conclusion, that labour requirements are higher for legume-based systems than cereal-based systems due to the fact that legumes are less mechanized and more labour is needed for weeding, as no effective post-emergence herbicide is available. They also show that sowing date has a strong effect on the efficiency of labour, for example spring-sown peas and chickpea may improve the efficiency of labour, by reducing the period of high requirement of labour as compared to cereals, which are mostly winter sown. This statement is supported by Nemecek and GL-Pro partners (2006), who found that in Saxony-Anhalt region (Germany), the cultivation of only winter rapeseed and winter cereals required a high number of labour in autumn for all agricultural operations such as tillage, seedbed preparation and sowing, which requires powerful and expensive mechanization. However, they found that it could be managed by integrating grain legumes into the rotation. For example, when a 500-ha farm introduces spring peas into a five-year rotation of rapeseed–wheat–wheat–wheat–barley more than 300 tractor hours/ha was saved between August and October. On the other hand, they found that only about 80 additional hours were required in spring. This indicates that machines and manpower were used more efficiently and the grain legume rotation allowed a larger cropped area to be managed.

5.4.3 Susceptibility to Pests and Diseases

There are two viewpoints. According to some researchers, inclusion of grain legumes with in continuous cereals rotation is helpful in reducing the pest and diseases due to 'break crop effect' (Robson et al. 2002; Prew and Dyke 1979; McEwen et al. 1989; Stevenson and van Kessel 1997). In addition, diversification of cereals rotation with grain legumes also reduced the application of pesticides and fungicides and hence protection of environment (von Richthofen et al. 2006). On the other hand, some scientists argue that the cost of protecting legumes against pest increases with the number of legumes in the system. It is considered that legumes are more susceptible to pests and diseases than cereals, especially in the tropics and sub- tropics (Beaver et al. 2003; Coyne et al. 2003).

5.5 Constraints for Grain Legumes Cultivation in Europe

Previous studies, surveys, farmer's interviews and special reports reported many constraints for the cultivation of grain legumes in Europe (Table 5.3). Some of these constraints are explained here.

Table 5.3 Major constraints identified by experts for grain legumes production in Europe

Main constraints	Adaptability and tolerance	Peas	Fababean	Lupins	Soyabean
Soil	Calcareous soils with $CaCO_3 > 2\%$	++	++	– –	++
	Shallow soils susceptible to drought	+	–	++	–
	Tolerance to waterlogged soil	+	++	+	++
Climate	Tolerance to high temperature	+	–	+	+++
	Tolerance to drought stress	+	–	++	–
	Frost resistance	++ to +++	+ to ++	Nd	– –
Technical and agronomic	Lodging problem	+	++	++	+
	Problem during sowing and harvesting (large seed size)	Nd	–	Nd	Nd
	Tolerance diseases	–	–	–	–
Economic (as compared to non-legume crops)	Premium	↘	↘	↘	↘
	Yield and sale price	↘	↘	↘	↘
	Price and yield variability	↗	↗	↗	↗
	Total cost	↗	↗	↗	↗

Mahmood (2011)
Tolerance sensitivity: +++ (perfect tolerant), ++ (good tolerant), + (moderate tolerant), – (low tolerant), – – (avoid), nd (not determind), ↗ or ↘ (high or low)

5.5.1 *Climate Constraints*

Regarding climate constraints, here is an example of pea and fababean. Pea is the main grain legume cultivated in the EU (GL-Pro partners 2007). One can find two types of peas according to their plantation timing, winter pea sown in October–November and spring pea sown in January–February. According to experts, peas are good cool-season alternative for regions not suited for growing soybeans, because they are comparatively less frost sensitive and may tolerate low temperatures during germination and growth. This is also confirmed by Miller et al. (2002). Experts further reported that most suitable planting period for peas is December and January. This is because of the chances of heavy frost in October and November. It also helps to reduce disease pressure and lodging problem, compared to October sowing and risk of yield loss due to high temperatures and drought during the grain formation stage compared to February sowing. Based on the plantation timing, fababean can also be classified into two types, spring fababean and winter fababean. According to experts, winter and spring fababean cannot tolerate the severe cold and frequent heat and drought conditions respectively. Thus the most suitable sowing period is December or January. This finding was also confirmed by Carrouée et al. (2003) and GL-Pro partners (2007), who suggested planting of fababean in mid December.

5.5.2 Soil Constraints

Pea and fababean are tolerant to calcareous soils with $CaCO_3 > 2\%$, whereas lupin is not suitable for such soils. They should not be grown on clay and limestone plateau regions with more than 2.5% limestone in the topsoil. In shallow soils, pea and lupin are more sensitive to drought as compared to fababean. Fababean is also more tolerant to waterlogged soils in winter, compared to pea and lupin.

5.5.3 Technical and Agronomic Constraints

A farmer survey in Belgium, Germany, Spain and Switzerland showed that the lack of competitiveness with cereals and alternative break crops (e.g. rapeseed) is the major obstacles for grain legume production (Von Richthofen and GL-Pro partners 2006). More technical skill and expertise are required for sowing and harvesting legumes, compared to cereals. For example, pea is characterized by a high tendency to lodging, so for sowing, it requires perfectly leveled soil with special equipment, which makes it costlier than other crops. Similarly, fababean seeds size is 2–3 times greater than peas seeds. This causes problems during drilling and harvesting making it difficult to adapt drills and combines. Carrouée et al. (2003) also reported the same technical problem faced by farmers during the drilling and harvesting of fababean due to the large seed size. Framers also mentioned the threshing problem of grain legumes especially the farmers of Barrrois in France (Von Richthofen and GL-Pro partners 2006). They also reported diseases as one of the major reasons, for the farmers' lack of interest in growing grain legumes in the region e.g. Anthracnose (lupin), Botrytis fabae and Ascochyta (winter fababean), rust (spring fababean) and root disease of Aphanomyces (pea) (Gueguen et al. 2008).

5.5.4 Economic Constraints

Market price, grain yield and the risk of yield fluctuations are also the major obstacles for cultivation of grain legumes in Europe. Farmers' survey in France also showed seed cost as an important constraint (Von Richthofen and GL-Pro partners 2006). The changes in agricultural policies (common agricultural policy – CAP – reforms) are one of the major factors for farmers 'lack of interest in cultivating grain legumes in Europe. According to a report of UNIP (2009), the impact of CAP reforms on the evolution of grain legume area and production can be analysed in two main phases of agricultural policy changes (UNIP 2009).

Developmental Phase Between 1981 and 1993 During this phase, the area under legumes grew very rapidly (Fig. 5.3 an example of France). The main driving force behind this growth was the establishment of an aid plan for the production of

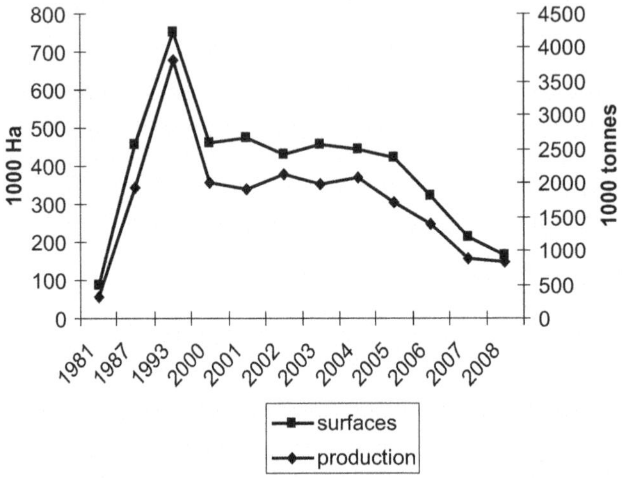

Fig. 5.3 Evolution of area and production of grain legumes in France between 1981 and 2008. (Source: UNIP 2009)

proteins intended to limit Europe's dependence on the major producers of soybean. The area of grain legumes peaked during this period at around 754,000 ha in 1993 (UNIP 2009). The main measures of this aid plan included the pro-active EU policy for protein and market standardization, i.e. (i) Provision of minimum price to farmers for growing peas, fababeans and lupins, and a subsidy for first users of these crop products in the animal feed supply chain. (ii) Provision of compensatory aid to adjust farmers' income, in case of fluctuating price of protein in the market.

Declining Phase Between 1993 and 2008 During this phase legume area began to decline slowly due to a price ratio, especially for compensatory payments. Although in 1998, a Maximum Guaranteed Quantity (MGQ) was fixed at 3.5 Mt. for grain legumes, the common agricultural policy (CAP) reform applied from 1st January 1993 (CAP reform 1992 called —Mac Sharry) changed the context. The guaranteed prices were reduced to bring them closer to market prices, especially for arable crops, and direct subsidies were applied with mandatory set-aside. Despite the aid, farmers 'interest in growing grain legumes and income related to grain legumes decreased strongly in this context. After the 2003 CAP reform, aid to protein crops was aligned with grain production rather than area, changing from 72.5 €/ha in 2000 to 63 €/t in 2004. In addition, grain legumes also got a standard decoupled aid of 55.57 €/ha (Table 5.4). For this reason, a slight recovery in legume cropping area was observed in 2001. However, this recovery was short-lived and cultivated area reached, in 2008, its lowest level since the 1980s with a 63% decrease between 2004 and 2008 (UNIP 2009).

In addition to policy changes of lower aid after CAP reforms, the experts identified lower yield and sale price, risk of fluctuating yield and prices and higher cost of

Table 5.4 Impact of evolution of the common agricultural policy (CAP) reforms on surface area of grain legumes in France

Evolution of CAP reforms	Year	Evolution of grain legumes area (1000 ha)
Granted prices + aide for produers	1978	101
Direct aid for farmers (78.45 €/T)	1993	754
Direct aid for farmers (72.5 €/T)	2000	461
Direct aid (63 €/T) + specific aid (55.57 €/T)	2004	445
Direct aid (63 €/T) + specific aid (205.57 €/T)	2010	165

Source: UNIP (2009)

Table 5.5 Comparison of wheat and pea crops for different variables observed at farm

	Crops	
Variables	Wheat	Peas
Premium (€/ha)	300	356
Sale price (€/t)	180	140
Yield (t/ha)	5	2.5
Total cost (€/ha)	459	481
Gross margin (price * yield) (€/ha)	900	350
Gross product (premium + gross margin) (€/ha)	1200	706
Total income (gross product – costs) (€/ha)	741	225

Chamber of Agriculture Ariege (2009)

seeds as main constraints for grain legumes production, especially in the presence of other more profitable crops such as wheat. Von Richthofen et al. (2006) after a survey of 533 farmers in Europe and France, reported that lower market price, grain yield and the risk of yield fluctuations is also one of the major obstacle of legume cultivation. According to Jeuffroy and Ney (1997), wheat (*Triticum aestivum*) yields increased by 120 kg ha^{-1} per year from 1981 to 1996, while for peas it increased by only 75 kg ha^{-1} per year over the same period. Schneider (2008) also reported the same trend of increasing yield gaps for wheat and pea crops in France for the same period. This fact can also be explained by an example of a farmer in the Ariege department of Midi-Pyrénées region (Chamber of Agriculture Ariege 2009). In 2009, that farmer received 300 €/ha of aid (CAP reforms 2003) for growing rainfed wheat and 356 €/ha for rainfed grain legumes (Chamber of Agriculture Ariege 2009). At harvest, he obtained yields of these crops as 5 and 2.5 t/ha for wheat and peas respectively. He sold the product (grains) at market price of 180 €/t for wheat and 140 €/t for peas. For growing these crops, he spent 459 and 481 €/ha for wheat and peas respectively (Table 5.5). At the end, he observed that wheat is more profitable than peas, with a difference in income of 516 €/ha (= 741–225). To make pea competitive with wheat, this 516 €/ha can be compensated by increasing the premium, sale price or crop yield of peas crop. It is estimated that peas can be competitive only, (i) if it receives a premium of 872 €/ha instead of 356 €/ha, (ii) market sale price must be increased from 140 to 346.5 €/t, (iii) peas yield should be 6.19 t ha^{-1} instead of 2.5 t ha^{-1}. Any of them could make the peas competitive with wheat but are not happening in the in any region of the Europe.

5.6 Strategies to Overcome Constraints

5.6.1 Technological Innovations

Introduction of new crop rotations with more proportion of grain legumes could be one of the solutions to increase the area of grain legumes in European Union. Generally, farmers show lack of interest in growing grain legumes due to many reasons as explained earlier. Biophysically suitable new rotations with more proportion of grain legumes can be introduced with the help of local experts throughout the Europe. However, it is always not true those rotations with grain legumes results in higher gross margin. Preissel et al. (2015) reported that out of 53 tested rotations only 27-grain legumes rotations showed the higher gross margin annually with 8 rotations where minor deficit in gross margin was observed. Therefore, overall 35 rotations showed competitiveness. Similarly, Von Richthofen et al. (2006) observed the higher gross margin for legumes based rotations as explained in Sect. 5.3.2. Table 5.6 showed the lower yield and gross margin of rotation including grain legumes as compared to rotation without grain legumes in some regions of the Europe. In some cases this difference is more than 50%. However, same table shows that some rotations also have grosser margin when grain legumes are included in main crop rotations (Legume Futures 2014). Mahmood (2011) also tested nine new rotations in rainfed and irrigated conditions for Midi-Pyrénées region of France. He observed no change in legumes area and gross margin of rotations with and without including grain legumes in cereal based rotations. So it can be concluded that depending on the biophysical conditions inclusion of grain legumes in main cereal rotations, sometimes resulted in profitability in term of gross margin and vice versa.

5.6.2 Provision of More Premiums to Grain Legumes

During the CAP reforms of 1992 and 2003, the potential of grain legumes was ignored leading to more premiums provided to non N-fixing crops (UNIP 2009; Von Richthofen et al. 2006). As a consequence, the area under legumes decreased drastically (Schneider 2008; UNIP 2009). It is assumed that provision of higher premiums for grain legumes would be the primary incentive for the adoption of these crops by farmers. In agreement with this argument, the EU commission projected a total of 40 million Euros per year between 2010 and 2012 to rapidly achieve a legume area of at least 400,000 ha in EU (Le syndicat Agricole 2009). This gives a premium per ha of legumes of: 150 €/ha in 2010 to achieve an area of 267,000 ha, 125 €/ha in 2011 to achieve an area of 320,000 ha, 100 €/ha in 2012 to achieve an area of 400,000 ha. Currently, experts from the Europe claimed that these amounts of premiums are insufficient for increasing significantly the grain legumes area in Europe. With their experience they acknowledged that peas can be more profitable than

Table 5.6 The economic performance of legume and non legume based rotations indifferent European regions

Region	Crop Rotation		Annual GM inclu.Precrop effect (€ ha^{-1})			
	Without Legume	With Legume	Without Legume	With Legume	Advantage legume rotation	Thereof per crop value
Romania, Sud Muntania	B-RS-W-M-SF	B-RS-W-P-SF	334	275	−59	−12
	W-SF-M	W-SF-M-P	319	314	−4	−4
		W-SF-M-SB		403	84	99
	B-SF-M	B-SF-M-SB	130	271	142	189
		B-SF-M-P		183	53	86
	W-RS-M-SF-M	W-P-M-RS-M	309	482	173	177
		W-RS-M-P-M		420	111	114
Sweden, Westren Sweden	R-O-R-RS-R-O	R-P-R-RS-R-O	486	482	−4	31
	O-W-RS-W-W	O-FB-W-RS-W-W	568	525	−43	25
	W-O-W	W-O-W-FB	459	422	−37	47
		W-O-W-W-O-W-P		444	−15	27
Italy, Calabria	O-B	O-B-FB	383	211	−172	9
		O-B-P		206	−177	9
Germany, Brandenburg (2)	RS-W-B	RS-W-P-W-B	161	120	−41	11
		RS-W-FB-W-B		97	−65	11
		RS-W-B-P-W-B		101	−60	0
Brandenburg (3)	RS-T-R-R	RS-T-P-R-R	91	54	−37	20
Brandenburg (1)	RS-W-W-R	RS-W-FB-R	308	198	−110	55
UK, Eastren Schotland	B-RS-W-W	B-RS-W-FB-W	799	757	−42	14
		B-RS-W-P-W		779	−20	14
	W-O-B-RS-Pot-sB	W-O-B-P-Pot-sB	1366	1318	−48	0
		W-O-B-FB-Pot-sB		1299	−66	0

(continued)

Table 5.6 (continued)

Region	Crop Rotation		Annual GM inclu.Precrop effect (€ ha^{-1})			
	Without Legume	With Legume	Without Legume	With Legume	Advantage legume rotation	Thereof per crop value
Average			439	425	−20	42
Range			91–1366	54–1318	−177–173	−12–189
Comparison						
Hayer et al. 2012[a]					−20 –33	
LMC International 2009[b]					−54–0	
Luetke-Entrup et al. 2006[c]					−31 – + 115	
Weitbrecht & Pahl 2000[d]					70–86	
Von Richthofen et al. 2006[e]					−181– + 7	−4–57

This table is taken from Legume Futures 2014. Legume-supported cropping systems for Europe. General project report. Available at www.legumefutures.de

Source: Data in upper part are based on own calculations, data provided by project partner

Crops: *W* Wheat, *B* Barley, *O* Oat, *M* Maize, *RS* Rapeseed, *SF* Sunflower, *Pot* Potato, *P* Pea, *FB* Faba bean, *SB* Soya bean

[a]Hayer et al. 2012 (France)

[b]LMC International 2009 (Germany UK, France, Spain, Considered Percrop effects: Yield effect on 1st subsequent crop, N fertilizer saving)

[c]Luetke-Entrup et al. 2006 (Germany, Ploughed and conservation tillage systems)

[d]Weitbrecht and Pahl 2000 (Germany, organic production system, high soys value partly for food use)

[e]Von Richthofen et al. 2006 (Switzerland, Germany, Denmark, France, Spain considered percrop effects: Yield effect on 1st subsequent crop, fertilizer saving, pesticide saving, reduce tillage)

wheat, only if it receives a higher premium. It is debatable that how much amount of premium should be given to increase an area of grain legumes so that Europe could fulfill their food and feed requirements rather importing soybean from USA. A study conducted by Mahmood (2011) showed that pea area and farm income in three farms types of Midi-Pyrénées region can be increased from 4 to 18 ha and 1–4% per farm by providing a premium of 400 €/ha. This was consistent with the finding of UNIP (2009) that provision of more premium to grain legumes could be one of the driving force for increasing grain legumes area and hence the farm income in Europe in order to make grain legumes more competitive than cereals.

5.6.3 Increase in Sale Price and Crop Yield

Farmers in EU believe that lower sale price and grain yields are two of the major obstacles for legume production (Von Richthofen et al. (2006). For example, Chamber of Agriculture Ariege (2009) reported that in rainfed conditions, average yields of wheat and peas are respectively 5 and 2.5 t ha^{-1}. On average, farmers sell the product (grains) at market price of 180 €/t for wheat and 140 €/t for peas. They spend almost the same amount of money to grow both crops: 460 and 480 €/ha respectively for wheat and peas. Obviously, this makes wheat more profitable than pea in these conditions, with a difference of gross margin of 516 €/ha (741–225). It is, therefore, assumed that an increase in sale price and/or crop yield would make grain legumes competitive compared to cereal. Similar findings were also reported by Mahmood (2011), 50% increase in sale price and crop yield did not significantly increased the pea area in 2 farms types of Midi- Pyrenees region. However, one 1-farm type it increases the pea area only by 2 ha with 2% increase in farm income. Schreuder and Visser (2014) also indicated that more than 50% increase in pea yield could only make it competitive to wheat and maize in Europe.

5.6.4 Decrease in Yield Variability

Farmers in Europe turn to cultivation of non-legumes crops like cereals, oilseeds and tubers. It is assumed that high inter-annual yield variability is one of the driving forces behind this diversion (Cernay et al. 2015). Von Richthofen et al. (2006) also reported the similar reason of yield instability for lower cultivation of grain legumes on European farms. Cernay et al. (2015) estimated the yield variability of major grain legume and non-legume crops in four European regions during the years 1961–2013. Overall, the results showed greater yield variability in grain legumes as compared to cereals across the four regions (Fig. 5.4).

It is assumed that a reduction in yield variability would make grain legumes more attractive to farmers of Europe. But how much amount of decrease in yield variability would be sufficient to make grain legumes more competitive? This

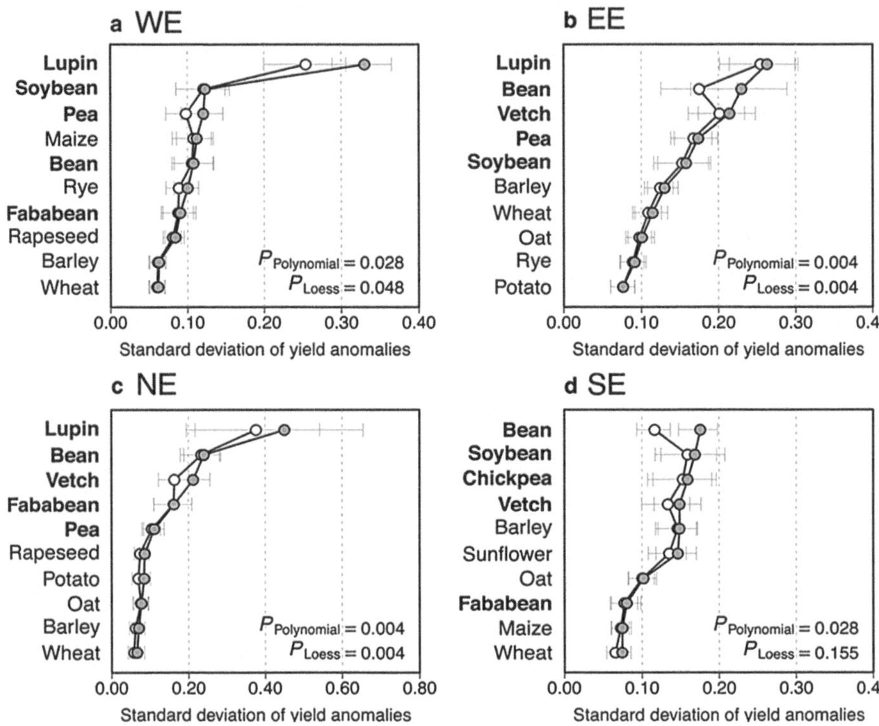

Fig. 5.4 Standard deviation of yield anomalies for 10 crops in Europe over 1961–2013. Standard deviation of yield anomalies for 10 crops in four European regions, i.e., Western Europe (WE), Eastern Europe (EE), Northern Europe (NE) and Southern Europe (SE). Figure is taken from Cernay et al. 2015

hypothesis was tested by Mahmood (2011) for three farms types of Midi-Pyrénées region of France. Results shows even with 50% decrease in yield variability could not make grain legumes more profitable than cereals.

5.6.5 *Use of Nutrient Policies e.g. Tax on Inorganic Fertilizers*

The EU directly or indirectly introduced several policies concerning the use of nutrients in agriculture e.g. Nitrate directives, Water Framework Directive and national regulations governing the use of nutrients. Such policies showed positive results e.g. Nitrate directive has increased the economic performance of white clover.

As we know, a huge amount of inorganic fertilizers are applied in agriculture annually as explained earlier, which resulted in negative environmental effects. Therefore, a tax on the use of mineral nitrogen can be applied as was applied in

Netherlands (Ondersteijn et al. 2002. Similarly, Sweden had also applied tax on mineral nitrogen from 1984 to 2010, but abolished it due to unsatisfactory competitive position of Swedish farm business as compared to other European countries. Such policies could be helpful in increasing the area of grain legumes in European farming systems due to N fixation ability of grain legumes (Bues et al. 2013).

5.7 Conclusion

Overall, it is concluded that the promotion of grain legumes in a context like Europe cannot be achieved in a realistic way by implementing individual above-mentioned strategies. These findings can explain the current low share of grain legume crops in the EU agricultural regions. It also explains why in some regions the implementation of only specific premium to promote grain legumes is insufficient (Schneider 2008). Most effective and realistic way to promote grain legumes on European farming systems is to implement combined agronomic and socio-economic strategies, like the ones used by Mahmood (2011). Results showed that by combining all these strategies grain legumes area can be increased significantly by 6% (of the total area of 111 ha), 32% (of the total of 107 ha) and 29% (of the total area of 110) respectively for cereal, cereals/follow and mixed farm types. Moreover, some more studies in order to confirm the findings of Mahmood (2011) should also be conducted throughout the Europe and then findings of all those studies should be communicated with researchers, farmers, stakeholders, agriculture research institutes and agricultural policy makers.

References

AEP (ed) (2004) Grain legumes and the environment: how to assess benefits and impacts? Zurich, November 18–19, 2004. AEP and FAL, pp 67–72

Alpmann D, Braun J, Schäfer BC, (2013) Analyse einer Befragung unter erfolgreichen Körnerleguminosenanbauern im konventionellen Landbau. Erste Ergebnisse aus dem Forschungsprojekt LeguAN. In: Wintertagung DLG (ed) Im Fokus: Heimische Körnerleguminosen vom Anbau bis zur Nutzung Berlin, p 20

Anonymous (1984) Legume inoculants and their use. FAO, Rome, pp 1–63

Bachinger J, Zander J (2007) ROTOR, a tool for generating and evaluating crop rotations for organic farming systems. Eur J Agron 26:130–143

Baddeley JA, Jones S, Topp CFE, Watson CA, Helming J, Stoddard FL (2014) Integrated analysis of biological nitrogen fixation (BNF) in Europe. Legume futures deliverable report 1.5 pp

Beaver JS, Rosas JC, Myers J, Acosta J, Kelly JD, Nchimbi-Msolla S, Misangu R, Bokosi J, Temple S, Arnaud-Santana E (2003) Contributions of the bean/cowpea CRSP program to cultivar and germplasm development in common bean. Field Crop Res 82:87–102

Bertoglio JC, Calvo MA, Hancke JL, Burgos RA, Riva A, Morazzoni P, Ponzone C, Magni C, Duranti M (2011) Hypoglycemic effect of lupin seed γ-conglutin in experimental animals and healthy human subjects. Fitoterapia 82:933–938

Bouwman AF (1996) Direct emission of nitrous oxide from agricultural soils. Nutr Cycl Agroecosyst 46:53–70

Bruce RR, Wilkinson SR, Langdale GW (1987) Legume effects on soil erosion and productivity. In: Power JF (ed) The role of legumes in conservation tillage systems. Soil Conservation Society of America, Ankeny, pp 127–138

Bues A, Preissel S, Reckling M, Zander P, Kuhlmann T, Topp K, Watson C, Lindström K, Stoddard FL, Murphy-Bokern D (2013) The environ- mental role of protein crops in the new common agricultural policy. In: directorate general for internal policies (ed) Agriculture and Rural Devel- opment, European Union Brussels, pp 113

Bulson HAJ, Snaydon RW, Stopes CE (1997) Effects of plant density on intercropped wheat and field beans in an organic farming system. J Agric Sci 128:59–71

Burris RH, Roberts GP (1993) Biological nitrogen fixation. Annu Rev Nutr 13:317–335

Byerlee D, White R (2000) Agricultural intensification and diversification through food legumes: technological and policy options. In: Knight R (ed) Linking Research and Marketing Opportunities for Pulses in the 21st Century. Kluwer Academic Publishers, Dordrecht, pp 31–46

Campbell BM, Costanza R, Van den Belt M (2000) Land use options in dry tropical woodland ecosystems in Zimbabwe: introduction, overview and synthesis. Ecol Econ 33:341–352

Carrouée B, Aveline A, Biarnes V, Charles R, Crozat Y, Jensen ES, Laurent F, Munier-Jolain N, Thévenet G, Viaux P (2002) Effets environnementaux des protéagineux dans les rotations de grandes cultures. Working document, UNIP, ITCF, INRA, ESA, Paris, p 32

Carrouée B, Crépon K, Peyronnet C (2003) Les protéagineux: intérêt dans les systèmes de production fourragers français et européens. Fourrages 174:163–182

Cassman KG, Dobermann A, Walters D (2002) Agroeco- systems, nitrogen-use efficiency, and nitrogen management. Ambio 31:132–140

Cernay C, Ben-ari T, Pelzer E, Meynard J, Makowski D (2015) Estimating variability in grain legume yields across Europe and the Americas. Sci Rep 5:11171

Chalk PM (1998) Dynamics of biologically fixed N in legume-cereal rotations: a review. Aust J Agric Res 49:303–316

Chamber of Agriculture Ariege (2009) Annual agricultural statistics region Midi-Pyrenees. Agreste in the region, France, p 09

Conrad R, Seiler W, Bunse G (1983) Factors influencing the loss of fertilizer nitrogen in the atmosphere as N2O. J Geophys Res 88:6709–6718

Coyne DP, Steadman JR, Godoy-Lutz G, Gilbertson R, Arnaud-Santana EA, Beaver JS, Myers JR (2003) Contributions of the bean/cowpea CRSP to the management of bean diseases. Field Crop Res 82:155–168

Crews TE, Peoples MB (2004) Legume versus fertilizer sources of nitrogen: ecological tradeoffs and human needs. Agric Ecosyst Environ 102:279–297

Dakora FD, Aboyinga RA, Mahama Y, Apaseku J (1987) Assessment of N2 fixation in groundnut (Arachis hypogea L.) and cowpea (Vigna unguiculata L. Walp.) and their relative N contribution to a succeeding maize crop in Northern Ghana. MIRCEN J 3:389–399

Dinnes DL, Karlen DL, Jaynes DB, Kaspar TC, Hatfield JL, Colvin TS, Cambardella CA (2002) Nitrogen management strategies to reduce nitrate leaching in tile-drained Midwestern soils. Agron J 94:153–171

Drinkwater LE, Wagoner P, Sarrantonio M (1998) Legume-based cropping systems have reduced carbon and nitrogen losses. Nature 396:262–264

Evans J, Scott G, Lemerle D, Kaiser A, Orchard B, Murray GM, Armstrong EL (2003) Impact of legume ‘break' crops on the yield and grain quality of wheat and relationship with soil mineral N and crop N content. Aust J Agric Res 54:777–788

Fillery IRP (2001) The fate of biologically fixed nitrogen in legume-based dry land farming systems: a review. Aust J Exp Agric 41:361–381

Giambalvo D, Stringi L, Durante G, Amato G, Frenda AS (2004) Nitro- gen efficiency component analysis in wheat under rainfed Mediter- ranean conditions: effects of crop rotation and nitrogen fertilisation. In: Cantero-Martínez C, Gabina D (eds) Mediterranean rainfed agricul- ture: strategies for sustainability. Mediterranean Agronomic Institute of Zaragoza, pp 169–173

Gierus M, Kleen J, Loges R, Taube F (2012) Forage legume species determine the nutritional quality of binary mixtures with perennial ryegrass in the first production year. Anim Feed Sci Technol 172:150–161

Giller KE (2001) Nitrogen fixation in tropical cropping systems. CABI Publishing, Wallingford

Glendining MJ, Powlson DS (1995) The effects of long continued applications of inorganic nitrogen fertilizer on soil organic nitrogen, a review. In: Lal R, Stewart BA (eds) Soil management, experimental basis for sustainability and environmental quality. CRC Press, Boca Raton, pp 385–446

GL-Pro partners (2007) Guidelines for growing grain legumes in Europe. GL-Pro concerted action, p 8

Graham PH, Vance CP (2003) Legumes: importance and constraints to greater utilization. Plant Physiol 131:872–877

Gueguen J, Duc G, Boutin JP, Dronne Y, Munier-Jolain N, Sève B, Tivoli B (2008) La filière protéagineuse, quels défis pour la recherche ? Rencontre au Salon International de l'Agriculture. INRA, Paris, p 6

Haque I, Powell JM, Ehui SK (1995) Improved crop-livestock production strategies for sustainable soil management in tropical Africa. In: Lal R, Stewart BA (eds) Soil management: experimental basis for sustainability and environmental quality. CRC Press, Boca Raton, pp 293–345

Harland JI, Haffner TA (2008) Systematic review, meta-analysis and regression of randomised controlled trials reporting an association between an intake of circa 25g soya protein per day and blood cholesterol. Atherosclerosis 200:13–27

Hayer F, Bonnin E, Carrouée B, Gaillard G, Nemecek T, Schneider A, Vivier C (2012) Designing sustainable crop rotations using life cycle assessment of crop sequences. In: Corson MS, van der Werf HMG (eds) Proceedings of the 8th International Conference on Life Cycle Assessment in the Agri-Food Sector, 1–4 October 2012. INRA, Saint Malo, pp 828–829

Jeanneret P, Baumgartner D, Freiermuth R, Gaillard G (2006) Méthod d'évaluation de l'impact des activités sur la biodiversité dans les bilans écologiqueses- SALCA-BD. Agroscope FAL Reck-enholz, p 67

Jensen ES (1997) The role of grain legume N2 fixation in the nitrogen cycling of temperate cropping systems. Risø National Laboratory, Roskilde, p 107

Jensen ES, Hauggaard-Nielsen H (2003) How can increased use of biological N2 fixation in agriculture benefit the environment? Plant Soil 252:177–186

Jensen CR, Joernsgaard B, Andersen MN, Christiansen JL, Mogensen VO, Friis P, Petersen CT (2004) The effect of lupins as compared with peas and oats on the yield of the subsequent winter barley crop. Eur J Agron 20:405–418

Jensen ES, Peoples MB, Boddey RM, Gresshoff PM, Hauggaard-Nielsen H, Bjr A, Morrison MJ (2011) Legumes for mitigation of climate change and the provision of feedstock for biofuels and biorefineries: a review. Agron Sustain Dev 32:329–364

Jeuffroy MH (2006) The ecological services of protein crops: the basis of sustainable farming systems. Meeting organized by the National institute of agricultural research at the international agricultural show, France. 6p

Jeuffroy MH, Ney B (1997) Crop physiology and productivity. Field Crop Res 53:3–16

Joshi PK, Birthal PS, Bourai VA (2002) Socioeconomic constraints and opportunities in rainfed rabi cropping in rice fallow areas of India. Patancheru 502 324, Andhra Pradesh, India: international crops research institute for the semi-arid tropics, p 58

Kabagambe EK, Baylin A, Ruiz-Narvarez E, Siles X, Campos H (2005) Decreased consumption of dried mature beans is positively associated with urbanization and nonfatal acute myocardial infarction. J Nutr 135:1770–1775

Kennedy IR, Cocking EC (1997) Biological nitrogen fixation: the global challenge and future needs. Rockefeller foundation Bellagio conference proceedings. SUN Fix Press, University of Sydney, Sydney, p 83

Kirkegaard JA, Christen O, Krupinsky J, Layzell D (2008) Break crop benefits in temperate wheat production. Field Crop Res 107:185–195

Lal R (2004) Carbon Emissions from Farm Operations. Environ Int 30:981–990

Lal R, Wilson GF, Okigbo BN (1978) No-till farming after various grasses and leguminous cover crops in tropical alfisol. i. crop performance. Field Crop Res 1:71–84

Le syndicat agricole (2009) Les nouvelles aides 2010. [Consulté en Janvier 2010]. http://www.syndicatagricole.com/actualites/vie-pratique-pac-les-nouvelles-aides-2010&fldSearch=:FXZ5S215.html

Legume Futures (2014) Legume-supported cropping systems for Europe. General project project report. Available at www.legumefutures.de

Liebman M, Dyck E (1993) Crop rotation and intercropping strategies for weed management. Ecol Appl 3:92–122

LMC International (2009) Evaluation of measures applied under the Common Agricultural Policy to the protein crop sector. Final report. http://ec.europa.eu/agriculture/eval/reports/protein_crops/(2013-02-25)

López-Fando C, Almendros G (1995) Interactive effects of tillage and crop Rota- tions on yield and chemical properties of soils in semi-arid Central Spain. Soil Tillage Res 36:45–57

Luetke-Entrup N, Schneider M, Stemann G, Gröblinghoff F-F, Heißenhuber A, Pahl H, Hülsbergen K-J, Maidl FX, Herr H, Sommer C, Korte K, Brunotte J, Kreye H, Lindwedel V, Ziesenіß H, Gienapp C, Schulz RR, Propp J (2006) Bewertung von neuen Systemen der Bodenbewirtschaftung in erweiterten Fruchtfolgen mit Körnerraps und Körnerleguminosen. Abschlussbericht über die Versuchsjahre 2001–2005. Fachhochschule Südwestfalen, Fachbereich Agrarwirtschaft, Soest

Magrini MB, Anton M, Cholez C, Corre-Hellou G, Duc G, Jeuffroy MH, Meynard JM, Pelzer E, Voisin AS, Walrand S (2016) Why are grain-legumes rarely present in cropping systems despite their environmental and nutritional benefits? Analyzing lock-in in the French agrifood system. Ecol Econ 126:152–162

Mahmood F (2011) Analysis of the conditions for the development of grain legumes in the midi-Pyrénées region (France), using the APES-FSSIM-indicators modeling chain. In: Systèmes Intégrés En Biologie, Agronomie, Géosciences, hydro- sciences et Environnement (SIBAGHE). Ecole National Supérieure Agronomique de de Montpellier, Montpellier, p 168

Marchesi C, Paradis P, Schiffrin EL (2008) Role of the renin–angiotensin system in vascular inflammation. Trends Pharmacol Sci 29:367–374

McEwen J, Darby RJ, Hewitt MV, Yeoman DP (1989) Effects of field beans, fallow, lupins, oats, oilseed rape, peas, ryegrass, sunflowers and wheat on nitrogen residues in the soil and on the growth of a subsequent wheat crop. J Agric Sci 115:209–219

Miller PR, McConkey BG, Clayton GW, Brandt SA, Staricka JA, Johnston AM, Lafond GP, Schatz BG, Baltensperger DD, Neill KE (2002) Pulse crop adaptation in the northern great plains. Agron J 94:261–272

MP3-Grain Legumes (2010) Enhanced food and feed security, nutritional balance, economic growth and soil health for smallholder farmers, CGIAR consortium board

Mudahar MS, Hignett TP (1987) Energy requirements, technology, and resources in the fertilizer sector. In: Helsel ZR (ed) Energy in plant nutrition and Pest control, vol 2. Elsevier, Amsterdam, pp 26–61

Munier-Jolain N, Collard A (2006) Grain legumes and weed management in crop rotations: opportunities and methodologies for reducing environmental impacts of weed control. In: AEP (ed) Grain legumes and the environment: how to assess benefits and impacts? Zurich, November 18–19, 2004. AEP and FAL, pp 67–72

Mvondo H, Owona S, Mvondo Ondoua J, Essono J (2007) Tectonic evolution of the Yaoundé segment of the Neoproterozoic Central African orogenic belt in southern Cameroon. Can J Earth Sci 44:433–444

Mwanamwenge J, Loss SP, Siddique KHM, Cocks PS (1998) Growth, seed yield and water use of faba bean (Vicia faba L.) in a shortseason Mediter- raneantype environment. Aust J Exp Agric 38:171–180

Ncube B, Dimes J, Vanwijk M, Twomlow S, Giller K (2009) Productivity and residual benefits of grain legumes to sorghum under semi-arid conditions in South-Western Zimbabwe: unraveling the effects of water and nitrogen using a simulation model. Field Crop Res 110:173–184

Nemecek T, Erzinger S (2005) Modelling representative life cycle inventories for Swiss arable crops. Int J LCA 10:68–76

Nemecek T, GL-Pro partners (2006) Economic and environmental value of European cropping systems that include grain legumes. Grain Legumes No. 45 – 1st quarter 2006

Nemecek T, von Richthofen JS, Dubois G, Casta P, Charles R, Pahl H (2008) Environmental impacts of introducing grain legumes into European crop rotations. Eur J Agron 28:380–393

Ondersteijn CJM, Harsh SB, Giesen GWJ, Beldman ACG, Huirne RBM (2002) Management strategies on Dutch dairy farms to meet environment regulations; a multi case study. Netherlands J Agric Sci 50:47–65

Owens LB, Edwards WM, Van Keuren RW (1994) Groundwater nitrate levels under fertilized grass and grass- legume pastures. J Environ Qual 23:752–758

Pappa VA, Rees RM, Walker RL, Baddeley JA, Watson CA (2012) Legumes intercropped with spring barley contribute to increased biomass production and carry-over effects. J Agric Sci 150:584–594

Paustian K, Andren O, Janzen HH, Lal R, Smith G, Tian H, Tiesen M, Noordwijk V, Woomer P (1997b) Agricultural soil as a sink to offset CO2 emissions. Soil Use Manag 13:230–244

Paustian K, Colins HP, Paul EA (1997c) Management controls on soil carbon. In: Paul EA, Paustian K, Elliott ET, Cole CV (eds) Soil organic matter in temperate agro-ecosystems: long-term experiments in North America. CRC Press, Boca Raton, pp 15–49

Peoples MB, Crasswell ET (1992) Biological nitrogen fixation: investments, expectations and actual contributions to agriculture. Plant Soil 141:13–39

Peoples MB, Herridge DF, Ladha JK (1995) Biological nitrogen fixation: an efficient source of nitrogen for sustainable agricultural production? Plant Soil 174:3–28

Peoples MB, Hauggaard-Nielsen H, Jensen ES (2009) The potential environmental benefits and risks derived from legumes in rotations. In: Nitrogen fixation in crop production. American Society of Agronomy, Crop Science Society of America, Soil Science Society of America, Madison, pp 349–385

Peoples M, Swan T, Goward L, Hunt J, Li G, Harris R, Ferrier D, Browne C, Craig S, van Rees H, Mwendwa J (2015) Legume effects on soil N dynamics-comparisons of crop response to legume and fertiliser N. Grains Research and Development Corporation, Government of Australia. Available at: http://grdc.com.au/Research-and-Development/GRDC-Update-Papers/2015/02/Legumeeffects-on-soil-N-dynamics%2D%2Dcomparisons-of-cropresponse-to-legume-and-fertiliser-N#sthash.9ncTiq1C.dpuf

Poss R, Saragoni H (1992) Leaching of nitrate. Fertil Res 33:123–133

Preissel S, Reckling M, Schläfke N, Zander P (2015) Field crops research magnitude and farm-economic value of grain legume pre-crop benefits in Europe: a review. Field Crop Res 175:64–79

Prew RD, Dyke GV (1979) Experiments comparing break crops as a preparation for winter-wheat followed by spring barley. J Agric Sci 92:189–201

Rao MR, Mathuva MN (1999) Legumes for improving maize yields and income in semi-arid Kenya. Agric Ecosyst Environ 78:123–137

Rao JK, Dart PJ, Sastry PV (1983) Residual effect of pigeonpea (Cajanus cajan) on yield and nitrogen response of maize. Exp Agric 19(2):131–141

Rao IM, Borrero V, Ricaurte J, Garcia R, Ayarza MA (1996) Adaptive attributes of tropical forage species to acid soils II. Differences in shoot and root growth responses to varying phosphorus supply and soil type. J Plant Nutr 19(2):323–352

Reckling M, Hecker JM, Schläfke N, Bachinger J, Zander P, Bergkvist G, Walker R, Maire J, Eory V, Topp CFA, Rees RA, Toncea I, Pristeri A, Stoddard FL (2014) Agronomic analysis of cropping strategies for each agroclimatic region. Legume Futures Report 1(4):75

Reckling M, Hecker JM, Bergkvist G, Watson CA, Zander P, Schläfke N, Stoddard FL, Eory V, Topp CF, Maire J, Bachinger J (2016) A cropping system assessment framework-evaluating effects of introducing legumes into crop rotations. Eur J Agron 76:186–197

Rego TJ, Seeling B (1996) Long-term effects of legume-based cropping systems on soil nitrogen status and mineralization in Vertisols. In: Ito et al (eds) Roots and nitrogen in cropping systems of the Semi-arid Tropics. JIRCAS, pp 469–479

Ribet J, Drevon JJ (1996) The phosphorus requirement of N2 fixing and urea-fed Acacia mangium. New Phytol 132:383–390

Roberts TL (2009) The role of fertilizer in growing the world's food. Better Crops 93(2):12–15

Robson MC, Fowler SM, Lampkin NH, Leifert C, Leitch M, Robinson D, Watson CA, Litterick AM (2002) The agronomic and economic potential of break crops for ley/arable rotations in temperate organic agriculture. Adv Agron 77:369–427

Rochester IJ, Peoples MB, Constable GA, Gault RR (1998) Fababeans and other legumes add nitrogen to irrigated cotton cropping systems. Aust J Exp Agric 38:253–260

Roman GV, Epure LI, Toader M, Lombardi AR (2016) Grain legumes - main source of vegetal proteins for European consumption. Agro Life Sci J 5:178–183

Salez P, Martin F (1992) Evolution de la production et de la fertilité du sol dans des rotations culturales incluant du sorgho, des légumineuses et du cotonnier. Document agronomique, n°3. p 17

Salisbury F, Ross C (1978) Nitrogen fixation. In: plant physiology, 2nd edn. W. Pub. Co., Ca., pp 195–198

Sanchez PA, Euhara G (1980) Management considerations for acid soils with high phosphorus fixation capacity. In: Khasawneh FE, Sample EC, Kamprath EJ (eds) The role of phosphorus in agriculture. American Society of Agronomy, Madison, pp 471–514

Schneider A (2008) The dynamics controlling the grain legume sector – analysis of past trends helps to focus on future challenges. (AEP European association for grain legume research, www.grainlegumes.com)

Schreuder R, De Visser C (2014) Raport EIP-AGRI focus group on protein crops, Bruxelles

Sinclair TR, Cassman KG (1999) Green revolution still too green. Nature 398:556

Sinclair TR, Muchow RC, Bennet JM, Hammond LC (1987) Relative sensitivity of nitrogen and biomass accumulation to drought in field-grown soyabean. Agron J 79:986–991

Singh RJ, Chung GH, Nelson RL (2007) Landmark research in legumes. Genome 50:525–537

Sirtori CR, Mombelli G, Triolo M, Laaksonen R (2012) Clinical response to statins: mechanism (s) of variable activity and adverse effects. Ann Med 44:419–432

Smil V (1999) Nitrogen in crop production. An account of global flows. Global Biogeochem Cycles 13:647–662

Smil V (2001) Enriching the earth. MIT Press, Cambridge, MA

Stevenson FC, van Kessel C (1997) Nitrogen contribution of pea residue in a hum- mocky terrain. Soil Sci Soc Am J 61:494–503

Tharanathan RN, Mahadevamma S (2003) Grain legumes a boon to human nutrition. Trends Food Sci Technol 14:507–518

UNIP (2009) Les chiffres clés: Protéagineux, France

UNIP/Arvalis-Institut du Végétal (2008) Se refaire un avis objectif sur le pois, la féverole et le lupin. UNIP, Paris

Unkovich MJ, Pate JS (2000) An appraisal of recent field measurements of symbiotic N2 fixation by annual legumes. Field Crop Res 65:211–222

Van Kessel C, Hartley C (2000) Agricultural management of grain legumes: has it led to an increase in nitrogen fixation? Field Crops Res 65:165–181

Von Richthofen JS, GL- Pro partner (2006) Economic and environmental value of European cropping systems that include grain legumes. Grain legumes No. 45 – 1st quarter 2006 Special report UNIP, France

Von Richthofen JS, Pahl H, Nemecek T, Odermatt O, Charles R, Casta P, Sombrero A, Lafarga A, Dubois G (2006) Economic interest of grain legumes in European crop rotations. GL-Pro Report, WP3. 58 pp

Wani SP, McGill WB, Haugen-Koyzra KL, Robertson JA, Thurstson JJ (1994) Improved soil quality and barley yields with faba-beans, manure, forages, and crop rotation on a gray luvisol. Can J Soil Sci 74:75–84

Wani SP, Rego TJ, Iot O, Lee KK (1996) Nitrogen budget in soil under different cropping systems. In: Ito et al. (eds) Roots and nitrogen in cropping systems of the semi-arid tropics. JIRCAS, pp 481–492

Wani SP, Pathak P, Jangawad LS, Eswaran H, Singh P (2003) Improved management of Vertisols in the semiarid tropics for increased productivity and soil carbon sequestration. Soil Use Manag 19(3):217–222

Weitbrecht B, Pahl H (2000) Lohnt sich der Anbau von Körnerleguminosen? Ökologie und Landbau 116:39–41

Wery J (1987) Relations entre la nutrition azotée et la production chez les légumineuses. In Nutrition azotée des légumineuses'. Les Colloques de l'INRA, Paris, 37:199–223

Wery J, Ahlawat IPS (2007) Analysing and improving the role of grain legumes in cropping system's sustainability: a system approach illustrated on chickpea in India and Europe. Proceeding of The Forth International Food Legumes Research Conference (IFLRC-IV), New Delhi, India

White RE (1988) Leaching. In: Wilson JR (ed) Advances in nitrogen cycling in agricultural ecosystems. CAB International, Wallingford, pp 193–211

Zander P, Amjath-Babu TS, Preissel S, Reckling M, Bues A, Schläfke N, Kuhlman T, Bachinger J, Uthes S, Stoddard F, Murphy-Bokern D (2016) Grain legume decline and potential recovery in European agriculture: a review. Agron Sustain Dev 36(2):26

Chapter 6
Nitrogen Management in the Rice–Wheat System of China and South Asia

Yingliang Yu, Linzhang Yang, Pengfu Hou, Lihong Xue, and Alfred Oduor Odindo

Abstract Nitrogen fertilization is one of the important agricultural practices for increasing crops production in modern farming. Excessive nitrogen fertilization without scientific guidance can also cause serious environmental problems. Therefore, the improvement of nitrogen management is critical for further sustainable agricultural development. In most areas of China and South Asia, the rice-wheat system is widely spread due to high precipitation. Since there are various lengths of rice flooding stages, nitrogen management is different compared with upland cultivation systems.

We review the characteristics of the general rice–wheat system, nitrogen transformation and existing techniques for nitrogen management. Less than 40% of the nitrogen applied is used by crops, with the other 60% being lost via denitrification (15–42% of total nitrogen application), ammonia volatilization (1–47%), runoff (5%) and leaching (2%). Thus, nitrogen transformations under actual soil conditions must be studied to improve nitrogen use efficiency and reduce losses.

Keywords Rice-wheat systems · Nitrogen management · China · South Asia · Nitrogen use efficiency · Nitrogen loss · Nitrogen transformation · Existing techniques · Sustainable · 4R technology

Y. Yu
Institute of Agricultural Resources and Environment, Jiangsu Academy of Agricultural Sciences, Nanjing, China

School of Agricultural, Earth & Environmental Sciences, University of KwaZulu-Natal, Scottsville, South Africa

L. Yang (✉) · P. Hou · L. Xue
Institute of Agricultural Resources and Environment, Jiangsu Academy of Agricultural Sciences, Nanjing, China
e-mail: lzyang@issas.ac.cn

A. O. Odindo
School of Agricultural, Earth & Environmental Sciences, University of KwaZulu-Natal, Scottsville, South Africa

E. Lichtfouse (ed.), *Sustainable Agriculture Reviews 32*, Sustainable Agriculture Reviews 32, https://doi.org/10.1007/978-3-319-98914-3_6

6.1 Introduction

Rice and wheat are the dominant cereal crops in the world. They provide 45% of the energy and 30% of the protein in people's diets (Evans 1993). Additionally, among cereals, rice and wheat are the highest in terms of yield and calories provided per hectare. Both crops have been cultivated for thousands of years. As of 2014, rice is grown in 150 countries and wheat in 128 (FAOSTAT 2016). Yields from 1 ha of rice and wheat could support 5.7 and 4.1 people for 1 year, respectively (Mahajan and Mahajan 2009).

In many countries, rice or wheat are grown for one season lasting 4–10 months per year (Fig. 6.1). However, they are widely and extensively grown in rotation in Asia, with China and South Asia employing a rice–wheat system in over 26.5 million ha. The nutrient environment for the growth and development of rice and wheat are quite different, and sophisticated technology is needed to construct rice and wheat in sequence in a double-cropping system. Therefore, the rice–wheat system has unique characteristics in its nutrient balance.

In recent years, crop production has had to keep increasing to cope with the food demand of the population. Unfortunately, land for agricultural cultivation is lost due to urbanization and soil degradation. Thus, the matter flows in the rice–wheat system have progressively increased and the system is now more fragile than ever before. Additionally, yields in the rice–wheat system have been difficult to increase or maintain (Timsina and Connor 2001) and inappropriate nutrient input has contributed to severe environmental degradation of soil (Zhu and Chen 2002), which seriously threatens food security. Nitrogen as an indispensable element for crop growth is also the vital crux of both grain production and soil sustainable development; however, many traditional practices of nitrogen application are not applicable to current soil conditions. Collecting and analyzing results from previous studies

Fig. 6.1 Farmers harvest rice at the end of October and then they sow wheat seeds as winter crop in the south-eastern of China

will help to understand nitrogen transformations under the frequent alternation between flooding and draining in the rice–wheat rotations and identify any problem of nitrogen cycling. A summary of the existing techniques provides new approaches and suggestions for optimizing nitrogen management based on nitrogen balance in a rice–wheat system.

6.2 Origin of the Rice–Wheat System

Rice is believed to have originated from China and India. Carbonized rice found in Yu Chanyan ruins in Hunan Province, China, is assumed to indicate rice cultivation 12,000 years ago. In comparison, wheat originated in West Asia and has fed more people over the last 8000 years (the Jordan Rift Valley) than any other crop. Wheat was introduced to China and the Indo-Gangetic Plains about 5000 years ago, but the two crops were cultivated and developed in different areas due to the contrasting environmental requirements.

The emergency of the rice-wheat system is a great innovation. The dominant factors in the expansion and development of the rotation system were (1) meeting the demand of human population growth and (2) identifying wheat varieties suited to growth in cold and short seasons.

Although a multi-cropping system began 1800 years ago, the earliest written record of the rice–wheat system is from only 1000 years ago in China (Zhou 2000). The Tang Dynasty, the ancient Chinese regime that developed trade from north to south, brought wheat to the south. Rice was preferred to wheat on dietary grounds and had become culturally significant in Chinese southern areas, while wheat was an addition to the cropping complex. Society was stable and rich that time. Agriculture technology allowed development of this unique double-cropping system to meet tremendous increases in food needs due to population growth (Li 1982). Since then, areas with the rice–wheat system have been considered as China's breadbasket. Rice and wheat together contribute 71–100% of the total cereal production in South and East Asia (Timsina and Connor 2001). Additionally, extension of the rice–wheat system is considered to show agricultural intensification. However, rice–wheat in South Asia is less historical and began with the introduction of rice into traditional wheat areas in India and Pakistan and wheat into traditional rice areas. It substantially increased in the twentieth century with the International Maize and Wheat Improvement Center introducing wheat varieties that could germinate at low temperature.

6.3 Distribution of the Rice–Wheat System

The rice–wheat system is located in subtropical to warm-temperate areas with appropriate air, moisture and thermal conditions for both crops during the annual cycle. It now spreads over 13 million ha in China (Zheng 2000) and another 13.5 million ha in South Asia (Yadav et al. 1998; Mahajan and Mahajan 2009) (Fig. 6.2). Other countries such as Bhutan, Iran, Japan, Korea, Mexico and Egypt also use this system. Most rice–wheat fields are converted from a traditional rice or wheat single-cropping system. The climate in winter and spring determines whether wheat can be grown; accumulated temperature, precipitation, frost-free days and sunshine hours are essential factors for wheat growth. In recent years, the rice–wheat system was introduced to Vietnam, the Philippines and Indonesia with a new wheat variety that can resist higher temperatures.

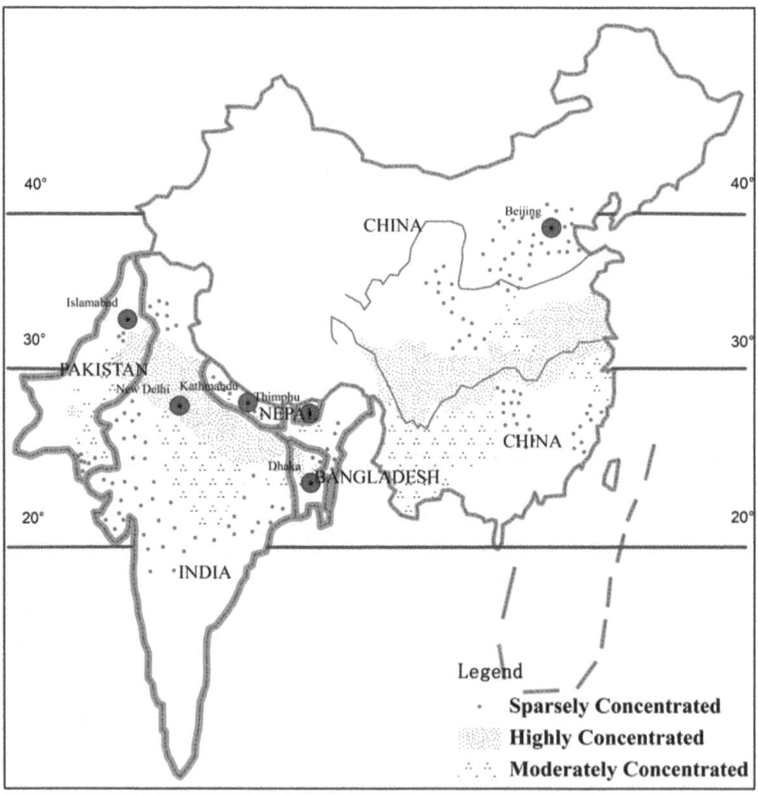

Fig. 6.2 The rice–wheat system mainly distributes between 20°N to 40°N latitude in India, China Bangladesh, Pakistan and Nepal. The rice–wheat fields distribute densely in Yangtze River Basin of China and in the northern of India. Proper precipitation in these areas is the predominant factor for the rice cultivation and low temperature determines the wheat cultivation

In China, the rice–wheat system is practiced in plains below 40°N latitude and in highlands below 28°N and is concentrated in the Yangtze River Basin including Jiangsu, Zhejiang and Anhui Provinces. This system accounts for 41.5% of the total rice area and 35% of total wheat area in China (calculated with data in National Bureau of Statistics 2012) and supplies a staple food for nearly 400 million people. Accumulated temperature and precipitation are the major restrictions for distribution of rice–wheat in China. In the north of the Yangtze River Basin, accumulated temperature is relatively low and the rice growing period is prolonged. Therefore, there is generally insufficient temperature and time for wheat growth. In the south of the basin, winter temperatures and water are sufficient to grow rice but the humid climate is unsuited to wheat. For these reasons, rice–rice is preferred to rice–wheat in the south.

In South Asia, the rice–wheat system occupies an area of 13.5 million ha: ten million ha in India, 2.2 million ha in Pakistan, 0.8 million ha in Bangladesh and 0.5 million ha in Nepal (Ladha et al. 2003). This represents about 33% of the total rice area and 42% of the total wheat area, and feeds more than 400 million people (Ladha et al. 2003). Rice–wheat is the most extensive cropping system in India and its production meets the cereal demand of more than 70% of Indian people (Minhas and Bajwa 2001). Additionally, the requirement of staple food from this system is still increasing due to population growth. However, some parts of these four countries are covered by mountains that can affect rainfall such that it can be inadequate for the water requirements of rice–wheat. In addition to precipitation rates, the topography controls water flow making it difficult to maintain flooding in some locations and in other places the soil drainage is too poor for wheat (Aslam and Prathapar 2001).

Since the 1960s, population growth has increased food demand, and the rice–wheat area was keeping growing until 1970s (Timsina and Connor 2001). However, most suitable land has already been used for multi-cropping systems. Increasing crop intensity alone is not a valid approach for increasing production and has raised concerns on how to improve soil fertility and promote cereal yield.

6.4 Characteristics of the Rice–Wheat System

6.4.1 Cultivation Characteristics

The rice–wheat system is one multi-cropping system that exploits differences in precipitation and temperature between summer and winter. As the world's primary cereal, rice has a unique ability to adapt and its efficiency of agricultural soil utilization has been greatly increased.

Generally, rice is grown during the rainy and warm season and wheat is then grown during the dry and cool season. The crop calendars for this system can differ greatly according to the various climates and agricultural and cultural requirements. The major rice–wheat sequences in China and South Asia are shown in Fig. 6.3.

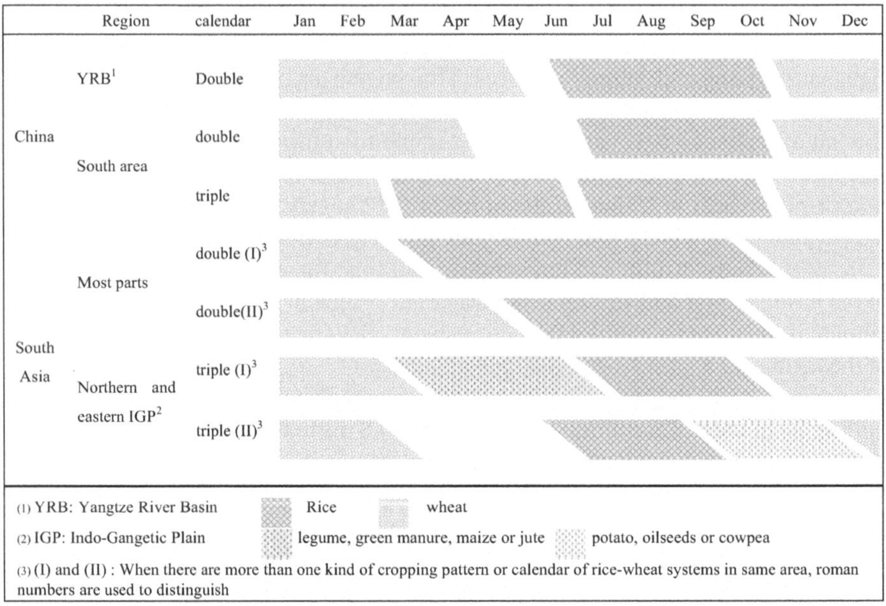

Fig. 6.3 There are double or triple growing seasons per year in rice-wheat systems. However, the cropping calendars are various due to the different climates such as temperature and rainfall in the area of rice-wheat systems distributing. In the Indo-Gangetic Plain, farmers usually plant other crops between rice and wheat seasons

In China, there are three main crop calendars (Zheng 2000). In the Yangtze River Basin, japonica rice and winter wheat are included: rice is transplanted in mid-June and harvested in late October; and wheat is grown from early November to late May. In subtropical areas, such as south, southeast and southwest China, where temperatures are higher, the system includes japonica/indica rice and winter/spring wheat (Crawford 2008). Both double- and triple-cropping systems are practiced. Rice is grown during July–October and wheat during November–April in double-cropping. Sometimes there is enough time and temperature for an additional crop after wheat and before rice, and for agricultural reasons another rice crop is a common option. Thus, rice is grown twice from mid-March to June and from July to November, respectively.

In most parts of South Asia (Timsina and Connor 2001), wheat cropping commences in October–November and is harvested in March of warmer areas, and in May of cooler parts in Pakistan. Basmati rice and Indica rice are grown to fill the time after wheat, and there is usually no time to include another crop. In the northern, northeastern and eastern Gangetic Plain, the wheat season is short. Thus, legumes, green manures, maize and jute are grown as an additional crop after wheat but before rice, when wheat is harvested in March and rice is transplanted in July. Less frequently, the rice harvest is advanced to September with potato, oilseeds and cowpea grown in October before the sowing of wheat. The sequences of the rice–wheat system are changed though the breeding and introduction of new crop varieties for higher productivity and economic income.

Fig. 6.4 Fields keep flooded in most time of rice season, while they keep drained in wheat season. The field ridges serve as levees once water flows into rice fields, while farmers break some ridges and ditch to ensure the drainage unhindered. The two pictures show different scenes of a rice–wheat system in the Yangtze River Basin, China

6.4.2 Environmental Characteristics

The predominant feature of the rice–wheat system is the annual conversion of soil water conditions (Fig. 6.4). Most of the time, rice grows in puddled and flooded soil conditions. Constant flooding at 3–8 cm depth is usually maintained by rainfall or irrigated water during the important rice growing periods. Furthermore, alternate wetting and drying is recommended at the rice tillering period. However, wheat is a crop suited to the lower temperature and precipitation in winter and following rice it is necessary to drain soil before wheat sowing to reduce the soil moisture. Good aeration is vital for maintaining the balance among air, moisture and thermal conditions. The water required for 1 kg of grain for wheat is only one-fifth of that for rice (IRRI 1995). Consequently, soil in a rice–wheat system is converted between aerobic and anaerobic conditions at least twice per year. This conversion has significant effects on the physical, chemical and biological properties of soil, which influences nutrient availability and transformation.

6.4.2.1 Physical Properties:

Advantages:

1. Soil conditions are beneficial for root growth under rice–wheat. Rice fields without rotation could lead to lower soil redox potential (usually reflect by soil Eh value) Eh during long-term flooding and could cause soil gleization, both of which have negative effects on rice root growth. In rice–wheat, the wheat rotation can alter soil granular structure, increase soil Eh and eliminate soil gleization (Xu et al. 1998; Ma 1999).

2. Pans prevent soil water loss and improve water and nutrient availability. Before the rice season begins, the field needs to be puddled, which breaks soil aggregates and reduces the void ratio, and these practices can reduce soil water leaching. After repetitive puddling, a pan is formed by clay particles settling at the base of the tilled layer. This enhances water and nutrient use efficiency through reduced water permeability and nutrient losses from leaching (Sharma and Datta 1986; Aggarwal et al. 1995; Singh et al. 2008).

Disadvantages:

1. Tillage before wheat sowing induces mass of clod in obtaining seedbed with tilth and could affect seed germination (Beyrouty et al. 1987).
2. Pans form physical resistance to wheat root penetration and nutrient uptake. With the commencing of the wheat season, the drying condition strengthens the puddled layer to a compacted pan, which is a dense zone at 20 cm deep that limits wheat root penetration (Gajri et al. 1999), although soil drainage may help roots extend. Destruction of soil structure is a major impediment to wheat growth (Oussible et al. 1992; Aggarwal et al. 1995).

6.4.2.2 Chemical Properties

Soil processes of oxidation and reduction are affected by aerobic and anaerobic conditions. They also affect nutrient availability and transformation. Moreover, nutrient availability and transformation are key factors for plant growth.

Advantages:

1. Flooding in the rice season weakens nitrogen nitrification, so ammonium nitrogen content is chemically increased (Swarup and Singh 1989; Dobermann et al. 2003). Rice growth is favored by ammonium compared with nitrate nitrogen.
2. Availability of phosphorus and potassium is increased by anaerobic conditions (Kirk et al. 1990).
3. Anaerobic conditions prevent the destruction of organic matter and so increase accumulation of soil organic matter (Mikha et al. 2005).
4. Good ventilation increases the soil content of nitrate nitrogen (Dobermann et al. 2003), which is the major nitrogen form for wheat absorption.

Disadvantages:

1. Manganese element changes from +4 valence to +2 under flooding conditions in the rice season. The +2 form is more easily lost with water leaching. In the wheat season, manganese is transformed to insoluble compounds, resulting in manganese deficiency in wheat (Lv and Zhang 1997).

2. Soil pH is changed by the carbon dioxide content in floodwater, which induces changes in valences of other nutrient elements.

6.4.2.3 Biological Properties

Advantages:

1. The conversion from anaerobic to aerobic conditions stimulates activity of soil micro community, increasing soil microbial biomass and the mineral carbon content. Mikha et al. (2005) confirmed that soil mineral nitrogen accumulation was significantly higher in rice–wheat growing system than in upland-upland.
2. Rice–wheat rotation can reduce pests and diseases in the wheat season. The flooding process for rice can reduce pathogens. The pathogen causing cercospora spot of wheat was found to die after several months of flooding (Fujisaka et al. 1994).

Disadvantages:

The alternate wetting–drying of soil enhances the activity of both soil nitrobacteria and denitrifying bacteria. Nitrous oxide (N_2O) emission under Eh value (indicating soil redox potential) +400 and 0 was induced by nitrification and denitrification processes, respectively, and resulted in substantial production and release of nitrous oxide (Aulakh et al., 2001).

Most physical changes are difficult to alter without special technology, while chemical changes are reversible through drainage. Biological changes can be regulated and exploited with significant nutrient management for the rice–wheat system.

6.5 Nitrogen Input, Transformation and Balance in the Rice–Wheat System

Nitrogen is an indispensable nutrient in the rice–wheat system and is the most active element in the soil system. Crops can use the available nitrogen in soil; therefore, there is a significant relationship between the mineral nitrogen content and productivity. However, nitrogen in the rice–wheat system is profoundly influenced by human activities, especially nitrogen fertilization, crop residue return and irrigation (Singh and Singh 2001). The nitrogen cycling in an agro-ecosystem reflects exchanges of the biosphere with the pedosphere, hydrosphere and atmosphere (Fig. 6.5). The complex system of nitrogen cycling in rice–wheat has significant implications for nitrogen management.

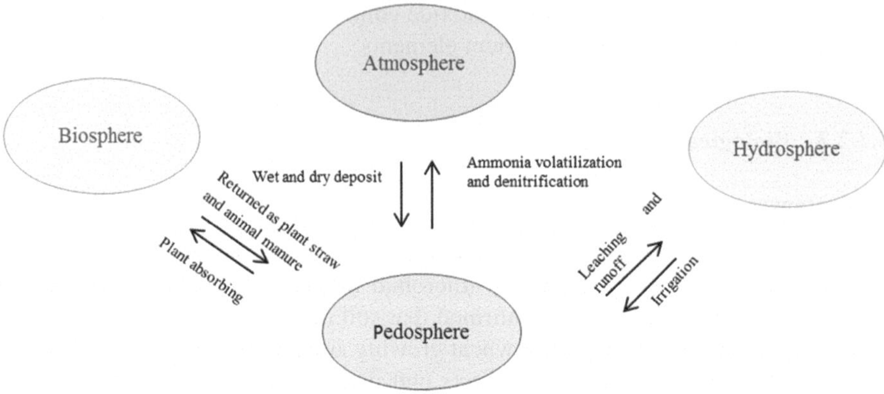

Fig. 6.5 Nitrogen is an active element. The nitrogen processes in agro-ecosystem reflect nitrogen exchanges of the pedosphere with biosphere, hydrosphere and atmosphere. Biosphere, hydrosphere and atmosphere input nitrogen to pedosphere and at the same time nitrogen in pedosphere is assimilated by plant or lost into hydrosphere and atmosphere

6.5.1 Input

6.5.1.1 Nitrogen Fertilization

Nitrogen Fertilizer Application

During 1960–2014, the amount of global nitrogen application increased from 11.6×10^6 t to 108.9×10^6 t, an average increase of up to 9.4 times (FAOSTAT 2016). However, in China, the nitrogen application amount increased by 45.9 times. About 20% of the world's rice fields are in China, and the annual consumption of nitrogen fertilizer in rice fields accounts for 37% of global consumption (Zhu 2000). Most of the rice–wheat area in China is in economically developed regions and this increases the amount of nitrogen application. In the city of Changshu, one of the main food production areas in the Yangtze River Basin, nitrogen application rates exceeded 100 kg nitrogen ha^{-1} in 1975 and increased by three times in the following 10 years (Fig. 6.6).

Although the nitrogen application rate is higher in the rice season, a survey showed that rice production was 50% more profitable with only 20% more fertilizer cost compared with wheat (Hofmeier et al. 2015). In 2009, nitrogen application rate in the rice season exceeded 300 kg nitrogen ha^{-1} in the Yangtze River Basin, and was nearly 250 kg nitrogen ha^{-1} in the wheat season (Wang et al. 2009).

In South Asia, nitrogen application rates are significantly lower than in China. A study on 30 long-term rice–wheat experiment sites showed that the nitrogen fertilizer application was in the range of 90–150 (average 115) kg nitrogen ha^{-1} in the rice season and 100–180 (average 123) kg nitrogen ha^{-1} in the wheat season (Ladha et al. 2003). The recommended nitrogen fertilization rates are 50–150 and 80–150 kg nitrogen ha^{-1} for rice and wheat seasons, respectively, depending on soil type and

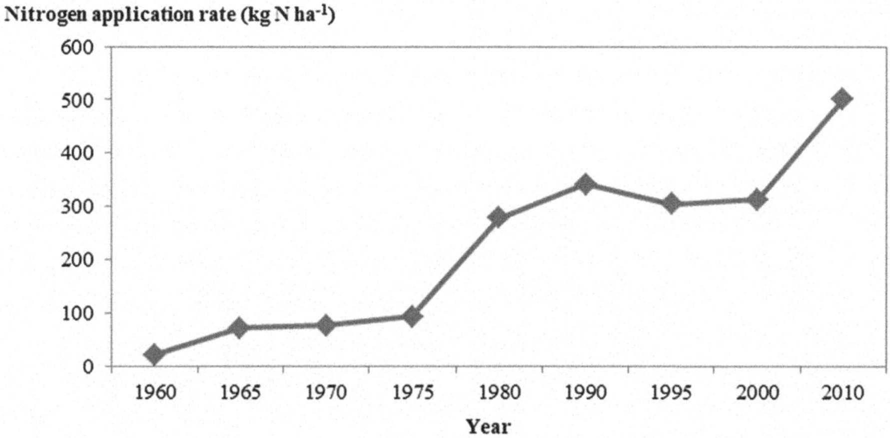

Fig. 6.6 Annual nitrogen application rates for 1960–2010 in the county of Changshu, Yangtze River Basin, China. The rate of annual nitrogen application has increased by 5 times during past 50 years

other characteristics in India and Bangladesh (Timsina et al. 2006; Bhaduri and Purakayastha 2014). The application of 120 kg nitrogen ha^{-1} has significant and economical responses in consideration of fertilizer cost and production value for most rice–wheat areas. However, farmers seldom adopt recommendations and much more nitrogen fertilizer has been applied to avoid yield decline with the spread of intensive rice–wheat in India (Singh et al. 2005).

Pattern of Nitrogen Fertilizer Use

The splitting of nitrogen fertilizer application is beneficial for plant growth. Rice takes up nearly half of its nitrogen during ear initiation, while wheat needs less nitrogen before the jointing stage and maintains high nitrogen absorption from jointing to grain filling. In most parts of South Asia, nearly 60% of nitrogen is applied as a mixture of diammonium phosphate and urea, and the other is top dressed in 2–3 applications (Singh and Singh 2001). In China, three applications of nitrogen are recommended for one growing season and in practice the nitrogen used is commonly 30:30:40 or 40:30:30 in the rice season and 50:25:25 in the wheat season for basal, tillering and panicle period fertilizer. Unfortunately, under the common split ratios of nitrogen, the utilization efficiency is less than 30% in the basal and tillering fertilizer periods. Xue et al. (2016) proposed that the split ratios of nitrogen fertilizer at different growth stages should be optimized according to soil fertility and found in a field study that nitrogen application ratios of 18:42:40 in medium and high soil fertility conditions and 25:25:50 in low soil fertility condition could result in the highest production. Usman et al. (2014) conducted a 2-year field experiment and showed that 200 kg nitrogen ha^{-1} in four equal splits enhanced wheat yield and nitrogen efficiency in a rice–wheat system.

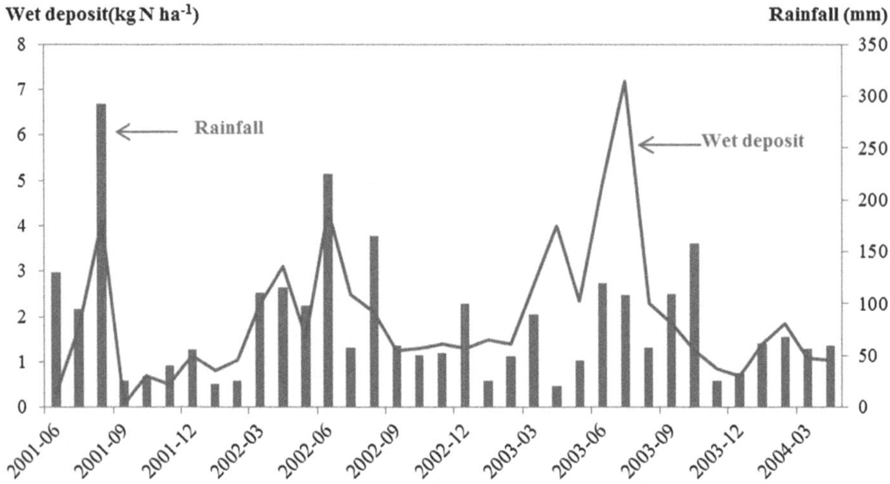

Fig. 6.7 Wet deposits and rainfall from June 2001 to May 2004 in the Yangtze River Basin, China. Wet deposits have a close relationship with rainfall. The period from June to September every year is the rainy season and rainfall takes 6-14 kg N ha^{-1} back to field in that period

6.5.1.2 Wet and Dry Deposition

Deposition is one stable nitrogen source in the rice–wheat system, and can offset nitrogen loss. Ammonium, nitrate and small amounts of soluble organic nitrogen are the major forms of wet deposition, while dry deposition consists of nitric oxide, ammonia and gaseous nitric acid. Regional wet and dry deposition has been determined by nitrogen emission rates in different areas. The deposition rates are strongly correlated with regional nitrogen application rates and precipitation rates. A 3-year field study in the Yangtze River Basin showed that rainfall could bring 17–26 kg nitrogen ha − 1 of wet deposition (from June 2001 to May 2004) (Fig. 6.7), with peak values in June–July for the rice season and November–December for the wheat season.

6.5.2 Nitrogen Used by Rice and Wheat

6.5.2.1 Yield

Crop production is the reason that people cultivate rice and wheat, and nitrogen fertilizer is applied for higher production. Nitrogen uptake by the crop is the proportion of nitrogen fertilizer called "effective", and crop yield is dependent on this.

The yields of rice and wheat in China and South Asia gradually increased during 1961–2014 (Fig. 6.8). The increases in China were large and indicated more intensive agricultural management. Interestingly, wheat yield increased rapidly resulting in higher yields compared with rice during 1986–2009 in China. However, the South Asia countries showed higher rice yield than wheat. The average yield data

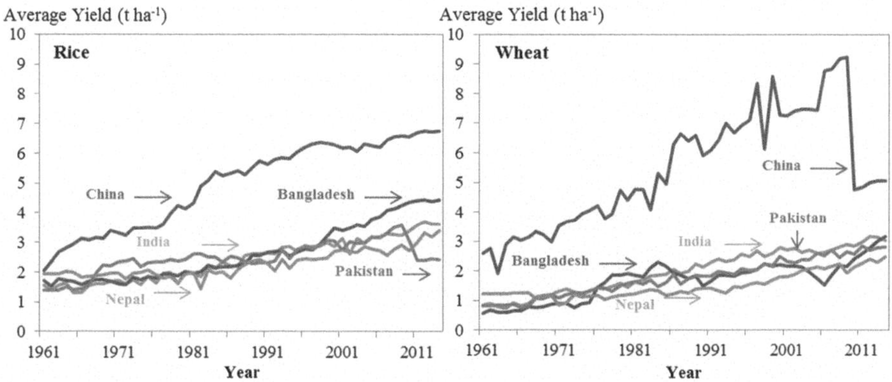

Fig. 6.8 National yield trends of rice and wheat in China, India, Bangladesh, Pakistan and Nepal (1961–2014). The increases in yields of rice and wheat in China were larger than those in other countries

included not only areas of rice–wheat but also all farmland for rice or wheat cultivation (FAOSTAT 2016). A long-term rice–wheat experiment in Asia showed that 22% and 6% of the sites had significant declining trends in rice and wheat yields, respectively (Ladha et al. 2003). In addition, rice yields have declined more rapidly than wheat. Timsina and Connor (2001) attributed the gap between actual and potential yield to severe biological or technological limitations. However, when the yield gap is narrow, crop yields will no longer rise with further increases in fertilizer application alone. Maintaining higher yield by greater fertilizer inputs may result in serious environmental degradation, which has been a major issue in China.

6.5.2.2 Nitrogen Taken Up by Crops

There are significant differences in crop nitrogen uptake in rice–wheat systems due to the variations in climate, nutrient management, soil and crop type between South Asia and China (Table 6.1). The rice yield and nitrogen uptake still have a good positive relationship with the amount of nitrogen applied, and China obtains high rice yields by high use of fertilizer (Xue et al. 2014a, b). Additionally, compared with South Asia, more rice is harvested per 100 kg of nitrogen fertilizer applied in China. This may be a consequence of higher ratios of nitrogen in grain to straw, indicating intensive and efficient agricultural cultivation in China.

However, the conditions for wheat are quite different. Nitrogen application rates show non-significant differences among different regions in South Asia, but wheat production is in the range of 1.88–4.8 Mg ha^{-1} (Aslam and Prathapar 2001; Ladha et al. 2003; Usman et al. 2013). Data from a long-term experiment showed higher production in the Indo-Gangetic Plain than other areas in India (Ladha et al. 2003). Wheat production varied in different areas of Pakistan, with low yield in Sindh attributed to inadequate levels of nutrient input and poor cultural practices, especial water management (Aslam and Prathapar 2001).

Table 6.1 Annual crop yields (t ha^{-1}), nitrogen input (kg nitrogen ha^{-1}) and uptake (kg nitrogen ha^{-1}) of rice–wheat systems

		Rice				Wheat			
		Fertilizer	Yield	Nitrogen uptake		Fertilizer	Yield	Nitrogen uptake	
Area				Grain	Straw			Grain	Straw
South Asia	IGP[a]	120[a]	4.65[a]	38–52 Mean 44.17[b]	17–59 Mean 33.17[b]	120[a]	3.64[a]	38–81 Mean 57.17[b]	9–20 Mean 14.11[b]
	Non-IGP in India	120[a]	4.11[a]			120[a]	2.73[a]		
	DeraIsmail Khan in Pakistan[c]	120	–[b]	–	–	120	4.80	94.82	29.34
	Sindh in Pakistan[d]		1.57				1.88		
China	Yangtze River Basin[e]	210	8.18	120.29	71.60	180	3.53	91.23	29.27
	Sichuan Province[f]	150	6.58	113.0	39.13	150	2.77	65.44	24.64

The rice yields in China is much higher than those in other countries due to high use of fertilizer. The highest wheat yield was found in Pakistan and its rate of fertilizer applied was 20%–33% lower than those in China

[a]IGP: Indo-Gangetic Plain

[b]–: data missing

[a]Ladha et al. (2003)

[b]Timsina et al. (2006)

[c]Usman et al. (2013)

[d]Aslam and Prathapar (2001)

[e]Xue et al. (2014b)

[f]Shi (2003)

6.5.3 *Transformations and Losses*

Nitrogen in soil is mainly in organic form from which it is continuously mineralized by microbial action. In addition, fertilizer is the dominant mineral nitrogen input as substrate for other nitrogen transformations in the rice–wheat system (Fig. 6.9). Nitrogen availability for crop growth depends on the form and amount in soil. Therefore, nitrogen transformations are the key to controlling nitrogen utilization efficiency by crops and nitrogen losses from the system.

6.5.3.1 Nitrification

Nitrification is the primary determinant of nitrogen loss into aquatic environments in the rice–wheat system. Urea, by far the most widely used nitrogen source, is first hydrolyzed to ammonia by the enzyme urease and then converted to ammonium. Hence, ammonium is the original ionic form of fertilizer. Ammonium nitrogen, which is positively charged, is not easily transported in water due to soil colloid absorption. However, once nitrification has transformed ammonium into nitrate, the nitrate can be carried away by water flow resulting in increasing losses via leaching and runoff. When nitrification rates are low, the retention period is lengthened as a consequence of lower leaching and runoff losses, but there is increased risk of ammonia volatilization in the rice season. Soil pH and aeration conditions are major factors determining nitrification. Evidence for the influence of soil pH on nitrification was presented by Nicol et al. (2008), who demonstrated a positive correlation between nitrobacteria abundance and soil pH. Nitrobacteria are the dominant ammonia oxidizers among soil microorganisms (Jia and Conrad 2009; Zhang et al.

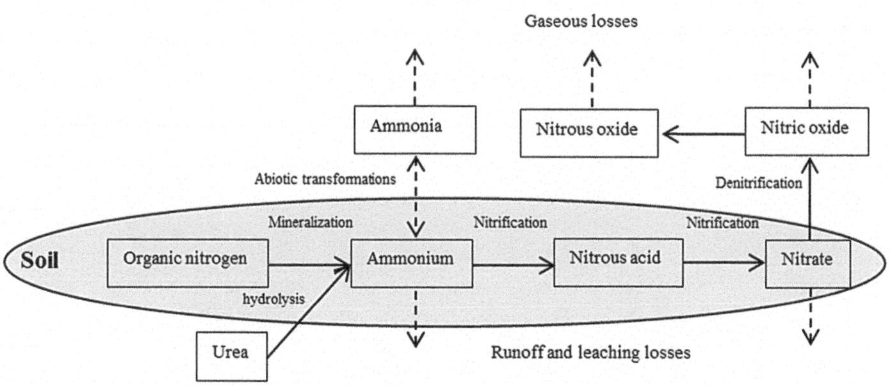

Fig. 6.9 Nitrogen in soil includes the processes of input and output. Nitrogen fertilizer is regarded as major nitrogen input from human being. Beside fertilizer input, soil mineral nitrogen is from the mineralization of organic nitrogen. The mineral nitrogen forms are changed by the processes of nitrification and denitrification. The major loss pathways in agricultural system are ammonia volatilization, denitrification, runoff and leaching

2012a). The lowest limits for nitrification are considered to be within pH 4.0–4.7 in soils (Persson and Wirén 1995; Hanan et al. 2016). However, soil nitrification occurs even under very acid conditions if ammonium content is sufficiently high as nitrifiers can adapt to these soil conditions (De Boer and Kowalchuk 2001). Nitrification as an aerobic process mainly occurs in oxidized or aerobic conditions and the rice–wheat system experiences conversions between aerobic and anaerobic seasons. Flooding in the rice season profoundly affects the nitrification rate; however, flooding is not maintained for the whole season. Nitrification occurs when flooding is removed, oxygen is introduced and thus nitrate is produced.

6.5.3.2 Runoff

Runoff formed by rainfall or excessive irrigation washes the soil surface and transports nitrogen into surface water. It is usually assumed that runoff is frequent in the rice season due to the rainy season and flooded conditions. However, runoff is more frequent during the wheat than the rice season; for example, during the 2009 rice season, there were three runoff events but seven events during the 2009–2010 wheat season (Fig. 6.10). This was attributed to the ridge of the rice field helping to prevent water overflow. Usually, the rand of a paddy field maintains the water table at around 10–15 cm, and this is raised over the rand only with heavy natural precipitation or irregular artificial draining. In contrast, considerable nitrogen runoff mainly occurs

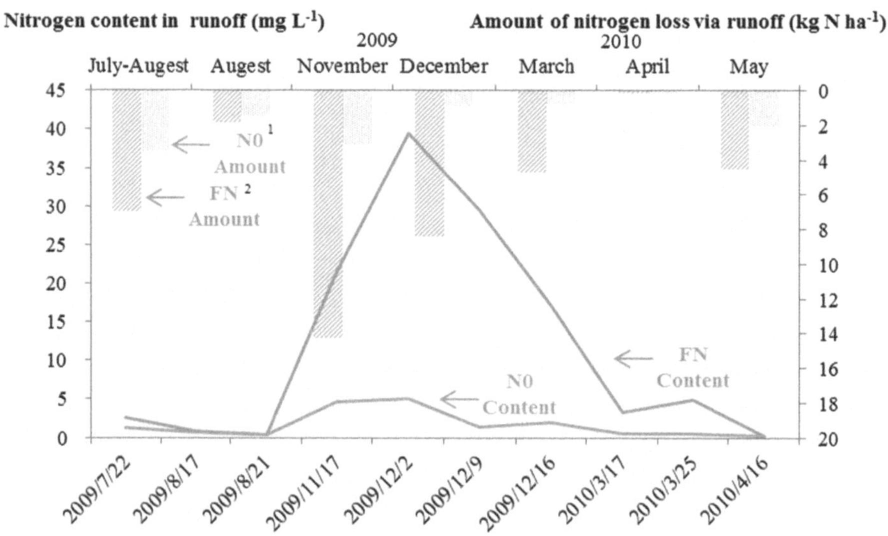

Fig. 6.10 Nitrogen loss via runoff (content and amount) from June 2009 to May 2010 (Modified after Xue et al. 2014b). The nitrogen content in runoff is determined by the timing of runoff. If runoff occurs during the week following fertilization, the nitrogen content in runoff could reach 30 mg L^{-1}. Compared to rice season, wheat season may loss more nitrogen via runoff. [1]N0: zero-nitrogen application. [2]FN: farmers' nitrogen application rate

with moderately heavy rainfall and accelerated water flow in existing drainage ditches in the wheat season. The nitrogen content of runoff water is determined by mineral nitrogen content in shallow soil layers. In Fig. 6.10 the nitrogen loss for FN treatment (farmers' nitrogen application rate) is much higher than that for the N0 treatment (zero-nitrogen application). Thus, if there is runoff during the week following fertilization, it will have very high nitrogen content. A 3-year field experiment in the Yangtze River Basin monitored the total nitrogen loss of rice–wheat into water systems and showed 82–93% was from runoff (Zhao et al. 2012). Zhu and Chen (2002) estimated that the average runoff loss of nitrogen represented 5% of nitrogen fertilizer application.

6.5.3.3 Leaching

It is generally though that nitrate is the main mineral nitrogen form lost via leaching and contributed to 64.5–82.9% to total nitrogen loss in the wheat season (Cao et al. 2014). However, nitrate and dissolved organic nitrogen were the predominant forms of nitrogen in leachate in the rice season, and contributed over 25% and 59% to total nitrogen for this season, respectively. During the flooding period, nitrification is very slow, leading to small amounts of nitrate produced, but downward percolation is constant due to concentrated flood irrigation and rainfall in summer. Fig. 6.11 shows that at 2 months after panicle fertilizer application for transplanted rice, the highest nitrogen content in leachate was at a depth of 40–60 cm, indicating a percolation rate of nitrogen of 6–10 mm d^{-1} (Yu et al. 2011). For the wheat season,

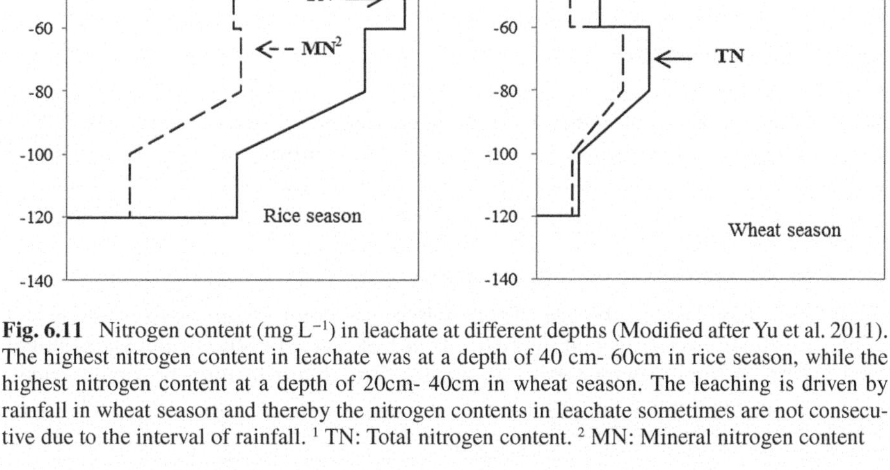

Fig. 6.11 Nitrogen content (mg L^{-1}) in leachate at different depths (Modified after Yu et al. 2011). The highest nitrogen content in leachate was at a depth of 40 cm- 60cm in rice season, while the highest nitrogen content at a depth of 20cm- 40cm in wheat season. The leaching is driven by rainfall in wheat season and thereby the nitrogen contents in leachate sometimes are not consecutive due to the interval of rainfall. [1] TN: Total nitrogen content. [2] MN: Mineral nitrogen content

leachate was collected 1 week after rainfall. The interval of rainfall is demonstrated by the higher nitrogen content in leachate at a depth of 60–80 cm than at 40–60 cm. According to previous data, the nitrogen loss via leaching was 6.75–27 kg nitrogen ha^{-1} y^{-1} (Xing and Zhu 2000) accounting for 2% of nitrogen fertilizer application rates (Zhu and Chen 2002).

6.5.3.4 Ammonia Volatilization

The amount of nitrogen loss via ammonia volatilization is substantial when urea is top dressed in alkaline soil, and this process occurs mainly at the water–air interface. Hence, ammonia is most likely to volatilize in the rice season with flooding. Ammonia volatilization is controlled by the ammonium phosphate slurry and wind speed at the water surface (Fillery and De Datta 1986). Soil characteristics, climatic conditions and agricultural practice all influence the kinetic processes of ammonia volatilization. Soil pH can affect the relative content of ammonia. When soil pH increased from 6 to 7, 8 and 9, the relative contents of ammonia increased from 0.1% to 1%, 10% and 50%, respectively (Tian et al. 2001). Other research also showed that soil ammonia volatilization rates in basic soil ($pH > 8.5$) were 39% higher than in neutral soil ($5.5 < pH < 7.3$) and 55% higher than in acid soil ($pH < 5.5$) (Cao 2006). Ammonia volatilization can cause nitrogen losses representing 1–47% of nitrogen fertilizer application (Tian and Cao 1998). Furthermore, along with the increased nitrogen input in agriculture, the proportion of nitrogen loss via ammonia volatilization is still increasing Zheng et al. (2002). Song et al. (2004) found that ammonia volatilization losses were mainly during the week following fertilization. Especially in the basal and tillering period, the amount nitrogen of ammonia volatilization accounted for 70% of the nitrogen losses in the rice season (Table 6.2).

Figure 6.12 (Modified after Yu et al. 2013) showed that the changes in ammonia volatilization rates were not related to the ammonium content of flooded water, indicating that ammonia volatilization was also affected by other climatic factors

6.5.3.5 Denitrification

Nitrate is the substrate of denitrification, and is converted from ammonium in both drained and flooded conditions and then reduced to nitrous acid, nitric oxide and nitrous oxide. With flooding in the rice season, denitrification is the main path of nitrate loss but is relatively weak in the wheat season. Alternating soil ventilation under rice can result in high nitrogen losses compared with maintaining continuous anaerobic conditions (Xu et al. 1998; Aulakh et al. 2001). Direct measurements are lacking for calculating gaseous nitrogen loss via denitrification. Zhu and Chen (2002) estimated that nitrogen denitrification loss via difference values accounted for 15–42% of the nitrogen fertilization rates. The denitrification process in rice–wheat has received much attention in recent years due to the considerable release of nitrous oxide, an important greenhouse gas.

Table 6.2 Volumes and ratios of cumulative ammonia volatilization in different fertilization periods of the rice season

Period	Ammonia volatilization (kg nitrogen ha^{-1})	Nitrogen fertilizer application (kg nitrogen ha^{-1})	Ratios (%)
Basal fertilizer period	28.0	81	34.6
Tillering fertilizer period	26.9	81	33.2
Panicle fertilizer period	20.9	108	19.4
Sum	75.9	270	28.1

Modified after Yu et al. (2013)
Compared to panicle fertilization period, more nitrogen lost via ammonia volatilization in basal and tillering fertilization period of rice season. The total nitrogen loss via ammonia volatilization could be over 25% of the nitrogen applied

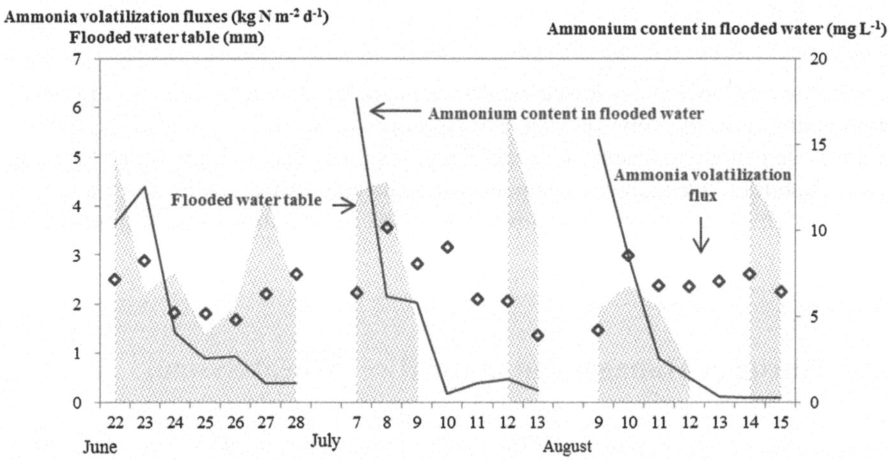

Fig. 6.12 Ammonia volatilization fluxes (kg nitrogen m^{-2} d^{-1}), ammonium content (mg L^{-1}) and water table (mm) in flooded water. The peak of ammonia volatilization flux appeared in the 2nd day after fertilizer application; however, no linear relation was found between ammonia volatilization flux and ammonium content in flooded water. Ammonia volatilization could still happen even there was no flooded water in field

6.5.4 *Nitrogen Balance in the Rice–Wheat System*

The annual nitrogen balance for each crop or the total system can be calculated from the difference between nitrogen input and the sum of nitrogen by crop removal and losses (Table 6.3). Timsina et al. (2006) noted that the nitrogen balance sheet will improve little with biological nitrogen fixation due to the dominant role of fertilizer, based on field experiments with different fertilizer rates that showed 30% losses in Bangladesh. Ju et al. (2009) computed an annual nitrogen surplus of 87 kg nitrogen ha^{-1} for current practices with large denitrification losses in the Yangtze River Basin.

Table 6.3 Nitrogen balance in the rice–wheat system (kg N ha^{-1})

Area	Period	Input			Removal		Balance[a]
		Fertilizer	Irrigation	Biological fixation	Crop	Loss	
Bangladesh[b]		120–380	–[c]	40–79	134–213	57–114	−37 to 64
Yangtze River basin, China[d]	Rice season	300	–	–	88	174	87
	Wheat season	250	–	–	46	155	

The rate of nitrogen fertilizer application in China was 1.3–4 times higher than that in Bangladesh, which resulting in lower nitrogen uptake by crop and higher nitrogen loss

[a]Balance is calculated by the difference between input and removal. The positive data indicates more nitrogen input than removal, while the negative data means less nitrogen input than removal

[b]Timsina et al. (2006)

[c]–: data missing

[d]Ju et al. (2009)

The positive balance in most rice–wheat results in a large proportion of nitrogen as surplus nitrate in the soil after harvest. Appropriate residuals could help maintain soil nitrogen supply capacity; however, large residuals lead to high nitrate leaching risks. Therefore, nitrogen management with reduced fertilizer application should be used to establish a temporary negative balance in consideration of the residual nitrogen in soil.

6.6 Nitrogen Management in the Rice–Wheat System

For most rice–wheat systems, less than 40% of nitrogen is used by crops, indicating that the remaining part (nearly 60%) is residual in soil or lost into the environment in different ways (Raun and Johnson 1999; Zhu and Chen 2002; Imran and Zed 2013). Accordingly, approaches to nitrogen management are needed to seek common goals that improve nitrogen utilization efficiency and reduce nitrogen losses.

6.6.1 Nitrogen Fertilizer Application

Fertilization is a direct way for humans to regulate soil mineral nitrogen content in rice–wheat, although a portion can be produced by mineralization of soil organic matter or acquired through irrigation and deposition.

1. Nitrogen fertilizer application rates should match crop demands. The over-use of nitrogen may decrease food production (Fig. 6.13) (Qiao et al. 2012). The theoretical nitrogen rate should be calculated based on the target yield and nitrogen requirement per unit grain (Ju et al. 2009). For fields in the Yangtze River Basin,

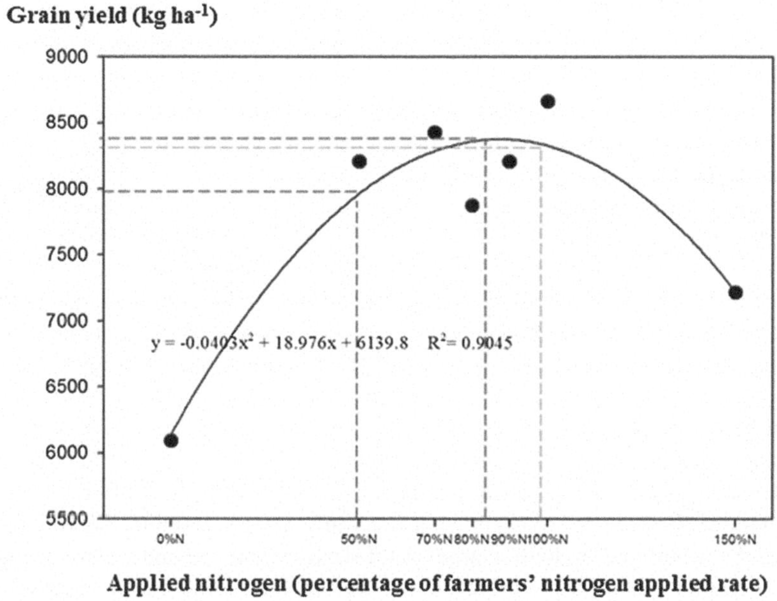

Fig. 6.13 Yields under different nitrogen rates in 2008. The rate of nitrogen application was reduced by 10%, 20%, 30% and 50% of farmers' nitrogen application. The highest yield was found with 80%-90% of farmers' nitrogen application indicating that excessive nitrogen may decrease yield. It was suggested that the rate of nitrogen applied could be reduced by 20% without yield decrease in experiment area.(Modified after Qiao et al. 2012)

the recommended optimum rate of nitrogen fertilizer is 200 kg nitrogen ha^{-1} for the rice season and 150 kg nitrogen ha^{-1} for the wheat season, which is 35% lower than farmers' practices. Based on field experiments and investigations, Hofmeier et al. (2015) concluded that nitrogen fertilizer application rates should be reduced by 15–25% for rice and by 20–25% for winter wheat. However, compared to farmers' practices no significant differences were found with reduced fertilizer application rate around 20%, indicating that similar high yields can be achieved with lower nitrogen fertilizer rates (Peng et al. 2006; Xue et al. 2014b).

2. The nitrogen form should fit the preference of crop utilization. Ammonium is preferable for rice, while both ammonium and nitrate can be used by wheat. Additionally, the flooded conditions of the rice season are helpful to decrease nitrate losses from nitrification. Urea is the dominant nitrogen fertilizer used in both China and South Asia; and ammonium chloride is also widely applied as a basal fertilizer in China and diammonium phosphate in South Asia. In recent years, organic manure has been recommended as a basal dressing along with chemical nitrogen fertilizer. Positive and significant trends in yield of rice have been observed in the Indo-Gangetic Plain with organic manure as a basal dressing (Yadav et al. 2000). After reducing nitrogen input by 22%, yield was still promoted by 4% with 30% organic fertilizer instead of all inorganic fertilizer in

a 3-year field study in the Yangtze River Basin (Xue et al. 2014b). Moreover, organic fertilizer also improves soil conditions for long-term agricultural cultivation.

3. The amount of nitrogen required by rice and wheat is substantial, so nitrogen fertilizer should be in split applications. Previous research showed that nitrogen loss in the basal and tiller fertilization period accounts for 55–70% of corresponding fertilizer rates, but is less than 20% in the panicle fertilization period for the rice season (Lin et al. 2014). Therefore, more nitrogen is lost if nitrogen fertilizer is in one application or in excess in the basal period.
4. Fertilization time and application rate should match the crop uptake pattern. An excessive fertilizer rate and incorrect timing can lead to half of the nitrogen being lost (Cassman et al. 1998). IRRI (International Rice Research Institute) suggests monitoring the plant nitrogen condition by testing the leaf SPAD (Soil and Plant Analyzer Development) value, and then determining the strategy for nitrogen fertilizer on this basis (Dobermann et al. 2002). The sufficiency index, calculated with an active sensor testing the canopy normalized difference vegetation index, is also used to establish a spectrally determined nitrogen topdressing model in combination with a target yield strategy and split-fertilization scheme. This results in similar or higher yields but with 20–40% lower nitrogen rates (Xue et al. 2014a).
5. The development of slow-release fertilizers may decrease the initial mineral nitrogen content after fertilization and improve nitrogen fertilizer performance in the long term (Ni et al. 2011).
6. Deep placement of nitrogen fertilizer into anaerobic soil could help prolong the existence of ammonium for rice utilization, reducing the ammonium content in floodwater, thus lowering ammonia volatilization. Fortunately, fertilizer deep placement can be performed at the time of rice transplantation using a machine, which reduces labor costs (Zhang et al. 2012a, b). Similar effects can be achieved by earthing (after fertilizer application, cover the soil surface with new soil) after nitrogen fertilization in the wheat growing season.

6.6.2 Tillage

1. Zero-tillage techniques have been widely applied for sowing wheat after rice (Hobbs and Giri 1997). Zero-tillage can prevent soil clodding, keep the soil capillary and pore system intact (Table 6.4) (Yonglu et al. 2009), and save water and labor (Bhushan et al. 2007). Thus, zero-tillage fields have better drainage, which provides appropriate air and moisture conditions for nitrogen transformation. Additionally, nitrogen loss via leaching may be weakened in the wheat season as zero-tillage soil has better water-holding capacity. Erenstein and Laxmi (2008) showed that the expansion of zero-tillage technology in the Indo-Gangetic Plains has enhanced income by about US\$100 ha^{-1} due to significant cost savings as well as potential gains in wheat yield through earlier planting of wheat.

Table 6.4 Effect of the zero-tillage technique on wheat yield and soil (0–7 cm) physical characteristics

Tillage practice	Yield (Mg ha^{-1})	Capillary porosity (%)	Non-capillary porosity (%)	Water content (%)
Plow	4.5	41.2	13.5	23.6
Zero-tillage	5.0	46.8	2.7	25.6

Plow will increase the soil non- capillary, while zero-tillage technique keeps the soil capillary and pore system intact and thereby the yield can be promoted

2. Straw retention can significantly reduce nitrogen extraction in the rice–wheat system. Traditionally, both grain and straw are harvested from the field, with large straw removed or burned in the field. This practice causes substantial nitrogen loss compared with straw retention (Table 6.1). In recent years, governments in China and South Asia have recommended straw retention. However, both rice and wheat straw are characterized by a high carbon to nitrogen ratio, which causes an initial nitrogen immobilization phase followed by a net re-mineralization phase (Müller et al. 1988). The use of crop straw as a nitrogen source depends on regulating the biological processes of soil to optimize nitrogen availability in response to plant demand (Singh and Sharma 2000). It is suggested that 8–10 weeks before rice transplanting is the appropriate time to incorporate wheat straw for alleviating the negative effect of wheat straw on rice growth owing to nitrogen immobilization. Mishra et al. (2001) and Nie et al. (2007) found increases in soil nitrogen, microbial biomass carbon and microbial biomass nitrogen by 14.8%, 12.7% and 15.1% compared with traditional practice, respectively.
3. Improving water use is one approach to keeping nitrogen in the rice–wheat system. Water plays a role as a nitrogen carrier and participates in nitrogen cycling in rice–wheat. Precipitation, irrigation water, leachate and runoff are all ways in which water is related to nitrogen cycling. Excess water, common in parts of southern China, India and Pakistani Punjab, may cause problems in both rice and wheat seasons (Minhas and Bajwa 2001; Zhao et al. 2012). Nitrogen fertilizer is transported into aquatic environments by flow losses in the rice season. Moreover, wheat production is affected by inadequate soil moisture or salinity. It is advisable to use water saving irrigation in the rice season to increase field capacity for rainfall and avoid runoff (Peng et al. 2007; Xu et al. 2012).

6.6.3 Additive

1. Nitrification inhibitors can mitigate the conversion of ammonium to nitrate. Compared with nitrate, ammonium is more easily held by soil colloids. The return to aerobic conditions before the wheat season results in rapid nitrification and the newly formed nitrate is susceptible to loss with drainage. The addition of nitrification inhibitors will contribute to limiting the loss of nitrate in this anaerobic–aerobic conversion period.

2. Use of surface films can reduce ammonia volatilization from flooded rice fields. These surface films are special emulsions made of water-soluble polymer, which rapidly spread as a close liquid molecular film on the surface of flooded water in rice fields. This physical barrier significantly lowers the ammonia volatilization. Moreover, the film can also improve rice growth by inhibiting algae for further weakening ammonia volatilization. Field experiments showed that surface films could help to reduce nitrogen fertilizer application by 25%, improve nitrogen utilization efficiency by 7.8–9.4% and increase rice yield by 6.5–7.9% (He et al. 2002).
3. Biochar is a new pathway for biomass residue. It has been shown to improve the structure and fertility of soils, thereby improving crop production. Biochar not only enhances the retention and therefore efficiency of fertilizers but, by the same mechanism, also decreases fertilizer runoff (Lehmann 2007).

6.6.4 Technological Integration

Many single nitrogen technologies have been proved effective in field experiments. Some technologies, influencing different stages of nitrogen cycling in the rice–wheat system, could be integrated. However, data on comprehensive effects of more than one technology on yield and nitrogen loss are limited. Integrated technologies require future study through both theory and practical application (Table 6.5).

6.7 Future Work

6.7.1 Encourage Nitrogen Management

The rice–wheat system is a fragile ecosystem with the large matter flow attributed to human activities such as harvest and fertilization. Many farmers still hold traditional opinions that higher crop yield will be obtained with more fertilizer. However, the proportionate increases of nitrogen fertilizer input and yield are different. Thus, excess nitrogen application with low utilization efficiency results in severe environmental degradation. Therefore, nitrogen management should be determined by accounting for nitrogen balance in the rice–wheat system.

1. Nitrogen fertilizer rate should be determined by accurate budgets for maintaining the balance between inputs (deposit and biologic fixation) and outputs (crop removal and loss).
2. Understanding the processes of nitrogen transformation is helpful to control forms of nitrogen in the rice–wheat system.

Table 6.5 The benefit of technological integration on crop yield or nitrogen loss

No.	Methods for technological integration	Benefits
1	Zero-tillage + residue returning in wheat season (Choudhury et al. 2014)	(1) Increasing yield by 8.3% (2) Improving soil aggregation by 53.8% and soil organic carbon sequestration by 33.6%
2	Nitrification inhibitors + zero-tillage in wheat season (Ma et al. 2013)	(1) Increasing yield by 0.9–6.9% (2) Reducing nitrous oxide by 18.1–44.6%
3	Reducing nitrogen application 25% + application in four splits: basal 50%, first topdressing 10%, and both the second and third topdressing 20% (Cao and Yin 2015)	(1) Increasing yield by 14% and straw biomass by 13% (2) Increasing plant nitrogen absorption 50% (48% of applied nitrogen versus 32% for conventional practice) (3) Reducing 52.1% ammonia volatilization loss (6.7% of applied nitrogen versus 14% for conventional practice)
4	Controlled released nitrogen fertilizer + non-flooding controlled irrigation (Xu et al. 2012)	Reducing ammonia volatilization loss by 81.1%
5	Site-specific nutrient management + non-flooding controlled irrigation (Xu et al. 2012; Dobermann et al. 2002)	Reducing ammonia volatilization loss by 70.0%
6[a]	Reducing nitrogen application 25.9% + 70% controlled released urea instead applied as basal fertilizer in rice season	(1) Increasing yield by 2.8% (2) Reducing nitrogen loss by 57.1% (3) Improving economic income by US$350 ha^{-1} in rice season
7[a]	Reducing nitrogen application 25.9% + 100% controlled released fertilizer instead deep placed as basal fertilizer + farm machinery under non-flooding controlled irrigation in rice season	(1) Increasing yield by 16.11% (2) Runoff loss significant (3) Improving economic income by US$340 ha^{-1} in rice season
8[a]	Reducing nitrogen application 25.9% in rice season + 30% organic fertilizer instead as basal fertilizer in rice season	(1) Increasing yield by 4.3% (2) Reducing nitrogen loss by 33% (3) Reducing irrigation water by 50% (4) Improving economic income US$540 ha^{-1} in rice season
9[a]	Reducing nitrogen application 25.9% in rice season + 70% mixed controlled released fertilizer instead + deep placed as basal fertilizer with farm machinery in rice season in rice season	(1) Increasing yield by 2.75% (2) Runoff loss significant (3) Improving economic income by US$340 ha^{-1} in rice season
10[a]	Reducing nitrogen application 25.9% + 100% mixed controlled released fertilizer instead + deep applied as basal fertilizer farm machinery in rice season	(1) Increasing yield by 9.77% (2) Runoff loss significant (3) Improving economic income by US$630 ha^{-1} in rice season

[a]The 6th–10th technological integration methods use the data unpublished by Lihong Xue

3. Any nitrogen management should be optimized by regional yield targets and soil conditions. In addition, a no-nitrogen treatment should be considered, which will provide some data for estimating nitrogen supplements from the soil organic matter pool.
4. Models estimating the track of nitrogen cycling could be used for predicting nitrogen losses in the rice–wheat system.
5. Long-term field experiments should be used to compute the feasibility and continuity of some methods of nitrogen management.

6.7.2 Regional Concept of Nitrogen Management in the Rice–Wheat System

The goal of nitrogen management in the rice–wheat system is to reduce the pathway of nitrogen transfer into the hydrosphere and atmosphere, but promoting the pathway of utilization in the biosphere. Much research only focuses on one nitrogen process, but nitrogen cycling occurs over space and time (Yang et al. 2013). For instance, a new technology that reduces nitrogen leaching could also raise the risk of ammonia volatilization. Additionally, nitrogen in runoff can be taken up by crops if runoff is used for irrigation. In a practical sense, what is needed is a workable nitrogen management strategy that considers the complete nitrogen cycling in the whole space and period of regional farmland. Yang et al. (2013) presented a '4R' technology to apply to regional rice–wheat fertilizer management (Fig. 6.14), which is explained in the following.

1. Reduce. Reducing chemical nitrogen fertilizer application rates could directly decrease nitrogen losses. The decrement of chemical nitrogen fertilizer should not affect grain production.
2. Retain. Provided by new agricultural technologies, biological materials are used to absorb nitrogen in water flow. Then, the controlled nitrogen is returned to the field with biological materials.
3. Reuse. Nitrogen loss from rice–wheat fields could be reused by eutrophic water irrigation. This has been adopted in Pakistan, where adjoining canals are built to use drainage water for water and nitrogen cycling (Aslam and Prathapar 2001).
4. Restore. Rice is the only crop that can be grown in flooded areas. Thus, rice could be used as nitrogen carrier for purifying polluted ponds and rivers.

The 4R technology is aimed to reduce nitrogen input and, by extending the nitrogen retention period in the field, to promote nitrogen utilization efficiency. The rice–wheat system acts as a purification unit at the regional scale. Yang et al. (2013) found that applying 4R technology reduced regional nitrogen loss by 47.5%.

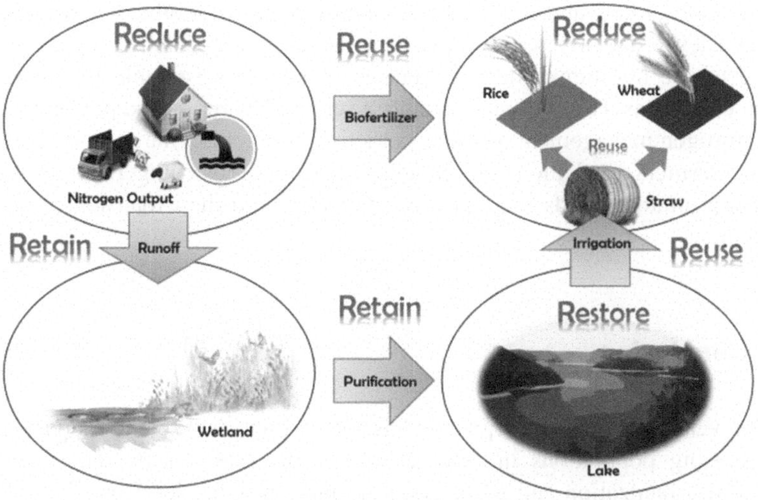

Fig. 6.14 Nitrogen cycling under '4R' technology. The concept of 'Reduce- Retain- Reuse-Restore' treats nitrogen as a cycling resource among biosphere, pedosphere and hydrosphere. With technology nitrogen will be removed from ground water and then used for crop growth again

6.7.3 Role of the Participants

There still a huge gap between theoretical conclusions and practical use in nitrogen management of the rice–wheat system. Much future work related to political decision-making, technology introduction and marketing dynamics is necessary for new nitrogen management to be accepted by farmers.

1. Government

The rice–wheat system is the main food source in both China and South Asia. Inappropriate nitrogen management is an increasing threat to these countries' food security and financial and political support are urgently required to optimize nitrogen use. For example, mechanization in most areas of China is inadequate to implement straw returning. Environmental standards and regulations should be set to guide farmers in nitrogen management and control the negative effects on environmental degradation. In addition, the lack of young and educated labor is a big issue for agricultural development; and the construction of water conserving facilities, such as channels and dams, are far beyond farmers' ability. Social capital should be attracted into the agricultural market to assist government plans.

2. Farmers

Farmers take an active part in every aspect of nitrogen management and so their participation is indispensable for sustainable agriculture. Owing to low profits in growing crops, young and educated people turn to cities for other employment. The older and less educated people, left in rural areas for farming work, are more inclined to use traditional methods that are generally unsuitable for present soil nutrient conditions. How to teach farmers a scientific strategy and persuade them to accept nitrogen management remain as practical problems. One recommendation is to invite farmers to demonstration fields instead of teaching them using data. Showing profitable yields may have a better effect on the expansion of scientific management.

6.8 Conclusion

For 1000 years, the rice–wheat system has had an important role in food production for increasing populations in Asia. However, the rice–wheat area is shrinking because of urbanization and more grain production is needed from the unit field. Therefore, massive amounts of fertilizer have been applied to increase yield in intensive agricultural production areas, which has led to significant environmental degradation. Nitrogen is the key factor to ensure food supply and security and imbalances in nitrogen application also limit the sustainable development of the rice–wheat system. Data on nitrogen utilization efficiency and loss describe the nitrogen movement in the rice–wheat system and show what is inappropriate under the present applications. Although there are many nitrogen technologies focusing on single nitrogen processes, studies on technological integration are lacking. Additionally, a general lack of historical soil samples is a problem for estimating the capacity of nitrogen supply. Nitrogen management should consist of technologies covering different nitrogen processes with definite yield targets based on the local soil environment. There should be continuing research on the performance of nitrogen cycling in the rice–wheat system to provide data for management decisions. Because it is the farmers that carry out field management, it would be helpful to expand better nitrogen management by improving agricultural equipment and educating farmers in environmental awareness.

Acknowledgments We thank Aijing Yin offering valuable suggestions that improved the manuscript. We gratefully acknowledge data support by Yixing Long-term Experiment Site that made this review possible. Data collection and analysis were undertaken as a part of research project funded by the Major State Basic Research Development Program of China (2016YFD0801101), the National Science Foundation for Distinguished Young Scholars of China (41501320) and the Special Fund for Agro-scientific Research in the Public Interest (201503106).

References

Aggarwal GC, Sidhu AS, Sekhon NK, Sandhu KS, Sur HS (1995) Puddling and N management effects on crop response in a rice-wheat cropping system. Soil Till Res 36:129–139. https://doi.org/10.1016/0167-1987(95)00504-8

Aslam M, Prathapar SA (2001) Water management in the rice-wheat cropping zone of Sindh, Pakistan. J Crop Prod 4:249–272. https://doi.org/10.1300/J144v04n01_07

Aulakh MS, Khera TS, Doran JW, Bronson KF (2001) Denitrification, N_2O and CO_2 fluxes in rice-wheat cropping system as affected by crop residues, fertilizer N and legume green manure. Biol Fert Soil 34:375–389. https://doi.org/10.1007/s003740100420

Beyrouty CA, Wells BR, Norman RJ, Marvel JN, Pillow JA Jr (1987) Characterization of rice roots using a minirhizotron technique. In: Minirhizotron observation tubes: Methods and applications for measuring rhizosphere dynamics. American Society of Agronomy, Madison, pp 99–108

Bhaduri D, Purakayastha TJ (2014) Long-term tillage, water and nutrient management in rice–wheat cropping system: Assessment and response of soil quality. Soil Till Res 144:83–95. https://doi.org/10.1016/j.still.2014.07.007

Bhushan L, Ladha JK, Gupta RK, Singh S, Tirol-Padre A, Saharawat YS, Pathak H (2007) Saving of water and labor in a rice–wheat system with no-tillage and direct seeding technologies. Agron J 99:1288–1296. https://doi.org/10.2134/agronj2006.0227

Cao ZH (2006) Material exchange and water environment quality between soil and water in Tai Lake region. Science Press, Beijing, pp 43–47

Cao Y, Yin B (2015) Effects of integrated high-efficiency practice versus conventional practice on rice yield and N fate. Agri Ecosys Environ 202:1–7. https://doi.org/10.1016/j.agee.2015.01.001

Cao Y, Tian Y, Yin B, Zhu Z (2014) Improving agronomic practices to reduce nitrate leaching from the rice–wheat rotation system. Agr Ecosyst Environ 195:61–67. https://doi.org/10.1016/j.agee.2014.05.020

Cassman KG, Peng S, Olk DC et al (1998) Opportunities for increased nitrogen-use efficiency from improved resource management in irrigated rice systems. Field Crops Research 56(1):7–39. https://doi.org/10.1016/S0378-4290(97)00140-8

Choudhury SG, Srivastava S, Singh R, Chaudhari SK, Sharma DK, Singh SK, Sarkar D (2014) Tillage and residue management effects on soil aggregation, organic carbon dynamics and yield attribute in rice–wheat cropping system under reclaimed sodic soil. Soil Till Res 136:76–83. https://doi.org/10.1016/j.still.2013.10.001

Crawford GW (2008) East Asian plant domestication. In: Archaeology of Asia. Blackwell, Malden, pp 77–95

De Boer W, Kowalchuk GA (2001) Nitrification in acid soils: micro-organisms and mechanisms. Soil Biol Biochem 33:853–866. https://doi.org/10.1016/S0038-0717(00)00247-9

Dobermann A, Witt C, Dawe D, Abdulrachman S, Gines HC, Nagarajan R, Chien NV (2002) Site-specific nutrient management for intensive rice cropping systems in Asia. Field Crop Res 108:37–66. https://doi.org/10.1016/S0378-4290(01)00197-6

Dobermann A, Witt C, Abdulrachman S, Gines HC, Nagarajan R, Son TT, Phung CV (2003) Soil fertility and indigenous nutrient supply in irrigated rice domains of Asia. Agron J 95:913–923. https://doi.org/10.2134/agronj2003.0913

Erenstein O, Laxmi V (2008) Zero tillage impacts in India's rice–wheat systems: A review. Soil Till Res 100:1–14. https://doi.org/10.1016/j.still.2008.05.001

Evans LT (1993) Crop evolution, adaptation and yield. Photosynthetica 34:56–56

Fillery IRP, De Datta SK (1986) Ammonia volatilization from nitrogen volatilization as a N loss mechanism in flooded rice fields. Fert Res 9:78–98. https://doi.org/10.1007/BF01048696

Fujisaka S, Harrington L, Hobbs P (1994) Rice-wheat in South Asia: Systems and long-term priorities established through diagnostic research. Agric Syst 46:169–187. https://doi.org/10.1016/0308-521X(94)90096-X

Gajri PR, Gill KS, Singh R, Gill BS (1999) Effect of pre-planting tillage on crop yields and weed biomass in a rice–wheat system on a sandy loam soil in Punjab. Soil Till Res 52:83–89. https://doi.org/10.1016/S0167-1987(99)00060-4

Hanan E J, Schimel J P, Dowdy K, D'Antonio CM (2016) Effects of substrate supply, pH, and char on net nitrogen mineralization and nitrification along a wildfire-structured age gradient in chaparral. Soil Biol Biochem 95: 87–99. doi:https://doi.org/10.1016/j.soilbio.2015.12.017

He WT, Sun Q, Zhu ZL, Liu SX, Kou DY (2002) Study on spraying film of liquid molecular in paddy field with techniques of both nitrogen fertilizer and water saving. Acta Pedo Sin 39:191–191

Hobbs PR, Giri GS (1997) Reduced and zero-tillage options for establishment of wheat after rice in South Asia. In: Wheat: Prospects for global improvement. Springer, Dordrecht, pp 455–465. https://doi.org/10.1007/978-94-011-4896-2_60

Hofmeier M, Roelcke M, Han Y, Lan T, Bergmann H, Böhm D, Nieder R (2015) Nitrogen management in a rice–wheat system in the Taihu region: Recommendations based on field experiments and surveys. Agriculture Ecosyst Environ 209:60–73. https://doi.org/10.1016/j.agee.2015.03.032

Imran MA, Zed R (2013) Physiology of nitrogen-use efficiency. In: Improving water and nutrient-use efficiency in food production systems. Wiley-Blackwell, Ames, pp 105–121

IRRI (1995) Water: A loaming crisis. International Rice Research Institute, Los Banos, p 90

Jia Z, Conrad R (2009) Bacteria rather than Archaea dominate microbial ammonia oxidation in an agricultural soil. Environ Microbiol 11:1658–1671. https://doi.org/10.1111/j.1462-2920.2009.01891.x

Ju XT, Xing GX, Chen XP, Zhang SL, Zhang LJ, Liu XJ, Zhang FS (2009) Reducing environmental risk by improving N management in intensive Chinese agricultural systems. Proc Natl Acad Sci U S A 106:3041–3046. https://doi.org/10.1073/pnas.0813417106

Kirk GJD, Ren YT, Choudhury FA (1990) Phosphorus chemistry in relation to water regime. Phosphorus Requirements for Sustainable Agriculture in Asia and Oceania: Proceedings of a Symposium. Rice Research Institute, p 211.

Ladha JK, Dawe D, Pathak H, Padre AT, Yadav RL, Singh B, Sakal R (2003) How extensive are yield declines in long-term rice–wheat experiments in Asia. Field Crop Res 81:159–180. https://doi.org/10.1016/S0378-4290(02)00219-8

Lehmann J (2007) A handful of carbon. Nature 447:143–144. https://doi.org/10.1038/447143a

Li BZ (1982) Rice-wheat system in China stared from the tang dynasty in Yangtze River Basin. Agr Archaeol 2.

Lin JJ, Li GH, Xue LH, Zhang WJ, Xu HG, Wang SH, Ding YF (2014) Subdivision of nitrogen use efficiency of rice based on ^{15}N tracer. Acta Agronomica Sinica 40:1418–1428. https://doi.org/10.3724/SP.J.1006.2014.01424

Lv SH, Zhang FS (1997) Ten-year retrospect and prospect of plant Mn in paddy-upland field. Agric Chem Bull Soil 1:1–7

Ma YJ (1999) Material change and ecological environment of rice soil. Science Press, Beijing, pp 1–8

Ma YC, Sun LY, Zhang XX, Yang B, Wang JY, Yin B, Yan XY, Xiong ZQ (2013) Mitigation of nitrous oxide emissions from paddy soil under conventional and no-till practices using nitrification inhibitors during the winter wheat-growing season. Biol Fert Soils 49:627–635. https://doi.org/10.1007/s00374-012-0753-7

Mahajan A, Mahajan A (2009) Chapter Seven: Integrated Nutrient Management (INM) in a sustainable rice-wheat cropping system. Springer, Dordrecht. https://doi.org/10.1007/978-1-4020-9875-8

Mikha MM, Rice CW, Milliken GA (2005) Carbon and nitrogen mineralization as affected by drying and wetting cycles. Soil Biol Biochem 37:339–347. https://doi.org/10.1016/j.soilbio.2004.08.003

Minhas PS, Bajwa MS (2001) Use and management of poor quality waters for the rice-wheat based production system. J Crop Prod 4:273–306. https://doi.org/10.1300/J144v04n01_08

Mishra B, Sharma PK, Bronson KF (2001) Decomposition of rice straw and mineralization of carbon, nitrogen, phosphorus and potassium in wheat field soil in western Uttar Pradesh. J Indian Soc Soil Sci 49:419–424

Müller MM, Sundman V, Soininvaara O, Meriläinen A (1988) Effect of chemical composition on the release of nitrogen from agricultural plant materials decomposing in soil under field conditions. Biol Fert Soil 6:78–83. https://doi.org/10.1007/BF00257926

National Bureau of Statistics (2012) Rural statistical yearbook of China. China Statistics Press, Beijing

Ni B, Liu M, Lü S, Xie L, Wang Y (2011) Environmentally friendly slow-release nitrogen fertilizer. J Agric Food Chem 59:10169–10175. https://doi.org/10.1021/jf202131z

Nicol GW, Leininger S, Schleper C, Prosser JI (2008) The influence of soil pH on the diversity, abundance and transcriptional activity of ammonia oxidizing archaea and bacteria. Environ Microbial 10:2966–2978

Nie J, Zhou JM, Wang HY, Chen XQ, Du CW (2007) Effect of long-term rice straw return on soil Glomalin, carbon and nitrogen. Pedosphere 17:295–302. https://doi.org/10.1016/S1002-0160(07)60036-8

Oussible M, Crookston RK, Larson WE (1992) Subsurface compaction reduces the root and shoot growth and grain yield of wheat. Agron J 84:34–38. https://doi.org/10.2134/agronj1992.00021962008400010008x

Peng S, Burcsh RJ, Huang J, Yang J, Zou Y, Zhong X, Zhang F (2006) Strategies for overcoming low agronomic nitrogen use efficiency in irrigated rice systems in China. Field Crop Res 96:37–47. https://doi.org/10.1016/j.fcr.2005.05.004

Peng S, Ding J, Mao Z, Xu JZ, Li DX (2007) Estimation and verification of crop coefficient for water saving irrigation of late rice using the FAO-56 method. T Chinese Soc Agric Eng 7

Persson T, Wirén A (1995) Nitrogen mineralization and potential nitrification at different depths in acid forest soils. Plant Soi 168–169:55–65. https://doi.org/10.1007/BF00029313

Raun WR, Johnson GV (1999) Improving Nitrogen Use Efficiency for Cereal Production. Agron J 91:357–363. https://doi.org/10.2134/agronj1999.00021962009100030001x

Sharma PK, Datta SKD (1986) Physical properties and processes of puddled rice soils, Advances in soil science. Springer, New York, pp 139–178. https://doi.org/10.1007/978-1-4613-8660-5_3

Shi XJ (2003) Characteristics of nutrient cycling in the rice-wheat rotation system. Doctoral Dissertation China Agricultural University, Beijing, China

Singh M, Sharma SN (2000) Effect of wheat residue management practices and nitrogen rates on productivity and nutrient uptake of rice (Oryza sativa)-wheat (Triticum aestivum) cropping system. Indian J Agric Sci 70:835–839

Singh Y, Singh B (2001) Efficient management of primary nutrients in the rice-wheat system. J Crop Prod 4:23–85. https://doi.org/10.1300/J144v04n01_02

Singh Y, Singh B, Timsina J (2005) Crop residue management for nutrient cycling and improving soil productivity in rice-based cropping systems in the tropics. Adv Agron 85:269–407. https://doi.org/10.1016/S0065-2113(04)85006-5

Singh VK, Dwivedi BS, Shukla AK, Mishra RP (2008) Effect of puddling and N management on productivity of rice (Oryza sativa L.)-wheat (Triticum aestivum L.) cropping system under sandy loam soils of Upper-Gangatic Plain. J Farm Syst Res Dev 14:12–19

Song YS, Fan XH, Lin DX (2004) Ammonia volatilation from paddy field in the Taihu Lake region and its influencing factors. Acta Pedologica Sinica 41:265–269

Swarup A, Singh KN (1989) Effect of 12 years' rice/wheat cropping sequence and fertilizer use on soil properties and crop yields in a sodic soil. Field Crop Res 21:277–287. https://doi.org/10.1016/0378-4290(89)90009-9

Tian GM, Cao JL (1998) Ammonia volatilization from winter wheat field top—dressed with urea. Pedosphere 8:331–336

Tian G, Cai Z, Cao J, Li X (2001) Factors affecting ammonia volatilisation from a rice–wheat rotation system. Chemosphere 42:123–129. https://doi.org/10.1016/S0045-6535(00)00117-X

Timsina J, Connor DJ (2001) Productivity and management of rice-wheat cropping systems: issues and challenges. Field Crop Res 69:93–132. https://doi.org/10.1016/S0378-4290(00)00143-X

Timsina J, Panaullah GM, Saleque MG, Ishaque M, Pathan ABMBU, Quayyum MA, Meisner CA (2006) Nutrient uptake and apparent balances for rice-wheat sequences. I. Nitrogen. J Plant Nutr 29:137–155. https://doi.org/10.1080/01904160500416539

Usman K, Khalil SKAU, Ghulam SGU (2013) Tillage and N Impact on Wheat Production and N Efficiency under Rice-Wheat Cropping System in Northwestern Pakistan. Philipp Agric Sci 95:33–42

Usman K, Khan EA, Yazdan F, Khan N, Rashid A, Din SU (2014) Short response of spring wheat to tillage, residue management and split nitrogen application in a rice-wheat system. J Integr Agric 13:2625–2633. https://doi.org/10.1016/S2095-3119(13)60737-6

Wang H, Xi YG, Chen RB (2009) Investigation on excessive application of fertilizer and pesticides in Taihu Lake region. Agro Environ Dev 26:10–15

Xing GX, Zhu ZL (2000) An assessment of N loss from agricultural fields to the environment in China. Nutr Cycl Agroecosy 57:67–73. https://doi.org/10.1023/A:1009717603427

Xu Q, Yang LZ, Dong YH, Xu H, Yan TM, Wang DJ, Lu YM (1998). Rice field ecosystem in China. Beijing: Chinese Agricultural Press pp1–4.

Xu J, Peng S, Yang S, Wang W (2012) Ammonia volatilization losses from a rice paddy with different irrigation and nitrogen managements. Agric Water Manage 104:184–192. https://doi.org/10.1016/j.agwat.2011.12.013

Xue LH, Li GH, Qin X, Yang LZ, Zhang H (2014a) Topdressing nitrogen recommendation for early rice with an active sensor in south China. Prec Agric 15:95–110. https://doi.org/10.1007/s11119-013-9326-5

Xue LH, Yu YL, Yang LZ (2014b) Maintaining yields and reducing nitrogen loss in rice–wheat rotation system in Taihu Lake region with proper fertilizer management. Environ Res Lett:9, 115010. https://doi.org/10.1088/1748-9326/9/11/115010

Xue LH, Li GH, Hou PF, Fan LH, Yang LZ (2016) Nitrogen reduction technique system for sustaining high yield of paddy fields. J Agro-Environ Sci 354:729–736

Yadav RL, Yadav DS, Singh RM, Kumar A (1998) Long term effects of inorganic fertilizer inputs on crop productivity in a rice-wheat cropping system. Nutr Cycl Agroecosy 51:193–200. https://doi.org/10.1023/A:1009744719420

Yadav RL, Dwivedi BS, Prasad K, Tomar OK, Shurpali NJ, Pandey PS (2000) Yield trends, and changes in soil organic-C and available NPK in a long-term rice-wheat system under integrated use of manures and fertilisers. Field Crop Res 68:219–246. https://doi.org/10.1016/S0378-4290(00)00126-X

Yang LZ, Shi WM, Xue LH, Song XF, Wang SQ, Chang ZZ, Li T (2013) Reduce-Retain-Reuse -Restore technology for the controlling the agricultural non-point source pollution in countryside in China: general countermeasures and technologies. J Agro Environ Sci 32: 1–8.

Yu YL, Xue LH, Yang LZ (2011) Effects of nitrogen management of nitrogenleaching of paddy soil in Taihu Lake Region. Acta Pedo Sin 48:98–103

Yu YL, Xue LH, Yang LZ (2013) Ammonia volatilization from paddy fields under different nitrogen schemes in Tai Lake Region. J Agro Environ Sci 32:1682–1689

Zhang AP, Liu RL, Yang SQ, Zhang QW, Li YH, Yang ZL (2012a) Effect of side bar fertilization technology based on slow-release fertilizer on rice yield and nitrogen losses. J Agro Environ Sci 31:555–562

Zhang J, Cai Z, Yang W, Zhu T, Yu Y, Yan X, Jia Z (2012b) Long-term field fertilization affects soil nitrogen transformations in a rice-wheat-rotation cropping system. J Plant Nutr Soil Sci 175:939–946. https://doi.org/10.1002/jpln.201200149

Zhao X, Zhou Y, Min J, Wang S, Shi W, Xing G (2012) Nitrogen runoff dominates water nitrogen pollution from rice-wheat rotation in the Taihu Lake region of China. Agric Ecosyst Environ 156:1–11. https://doi.org/10.1016/j.agee.2012.04.024

Zheng JG (2000) Rice-wheat cropping system in China. In: Hobbs PR, Gupta RKO (eds) Soil and crop management practices for enhanced productivity of the rice-wheat cropping system in the Sichuan Province of China', Rice-wheat consortium paper series 9. RWC-CIMMYT, New Delhi, pp 1–10

Zheng X, Fu C, Xu X, Yan X, Huang Y, Han S, Chen G (2002) The Asian nitrogen cycle case study. Ambio: J Human Environ 31:79–87. https://doi.org/10.1579/0044-7447-31.2.79

Zhou LS (2000) Agro-geography of China. Science Press, Beijing

Zhu ZL (2000) Loss of fertilizer N from plants-soil system and the strategies and techniques for its reduction. Soil Environ Sci 9:1–6

Zhu ZL, Chen DL (2002) Nitrogen fertilizer use in China – contributions to food production, impacts on the environment and best management strategies. Nutr Cycl Agroecosy 63:117–127

Chapter 7
Oilseed Rape Crop Residues: Decomposition, Properties and Allelopathic Effects

Zita Kriaučiūnienė, Rita Čepulienė, Rimantas Velička, Aušra Marcinkevičienė, Kristina Lekavičienė, and Egidijus Šarauskis

Abstract Oilseed rape, *Brassica napus* L., is an important crop for producing edible oil, biofuel and fodder. Oilseed rape is not only useful as a commercial plant, but it also improves the structure of the heavy granulometric soil fraction, activates microbiological processes, and reduces weed growth and disease rates. Oilseed rape belongs to the oil crop group, which increases soil fertility during crop rotation. Maintenance and restoration of soil fertility is one of the most relevant topics in agronomy science. One way to increase soil fertility is to increase organic matter content. In agriculture, the amount of organic matter in the soil depends on the crops grown and the residues that they leave. Crop residues of the previous harvest left on the soil surface interfere with soil tillage and sowing operations and can cause problems during application of environmentally sustainable tillage technologies.

We review here the properties of oilseed rape residues, the effects on agrocenoses and sustainable tillage technologies. The improvement in soil quality by growing oilseed rape can be determined by measuring the amount of remaining organic matter, the chemical composition and the intensity of mineralisation and humification, which depends on the carbon and nitrogen (C: N) ratio and lignin content. The C:N ratio, of 39–55, of oilseed rape residues is favourable for decomposition, but higher than that in most agricultural plants, as the higher lignin content, of 89.5–155.6 g kg^{-1}, slows decomposition, so the impact on the soil and plants continues longer. Oilseed rape synthesises allelochemicals such as glucosinolates and phenolic compounds that are released through the leaves and roots and penetrate the soil as the rape residues decompose. Comprehensive studies have revealed the allelopathic effects of oilseed rape. Moreover, the period from the end of harvest of one crop to the beginning of soil tillage and sowing of the new crops is very short. Crop residues do not lose their physical-mechanical properties within this short period,

Z. Kriaučiūnienė (✉) · R. Čepulienė · R. Velička · A. Marcinkevičienė
K. Lekavičienė · E. Šarauskis
Aleksandras Stulginskis University, Kaunas, Lithuania
e-mail: zita.kriauciuniene@asu.lt

E. Lichtfouse (ed.), *Sustainable Agriculture Reviews 32*, Sustainable Agriculture Reviews 32, https://doi.org/10.1007/978-3-319-98914-3_7

which exerts negative effects on the soil tillage and sowing. Experiments showed that biological preparations significantly reduce the breaking force required for spring oilseed rape residues by 39% and cutting force by 42% during the first week after harvesting.

Keywords Crop residues · Decomposition · Chemical composition · Physical-mechanical properties · Allelochemicals · Phytosanitary effect

7.1 Introduction

Soil erosion has accelerated on Earth since the advent of agriculture, and the intensification of agriculture has further sped up this process. Eroding soil is becoming increasingly more serious; however, attention has been drawn to this issue only in the last century. One of the causes of soil degradation is that no investments are made in soil restoration, no erosion prevention measures are considered and focus is placed on achieving the highest possible yield, regardless of the fertilizer input level or agronomic practices employed. Although soil quality depends on physical, chemical and biological properties influenced by the climate and ecosystem, the decisive factor is the land user. Both soil and crop productivity depend on soil quality including soil structure, water regime, chemical and biological composition. Many of soil's biological, physical and chemical properties are determined by the soil organic matter content. Soil microorganisms absorb nutrients and energy and produce biomass and soil organic matter while decomposing plant residues (Lal 1999). Crop residues help maintain the balance of biogenic elements in arable soils. Part of the nutrients are returned to the soil with the crop residues. The mineral matter that forms as organic matter decomposes is utilised by plants for nutrition. The value of crop residues in enriching the soil with nutrients depends on their chemical composition and the quantity incorporated into the soil (Tripolskaja 2005) (Fig. 7.1).

7.2 Effect of Chemical Composition of Crop Residues on Decomposition Rate

Winter oilseed rape produce a seed yield of 3.0 t ha^{-1} and in soil leaves 5.0 t ha^{-1} crop residues, including 1.9 t ha^{-1} of roots, 1.3 t ha^{-1} of stubble and 1.8 t ha^{-1} of other residues. Wager et al. (1998) reported that crop residues from hybrid oilseed rape varieties total 5–10 t ha^{-1}. However, when reduced crop and soil management practices were used, it dropped to 2.5–5.0 t ha^{-1} oilseed rape residues in the field (Hatfield and Stewart 1994).

Fig. 7.1 Oilseed rape residues left after harvesting has influence on tillage, sowing, soil, following crops and weeds. (Photos by Kriauciuniene and Sarauskis)

The decomposition rate of plants in different taxonomic classes differs; carbon is decomposed more intensively from the residues of Magnoliaceae plants at the beginning of decomposition, compared to those of Liliaceous plants (Magid et al. 2004). Above-ground plant residues decompose more rapidly than roots (Aleksandrova 1980; Teit 1991). Leaves and vines decompose at a higher rate than stems and roots (Wolf and Snyder 2003). Young, succulent plant parts decompose at the highest rate, while more mature parts are decomposed at a lower rate because young tissue contains more readily decomposable substances, such as sugars, starch, amino acids and protein, while older tissue contains higher concentrations of slowly decomposable hemicellulose and lignin (Wolf and Snyder 2003). Based on the C:N, polyphenol and lignin content, leaf and stem size and thickness and material age, some authors have classified plant residues based on their persistence into fragile (decompose rapidly) and nonfragile parts (resistant to decomposition). According to this classification, oilseed rape residues are fragile (Farnsworth et al. 1993).

The direction and rate of straw decomposition in soil largely depend on chop length of the straw particles. The finer and more crushed the straw is the higher the decomposition rate (Tripolskaja 2005). Residue particle size and their distribution in soil depend on tillage and harvesting processes. Finer residues are better mixed with the soil and are more readily decomposed by microorganisms (Singh et al. 2004). In addition, the chemical composition of the residues is important (Tripolskaja 2005). Plant chemical composition depends on the species and decomposition rate, as many authors have suggested, and it is determined by the environmental conditions, C: N and lignin content in plant residues (Jenkinson 1965, 1966, 1968a, b, 1977a, b; Jenkinson and Rayner 1977, 2006; Jenkinson et al. 1987, 1990; Teit 1991; Moran et al. 2005).

7.2.1 Nitrogen

Plant residue decomposition rate depends on its nitrogen content (Ågren et al. 2001; Adams 2003). The amount of nitrogen in plant residues is based on species, age, morphological part, environmental conditions and fertilisation. Nitrogen bound in organic compounds becomes available to plants only after mineralisation. Organic nitrogen mineralisation in the soil occurs due to the ammonification and nitrification processes. Ammonia is released as organic matter decomposes and is quickly engaged in the second mineralisation cycle called nitrification. Nitrate-nitrogen present in the soil is reduced to ammonia again or denitrified to gaseous nitrogen (Tripolskaja 2005). Carbon dioxide, mineral matter (mineralisation) and formation of stable humic substances (humification) occur as plant residues decompose in soil (Aleksandrova 1980; Teit 1991).

Organic matter content is a primary factor in the formation of soil. Organic matter is produced during plant photosynthesis, and it decomposes in soil due to biotic energetic activity (Kononova 1963; Aleksandrova 1980; Teit 1991; Eidukeviciene 2001). This is the circle of life on Earth, which is not possible without solar radiation and precipitation. Organic matter is produced in the soil and chemical elements necessary for the development of new organics are released (Eitminaviciute 1994). As a result, soil productivity depends on the organic carbon content left by preceding crops and its oxidation rate (Teit 1991). These processes are determined by different biological properties of plants, their chemical composition, soil genetic type as well as physical, chemical and biological properties, meteorological conditions and agronomic practices (Johnston 1986; Smukalski 1988; Teit 1991; Titova and Kogum 1991; Larson et al. 1994). Long-term research done at the Rothamsted Experimental Station (UK) has shown that the organic carbon content in the soil changes very slowly (Jenkinson and Rayner 2006). The biological activity of moderately heavy loam does not decline when is continually supplemented with fresh biomass, and a stable energy potential is maintained (Marcinkeviciene 2003; Pupaliene 2004). However, data indicate that only 38–43% of the phytomass accumulating in the agrocoenosis is left annually in light-textured soils (Janusiene 1994).

7.2.2 Carbon and Nitrogen Ratio

The most important factors influencing the duration and intensity of plant residue decomposition are the C:N ratio and lignin content (Teit 1991; Flaig 1969; Aleksandrova 1980; Janusiene 2002). When the C:N ratio in plant organic matter is high, the activities of microorganisms decreases rapidly during carbon decomposition due to a nitrogen shortage (Klimanek 1988). The C:N ratio in young succulent tissue is close to that in microorganisms (10:1); therefore, nitrogen is present in a sufficient amount and organic matter decomposes rapidly (Wolf and Snyder 2003). A crop whose phytomass more rapidly decomposes and leaves less non-humified plant residues for subsequent crops is the best for agricultural production (Robles and Burke 1997; Wager et al. 1998). The optimal C:N ratio for organic matter is 30–50:1, and that which is favourable for organic matter humification is 10–20:1 (Aleksandrova 1980; Wager et al. 1998). Oilseed rape residues decompose slowly due to their rather high C:N ratio of 50–70:1. Research done at the Rothamsted Experimental Station (UK) indicates that 13% of winter oilseed rape organic matter is decomposed within 5 months under field conditions (Watkins and Barraclough 1996). Data obtained in France show that 50% of winter rape carbon is decomposed within 2 months (Trinsoutrot et al. 2000). Plant residues with low (about 10:1) C:N ratios decompose rapidly. For example, green manure plant ploughed in at the early growth stage rapidly decomposes without leaving organic carbon in the soil and promotes decomposition of humus due to higher microorganismal activities; thus, reducing soil stocks (Stancevicius and Boguzas 1995). Oilseed rape straw, characterised by a high C:N (80–100:1), mineralises slowly (Stanceviius and Pranaitiene 1982). Slow mineralisation occurs particularly in roots as they have higher concentrations of lignin and cellulose compounds as well as fibrous tissue (Aleksandrova 1980). Research done in Lithuania shows that the C:N ratio in winter oilseed rape roots is 133:1, whereas that in stubble is 114:1. The coefficient of variation of this ratio depends on the meteorological conditions. The C:N ratio in winter and spring oilseed rape residues differs and it depends on morphological part of the plant. After harvesting in winter oilseed rape threshing remains C:N ratio was 48, in stubble 59 and in roots 55, in spring oilseed rape was accordingly 48, 67 and 67 (Kriauciuniene et al. 2012).

The plant residue decomposition rate is determined by soil and climate conditions as well as crop cultivation technologies (Aleksandrova 1980; Stanceviius and Pranaitiene 1982; Beyer 1996; Zekoniene and Janusiene 1997; Velicka 2002). Humification, mineralisation, fixation and migration are balanced once these soil and climatic factors have reached equilibrium (Aleksandrova 1980). These processes are stable and well-balanced in natural biocenoses, whereas they are disturbed in the agrocenose by intensive anthropogenic activity. A large part of solar energy that accumulates in humus is removed with yield, so additional organic matter is needed to restore the soil. Using different crops in different areas of a crop rotation and extending their cultivation duration also largely influence humus content and quality (Velicka 2002).

7.2.3 Lignin

Organic matter decomposition rate is determined by the concentration of readily decomposable components and lignin and their ratios in plant residues (Sollins et al. 1996; Kalbitz et al. 2000). Lignin is a key component in the formation of humus and maintaining soil productivity (Kononova 1963; Teit 1991). Lignin (lot. *Lignum*–a tree) is a phenolic polymer that forms the cell walls of plants with cellulose and hemicellulose (Kögel-Knaber 2002; Boerjan et al. 2003; Grabber et al. 2004; Grabber 2005). Lignin is very resistant to external factors and protects the cell cytoplasm from decomposition (Sollins et al. 1996; Blanchette 2000; Boerjan et al. 2003). However, some soil fungi are capable of using lignin as carbon source (Lugauskas 1997). Lignins are complex, polymeric hard-to-decompose substances of organic origin that have accumulated large reserves of photosynthetic energy

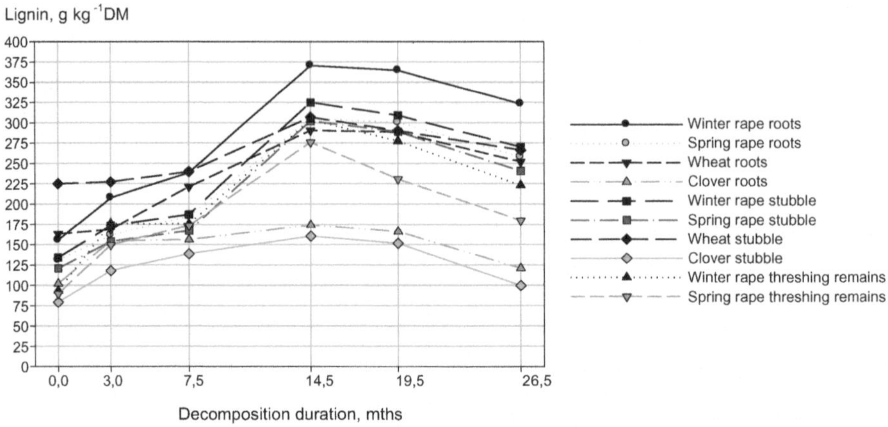

Fig. 7.2 Lignin content in crop residues increases for 14.5 of a 26.5-month period of crop residue decomposition in soil. It subsequently started to decrease, but a significant decrease in lignin contents in all investigated crop residues occurred after 26.5 months in the soil. *DM* dry matter; $LSD_{05} = 22.04$. (From: Kriauciuniene 2008)

(Blanchette 2000; Varnaite 2004). Lignin content and its chemical composition depend on the type of cell wall and structure (Donaldson 2001; Grabber et al. 2004). The composition of lignin markedly differs between different species of field crops. The lignin in oilseed rape stems is typical of angiosperm wood, as the syringyl:guaiacyl ratio varies slightly but increases with lignin content (Evansa et al. 2003). The highest concentration of lignin is found in the middle cell wall layer (primary wall and middle lamella) and cell corners (Blanchette 2000; Boerjan et al. 2003). Macromolecular polysaccharides, which are the products of monosaccharide condensation, account for 30–50% of lignin (Teit 1991).

It was established that the lignin content in crop residues increases for 14.5 of a 26.5-month period of crop residue decomposition (Fig. 7.2). It subsequently started to decrease, but a significant decrease occurred after 19.5 months only in winter and spring oilseed rape threshing remains. Significant decreases in lignin contents in all investigated crop residues occurred after 26.5 months in the soil. Lignin decomposed the most from its peak in red clover stubble (38%) and the least in winter oilseed rape roots (13%). The C:N ratio of the winter oilseed rape residues during dry matter and lignin decomposition decreased slower than that in the spring oilseed rape residues. Strong negative correlations were detected between lignin content and dry matter and between lignin content and the C:N ratio in decomposing winter and spring oilseed rape roots (r = from–0.86** to–0.93**), stubble (r = from–0.82** to–0.93**) and threshing remains (r = from–0.78** to–0.87**).

7.2.4 Nitrogen, Phosphorus and Sulphur

Dry matter content decreases and the concentrations of nitrogen, phosphorus and sulphur increase during microbial decomposition of plant residues (Ågren et al. 2001; Salas et al. 2003; Eriksen 2005). Maximum nutrient mineralisation occurs in the summer and warmer part of autumn (Orlova et al. 2002). Mineralisation of sulphur during plant residue decomposition is closely related to changes in nitrogen (Teit 1991). Decomposition of nitrogen- and sulphur-containing compounds is part of organic matter synthesis and a shortage in either element inhibits these processes (Aleksandrova 1980; Kazuki 2004; Eriksen 2005).

Various chemical elements are returned to the soil during decomposition. Biogenic elements returned to the soil with crop by-products do not increase the total content of elements in the soil but may affect their available forms. The nutrients utilised to enhance crop yield remain in the field when crop residues are ploughed in and returned to the soil, which improves the overall balance, and less mineral fertilizer is needed for the next year's crop (Tripolskaja 2005). The soil in a

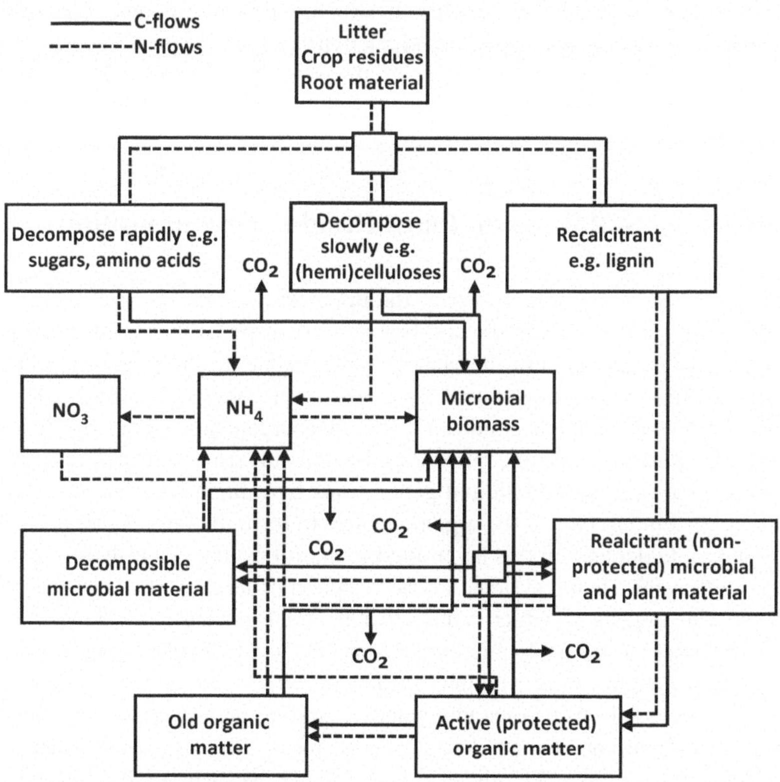

Fig. 7.3 Nitrogen mineralisation-immobilisation pools and fluxes during degradation of plant residues and formation of soil organic matter. Flows: — carbon; - - - nitrogen. (Modified after Paul and Clark (1989))

winter oilseed rape field of light loamy *Cambisol* was 1.7-, 2.5- and 2.9-times more enriched with nitrogen, phosphorus and potassium, respectively, compared with cereal residues; however, nitrogen and phosphorus contents were 1.9- and 1.3-times less compared with that in perennial grasses residues after oilseed rape cultivation, respectively (Magyla et al. 1997). Sidlauskas (2002) reported that spring rape dry matter residues contain 0.63% N, 0.05% P and 1.61% K. Shkarda (Shkarda 1985) reported that rape straw contains 85% dry matter, including 0.53% N, 0.11% P and 0.85% K, and the C:N ratio was 60–70:1.

The model (Fig. 7.3) demonstrates the changes occurring during decomposition of such complex substrates as plant residues and the direction of carbon and nitrogen flow. It is very important to consider microbial biomass production when estimating carbon and nitrogen contents. The three constituents of plant residues are readily decomposable sugars and amino acids, slowly decomposable cellulose and hemicelluloses and hardly decomposable lignin.

The largest amount of CO_2 is returned to the atmosphere by decomposers when they decompose plant and animal residues, which impacts the greenhouse effect. Carbon dioxide present in the atmosphere is in dynamic balance with that used for photosynthesis and that present in water and humus. More CO_2 will be released from decomposing humus, as less vegetation is established, which will result in higher concentration of CO_2 in the atmosphere and increased productivity and carbon mass in humus. In this way, a new equilibrium is reached; however, the ecosystem is only stable up to a particular CO_2 concentration (Eitminaviciute 1994; Teit 1991; Ryden et al. 2003).

7.3 Effect of Soil Biota on Plant Residue Decomposition

The accumulation of organic matter in the upper horizon of the soil is a complex biochemical process involving the soil biota and is responsible for nutrient cycling and stability of the ecosystem. In each system, species in the biota are related by close trophic links (Eitminaviciute 1994). Soil invertebrates chop and crush plant residues and create suitable conditions for microorganismal activities. Microbes decompose organic matter concentrated at the soil surface and provide good living conditions for new plant and animal generations by mineralising the organic content. Microbes bind to part of the organic matter to form the humus soil layer. Plant growth and yield depend on this layer and the microbes that occur there.

Soil is a dynamic, living system, as one teaspoon of fertile soil (10 g) can contain >9 billion living organisms, which is more than the total world's population (Lal 1999). The main function of soil microorganisms is to decompose organic matter. Microorganisms are involved in all intermediate processes of plant residue decomposition until cellulose is broken down into glucose; proteins are metabolised to individual amino acids, or ammonia and amides; and lignin forms aromatic compounds, such as syringaldehydes, vanillin and ferulic acid (Teit 1991; Kögel-Knaber 2002; Raudoniene 2003).

7.3.1 Micromycetes

Micromycetes breakdown lignin, actinomycetes breakdown pectin, cellulose, and fatty acids and bacteria breakdown carbohydrates and fibre (Lugauskas 1997; Henriksen and Breland 1999; Tripolskaja 2005). Micromycetes are highly adaptable organisms involved in various energy metabolic processes ranging from decomposition of complex chemical compounds to biosynthesis of new biological matter (Raudoniene 2003)., About 60% of decomposed carbon and nitrogen are immobilised in the organisms themselves during micromycetes nutrient metabolism, which temporarily retains the nutrients in the soil (Balota et al. 2003).

Plant residues with a high C:N ratio (>25:1) do not supply adequate nitrogen for microorganisms, so their activity decreases. Microorganisms utilise more carbon for food when readily metabolisable nitrogen compounds are depleted, and this increases the energy expenditure (Teit 1991; Hadas et al. 2004). Microorganisms breakdown plant residues into simpler compounds, part of which are mineralised, part become an energy source for microorganisms and part are humified (Teit 1991; Moran et al. 2005). The main enzymes involved in carbon decomposition are invertases and xylanases (Luxhøi et al. 2002). Nitrogen bound to the organic compounds of plant residues mineralises due to ammonification and nitrification. Ammonification is microbiological decomposition of nitrogen-containing organic compounds and conversion to ammonium ions (NH_4^+). This process is performed by a multitude of soil microorganisms, including aerobic and anaerobic bacteria, bacilli, mould fungi and actinomycetes (Tripolskaja 2005).

7.3.2 Fungi

Two stages are distinguished during phytomass decomposition. Mould fungi are involved in the first stage, as they damage plant cell walls and create conditions for decomposition (Kononova 1963; Teit 1991; Eitminaviciute 1994). The first organic compounds to be decomposed are those that are readily decomposable, such as carbohydrates, protein and water soluble organic compounds (Kononova 1963; Freytag and Luttich 1988). Furthermore, the rate of transformation of organic compounds depends on the relationship between the biochemical components of organic matter that are readily and not readily available to microorganisms. For example, fibre bound with lignin decomposes at a considerably lower rate than free fibre and protein and complexes with lignin (Cramer 1990; Teit 1991). These compounds as well as complex organic substances that compose cell walls of plant residues, including cellulose, hemicellulose and lignin are very slowly decomposed during the second stage of phytomass decomposition.

Lignin is depolymerised as plant residue is decomposed; inter-monomeric linkages break and aromatic compounds are condensed (Sjöberg 2003). Phenoloxidases are lignolytic enzymes that decompose lignin to phenolic alcohols and aldehydes

(Gisi 1990). Fungi are the main decomposers of lignin (Sollins et al. 1996; Varnaite 2004). Lignin polymers decomposed by microorganisms become biopolymers that bind to amino acids or proteins by forming primary humic substances (Kononova 1963; Teit 1991). The longer the polymers decompose in the soil, the greater the probability that stable humic compounds will form (Teit 1991).

The appropriate nutrient substrate and environmental conditions on which plant residue mineralisation and humification depend are essential for the functioning of each microorganismal species. The main environmental factors determining the course of these processes are climate, primary rock composition, soil type, acidity, moisture, oxygen content, phytocenose type, structure of fauna and other factors (Lugauskas 1997). Plant residues of different chemical compositions and soil tillage play significant roles under these conditions (Cesevicius 2007; Lejon et al. 2007). Some studies have reported that in *Endocalcari-Epihypogleyic Cambisol* ploughed in straw increases micromycetes populations by 24–40%, whereas they increase by 11% after conventional deep ploughing (Cesevicius 2007). However, the course of mineralisation and humification depends less on the quantity of microorganisms and more on the physiological characteristics of the species. *Penicillium* are less common in soil than *Mucor, Trichoderma* and *Fusarium* fungi. *Mucor* and *Trichoderma* micromycetes are more prevalent in cultivated soils rich in organic matter and those of the *Fusarium* genus are more frequently found in the soils under grass vegetation (Arlauskiene 2001). Soil biology research has become particularly relevant with the increase in anthropogenic effects on agrocenoses and cultivation of high yield potential crops.

7.4 Effects of Environmental Conditions on Plant Residue Decomposition

Decomposition of plant residues is determined by many factors, including the biological properties of the plants and their chemical composition, soil type, soil physical, chemical and biological properties, as well as soil texture and meteorological and agronomic conditions (Larson et al. 1994; Wolf and Snyder 2003). Hydrothermal conditions play a significant role in organic residue decomposition as well. Decomposition of organic matter occurs under aerobic or anaerobic conditions depending on moisture conditions. More intensive decomposition occurs at a soil temperature of 30 °C and soil moisture of 60–80% from total field moisture capacity. Excess moisture slows down microorganismal activity and organic matter decomposition and shifts destructive processes towards humification. As a result, greater concentrations of humic substances accumulate in waterlogged soils (Tripolskaja 2005). Soil oxygen content is another important factor affecting the decomposition rate of organic matter and soil organic matter. Wet and compacted soils contain less oxygen, leading to slower rates of organic matter decomposition (Wolf and Snyder 2003).

7.4.1 Temperature

Kononova (1963) suggested that organic matter decomposition begins at 0 °C. The decomposition rate increases as temperature increases, whereas moisture has less of an effect on this process. The findings obtained under simulated conditions indicate that organic nitrogen mineralisation begins at 1 °C, markedly intensifies with an increase in temperature to 10 °C and increases 2.5–3-fold from 10 to 30 °C. Organic nitrogen mineralisation occurs at rather different soil moisture concentrations of 20–90% maximum field moisture capacity (Hanschmann 1983). Field experiments and simulations show that 20% of total organic nitrogen in the catch crop material is mineralised after 10 weeks at 1 °C, indicating that mineralisation at low temperatures is not negligible. Maximum mineralisation occurs after 10 weeks at 15 °C and was 39% of total applied organic nitrogen was from shoots and 35% was from roots (Van Scholl et al. 1997). Research done on residue decomposition in Denmark suggests that 3 and 9 °C have no effect on nitrogen mineralisation, but organic carbon decomposition differed significantly. The C:N ratio of plant residues decomposed at 3 °C is higher than those decomposed at 9 °C. The growth of microorganism populations and decomposition of macropolymers at the same time slows down at a lower temperature. Decomposition of readily decomposable components and macro-polymers differs; these processes occur at the same rate at higher temperatures but the rate differs at lower temperatures (Magid et al. 2004). Plant residue decomposition experiments conducted in boreal soils show that whether the C:N and C:P ratios of plant residues increase or decrease depends on the temperature. Soil structure and the ecosystem affect these changes (Nadelhoffer et al. 1991).

7.4.2 Soil

The microbiological decomposition rate of organic matter also depends on soil chemical composition. Decomposition of organic residues occurs more slowly in soils with a high secondary mineral content (Tripolskaja 2005). Kaboneka et al. (2004) reported that various mineral fertilizers must be incorporated into acid, nutrient-poor soils to activate straw decomposition processes. They also found that incorporating nitrogen and phosphorus accelerates straw decomposition and increases available nutrient stocks in the soil. Mineral fertilizers are usually incorporated with straw to accelerate straw decomposition at a rate of 7–10 kg N 1 t straw. In this way, the nitrogen shortage is compensated for and biological nitrogen mobilisation from soil organic matter is prevented (Tripolskaja 2005). The soil is not enriched with nitrogen when agricultural crop residues containing 40% carbon and 1.6% nitrogen (C:N = 25:1) decompose. If the C:N ratio is higher, the nitrogen necessary for residue decomposition will be taken up from the soil or dead microorganisms and the decomposition rate of the residues will slow down. This explains the empirical formula in which 100 kg of straw requires incorporation of 1 kg of additional N. Mineral N must be incorporated into the soil to prevent soil N stocks from being used for decomposition (Paul and Clark 1989).

The duration of plant residue decomposition depends on soil structure and organic matter distribution in the soil (Golchin et al. 1994). Residues of plant roots or above-ground plant parts that are incorporated into the soil are first colonised by microorganisms and absorb soil mineral particles. Some of the plant residues with adhering clay particles becomes the centre of water stable aggregates and does not decompose quickly. These aggregates are very stable during the initial plant residue decomposition stages because metabolites released by microorganisms strengthen them. After the microorganisms have decomposed the easily decomposable substances, organic residues turn into more stable organic structures containing high concentrations of aromatic and alkyl carbon (Stott and Martin 1990). Aggregate stability becomes weaker when microorganismal metabolites decompose, as they serve as a binding material (Lal 1999).

Many changes occur in the soil that affects microorganism populations during the transition from intensive to ploughless agricultural system. Plant residues from intensive agriculture are either ploughed in or incorporated in the plough layer depending on the soil tillage machinery used. Plant residues are left on the soil surface in ploughless agriculture. The decomposition rate of residues depends on their placement (Lal 1999). The numbers of aerobic microorganisms in intensively tilled soils, particularly that of fungi and actinomycetes, increase and organic matter mineralisation intensifies (Tripolskaja 2005). Plant residues left on the soil surface decompose at a slower rate than those incorporated into the soil (Wolf and Snyder 2003).

7.5 Physical-Mechanical Properties of Oilseed Rape Residues

Theoretical and experimental research on the physicomechanical characteristics of plant residues is relevant to enhance the quality of tillage and drilling processes, manufacture of agricultural machinery and selection of working parts (Hemmatian et al. 2012). Problems caused by plant residues present on the soil surface are becoming increasingly more common with the increasing popularity of ploughless soil tillage. Due to physical-mechanical, biological and other properties, plant residues deteriorate the performance of soil tillage and sowing machinery by interfere with operations. When soils are heavily covered with plant residues, tillage and drilling machinery with regular coulters clogs up; therefore, the quality of seedbed preparation decreases (Arvidsson 2010; Sarauskis et al. 2012). Machinery with a working disc is commonly used in no-till when abundant plant residues are on the surface (Magalhaes et al. 2007; Sarauskis et al. 2013). Disc coulters with serrated cutting edges cut plant residues better than disc coulters with smooth cutting edges (Bianchini and Magalhaes 2008; Sarauskis et al. 2013a).

The effect of a disc coulter depends on a good interaction among the coulter, the plant residues and the soil. If the characteristics of at least one of the components change, tillage and drilling can improve or worsen significantly. Sarauskis et al. (2005) established that both single-disc and double-disc coulters cut or press plant residues into notches made by the coulter while penetrating the soil. Notably, coulters consisting of two discs cut the soil and plant residues on the soil surface at the same time in two different sections. This means that under certain climatic conditions, when penetration resistance of the upper soil layer is insufficient to ensure cutting through the plant residues, they are pressed into the soil in two sections and dragged along the soil surface curved outwards between the discs. Therefore, the plant residues are both cut and broken. If the soil resists penetration of the disc coulters and plant residues, the residues are broken up only if the stress on the plant residues is exceeded. Thus, penetration resistance has to be higher than the normal stress of plant residues to cut through plant residues on the soil surface (Sarauskis et al. 2005; Sarauskis et al. 2013).

Researchers from different countries (Linke 1998; Tavakoli et al. 2009; Hemmatian et al. 2012) have conducted investigations and observed that the force required to cut or break plant residues depends on the plant species, stem diameter, plant length, moisture, cell structure and elasticity. The design and technological parameters of the working parts have significant effects on the ability to cut or breakthrough plant residues (Liu et al. 2010). Canadian researchers (Kushwaha et al. 1983) investigated cutting of unchopped straw (5 t·ha^{-1}) with coulters containing disks of different diameters. They found that 360 mm diameter and 2 mm thick coulter disks cut approximately 80% of the straw on the soil surface and the remaining 20% of the straw was pressed into the soil. The disks cut approximately 95% of straw when they were 460 mm in diameter and 4 mm thick. Even larger disks (600 mm diameter and 4.5 mm thickness) cut only approximately 20% of the straw.

The straws and stems of different plants have different physical and mechanical characteristics. Investigations into the influence of sunflower stems on the operation of disc coulters demonstrated that the force required to bend a sunflower stem was 34–47 N mm^{-2} at 80% moisture and 41–64 N mm^{-2} at 55% moisture content. Sunflower stems are fibrous with a tubular profile. Due to these attributes, sunflower stems are crushed prior to cutting, which influences cutting. The force required to cut through a sunflower stem is 10–95 N mm^{-2} at 80% stem moisture content (Ince et al. 2005). Liu et al. (2007) investigated the cutting force required to cut through a sugar cane stem. The cutting force required to cut through a sugar cane stem increased as cutting speed was increased. Hemmatian et al. (2012) conducted similar cutting force experiments with sugar cane stems and established that reducing plant moisture content from 78 to 46% resulted in a decrease in cutting force of approximately 16.3%. A cutting speed increase from 5 to 15 mm·min^{-1} increases the cutting force required by approximately 3.2%. Nazari et al. (2008) investigated the physicomechanical characteristics of alfalfa (*Medicago saliva* L.) and proposed that less force is required to cut an alfalfa stem as its moisture content decreases. These investigations also established that a decrease in plant stem diameter decreases the force required for cutting.

Tavakoli et al. (2009) investigated the effect of barley straw moisture content, length of the internode and loading rate on straw bending characteristics. It was established that an increase in straw moisture content and bending rate results in a decrease in the relative force required for bending and that this force was 6.32–12.41 N mm^{-2}. Cakir et al. (1994) proposed that the relative force required by a coulter disc to cut through wheat straw is 2.8–6.4 N mm^{-2}, that for a maize stem is 0.75–1.65 N mm^{-2}, that for a soya straw is 3.8–5.8 N mm^{-2} and that for cotton stem is 6.0–10.0 N mm^{-2}. Chen et al. (2004) reported that the mean force required to cut through hemp straw is approximately 243 N.

The mechanical characteristics of plant residues depend on how long the residues are left on the soil surface after harvesting. The forces required to cut or break through fresh or overwintered plant residues are very different (Linke 1998). For example, the force required to break overwintered winter wheat straw is about 3.2-fold lower than the force required to break fresh winter wheat straw harvested in autumn. Investigations with spring barley showed that the breaking force for overwintered spring barley straw is approximately 34% lower than that required for autumn straw (Sarauskis et al. 2013). Long periods deteriorate the mechanical strength of plant residues. However, modern agricultural technologies do not always provide an opportunity for waiting until the mechanical characteristics of the plant residue are weakened under natural conditions, as new plants are very often being drilled several weeks after crop harvest. The applications of no-till and strip-till result in leaving the crop residues from the previous harvest on the soil surface, which directly influences the drilling coulter operation. Because plant residues left on the surface for a short time can maintain strong mechanical characteristics, disc coulters may fail to cut through or break them, and plant residues will be pressed into the notches. Therefore, it is necessary to speed up the processes of plant residue mineralisation and associated mechanical weakening. Different biological preparations are being increasingly used to activate such processes. Biological preparations create a distinct culture and ensure long-term and stable fertility of field plants, while maintaining a clean environment without causing damage to people (Ahmadi 2010). Biological preparations are most often used as soil and plant nutrients. Plants sprayed with a solution of such a preparation assimilate mineral nutrients much better and grow more vigorously, so plant productivity increases. Some biological preparations consist of nitrogen-fixing stem bacteria and biologically active materials that affect the structure of plant residues. Therefore, mineralisation of the plant residues is activated on the soil surface, and the nitrogen-fixing bacteria perform the function of speeding up the processes of plant residue decomposition and weakening of the mechanical characteristics of the residues at the same time (Jakiene 2011; Holtze et al. 2008; Ahmadi 2010).

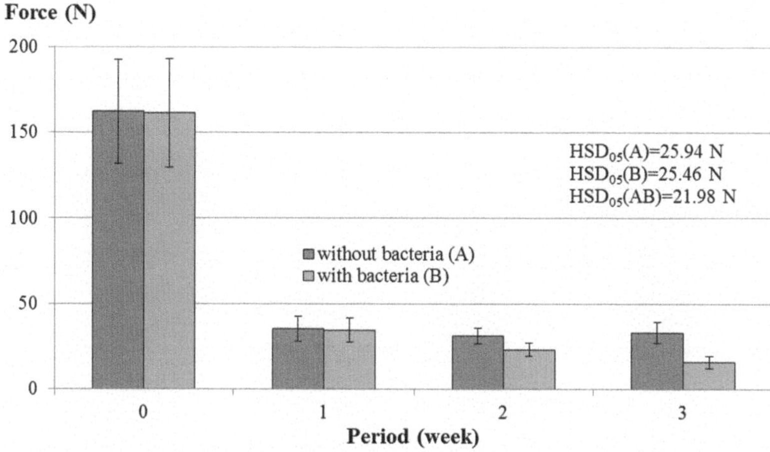

Fig. 7.4 Influence of biological preparation with free-living nitrogen-fixing bacteria on the breaking force of winter oilseed rape over time. Breaking force required for winter oilseed rape residues decreased significantly (4.7 times) during the first week after harvesting. The significant effect of biological preparation was estimated only after 3 weeks after harvesting: breaking force required for residues treated with bio-preparation was two times lower to compare with untreated residues. (From: Vaitauskiene et al. 2015)

7.5.1 Breaking Force of Oilseed Rape Residue

Breaking of winter rape residues decreased significantly within the first 7 days of the experiment. After the first week, the force required to break winter oilseed rape residues decreased approximately 4.7-fold (Fig. 7.4). Maintaining winter oilseed rape plant residues under natural climatic conditions for a longer period did not have any significant effect. Breaking forces varied depending on the moisture content of the winter oilseed rape residues. The effect of the biological preparation with free-living nitrogen-fixing bacteria on the mechanical characteristics of plant residues was slightly different. At the beginning of the experiment (after 1 week), the biological preparation had no significant effect on breaking the residues, as the force required to break the residues was similar to that required to break the plant residues that were not treated with the biological preparation. Significant differences were detected only after 3 weeks. Notably, mechanical resistance to breaking of winter rape residues treated with the preparation decreased significantly after each week of the experiment, unlike that of naturally maintained winter rape residues.

The tentative findings on the mechanical breaking characteristics of spring oilseed rape residues suggested that 2.6 times lower force is required to break spring rape residues after harvesting compared with that required for winter rape residues. However, no such differences were observed at later periods of the experiment. The effect of the biological preparation on the breaking characteristics of spring rape residues was significant during the entire experiment. A comparison of the forces required to break treated and untreated spring rape residues with the biological

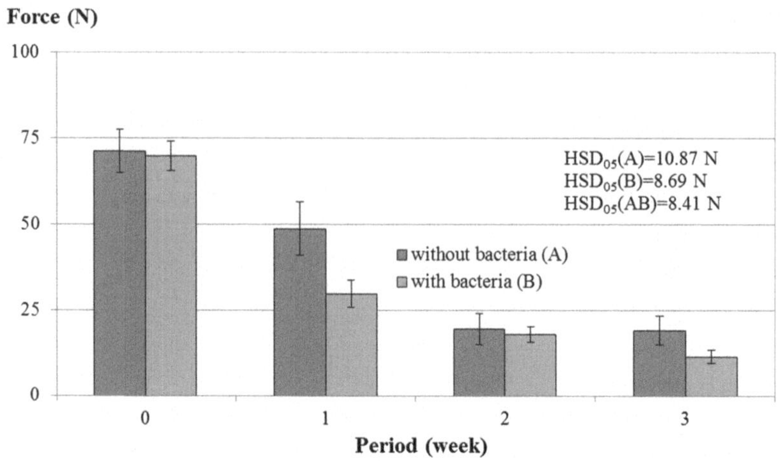

Fig. 7.5 Influence of the biological preparation with free-living nitrogen-fixing bacteria on the breaking force of spring oilseed rape residues over time. The significant effect of biological preparation on spring oilseed rape residues was estimated during the first week after harvesting: breaking force required for residues treated with bio-preparation was 39% lower to compare with untreated residues. (From: Vaitauskiene et al. 2015)

preparation revealed significant differences after weeks 1 and 3 of the experiment (Fig. 7.5). The mechanical characteristics of spring rape residues kept under natural climatic conditions on the soil surface significantly weakened within the first 2 weeks. However, no significant difference was observed after week 3.

7.5.2 *Oilseed Rape Residue Cutting Force*

The mechanical cutting characteristics of winter rape residues kept under natural climatic conditions and those treated with the biological preparation significantly weakened within the first 2 weeks. Maintaining the winter rape residue on the soil surface for a longer period did not have any significant effect on the cutting characteristics. The assessment showed that the greatest significant effect was noted after week 1 of the experiment, regardless of the angle of the cutting knife. Later in the experiment, the biological preparation tended to lower the cutting force required for treated winter rape residues compared with that of the untreated residues; however, the difference was not significant.

The success of cutting winter rape residue was highly dependent on the process used for cutting, as a vertical knife (Fig. 7.6a) and an angled knife (Fig. 7.6b) differed.

The vertical knife required twice the force to cut the winter rape residues compared to the angled knife, regardless of whether the plants were treated with the biological preparation or not. This difference can be explained by the fact that a

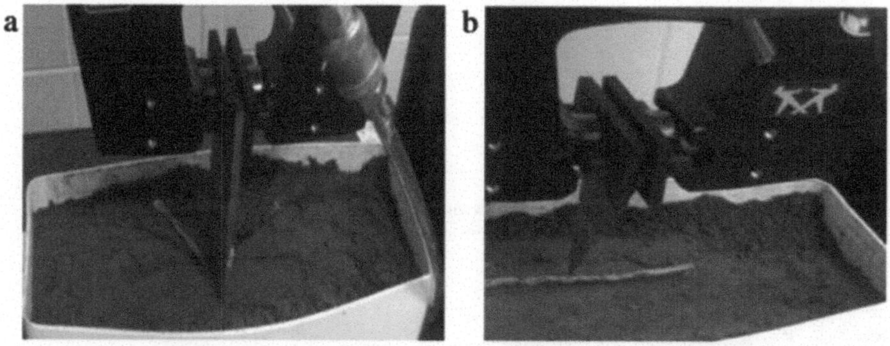

Fig. 7.6 Cutting of plant residues with different knives: (**a**) cutting with a smooth knife simulating a disc coulter with a smooth cutting edge; (**b**) slide-cutting simulating a disc coulter with a serrated cutting edge. (From: Vaitauskiene et al. 2015)

slide-cutting process occurs when cutting with an angled knife. The soil surface serves as a carrier, and the knife penetrates the soil, touches the plant residues and slides through them to cut them. Such a cutting process is seen in no-till and strip-till technologies, when a disk coulter with serrated blades or knives is used. Plant residues placed in the notch made by the blades of a disk coulter are cut by slide cutting. This is a complex process that depends on the shape and geometry of the disk coulter (Kushwaha et al. 1986) as well as on the depth and radius of the notch (Sarauskis et al. 2013), cutting speed (Liu et al. 2007), moisture content of the plant residues and other factors (Ince et al. 2005; Nazari et al. 2008). No sliding occurs when cutting with a vertical knife; therefore, a higher force is required to cut the winter rape residues. Slide cutting with a vertical knife can only be achieved if the disk coulter is forcibly rotated.

Similar experiments early in the study period showed that approximately 62 N force was required to cut through spring rape residues with a vertical knife, and similar results occurred with the angled knife (Fig. 7.7). However, later in the study period, the differences were significant. Spring rape residues maintained on the soil surface under natural climatic conditions required a mean of 1.7–2.0-times more force to cut with a vertical knife after weeks 1, 2 and 3 compared with that required for cutting with an angled knife. Estimates of the effects of the biological preparation revealed that spring rape residues treated with the biological preparation required less cutting force, regardless of knife type. Significant differences were established after weeks 1 and 3 when cutting with the vertical knife, and significant differences were noted during all experimental periods, i.e. after weeks 1, 2 and 3, when cutting with the angled knife.

The spring rape residue required 1.7-times less force when cutting with a vertical knife compared with that required to cut winter rape residue. However, no difference was found when cutting with the angled knife.

Summarising the experimental results, the mechanical cutting and breaking characteristics of the plant residues were affected by the biological preparation.

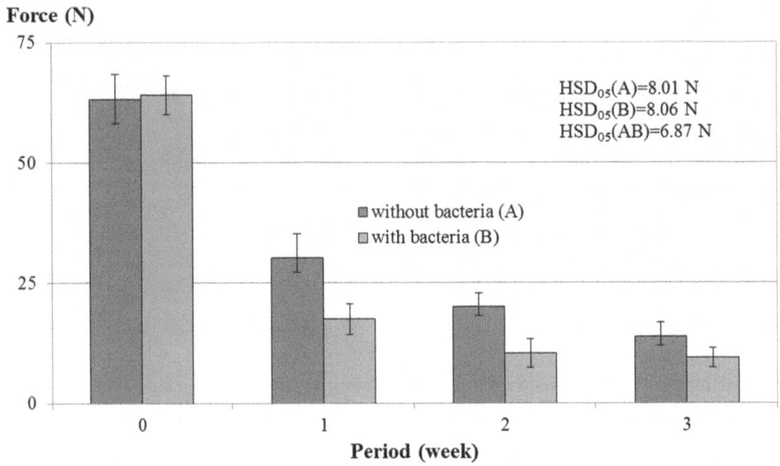

Fig. 7.7 Influence of biological preparation with free-living nitrogen-fixing bacteria on the cutting force of spring oilseed rape residues over time when using an angled knife and a sliding motion. The significant effect of biological preparation on spring oilseed rape residues was estimated during the first week after harvesting: cutting force required for residues treated with bio-preparation was 42% lower to compare with untreated residues. (From: Vaitauskiene et al. 2015)

This must be emphasised, as use of a biological preparation cannot only intensify residue decomposition but also weaken mechanical characteristics. These processes depend on the species of plant residue, chemical composition, the C:N ratio, lignin concentration and other factors (Grabber 2005; Kriauciuniene et al. 2012). The rapid weakening of mechanical characteristics of plant residues is particularly relevant when applying no-till and strip-till technologies, where plant residues play a major role in drilling quality. If plant residues are not removed from the seed placement row, the seeds might not be inserted properly into the soil or are incorporated into the plant residues. This results in lower rates of seed emergence and plant development. It is very important that the appropriate construction and technological parameters of the no-till and strip-till machinery are selected to minimise the problems associated with plant residues, as a smooth process is highly dependent on them. Our experimental results suggest that significantly less force is required to cut plant residues when slide cutting is ensured.

7.6 Accumulation of Allelochemical Compounds in Oilseed Rape and Effect on Agricultural Crops and Weeds

7.6.1 Plant Allelopathy and Factors Governing These Properties

Allelopathy is any process involving secondary metabolites produced by plants, algae, bacteria and fungi that affect the growth and development of agricultural and biological systems. This definition of allelopathy was adopted by the International Allelopathy Society after the first World Congress on Allelopathy in Spain in 1996. The effects can be positive, negative or neutral. The active compounds in this process are called alellochemicals. Allelochemical interactions among plants have been recognised as a key factor for modelling crop and weed growth, cropping systems and technologies (Rice 1984).

Particular species of agricultural plants and weeds can be used to control other weeds, diseases and pests in traditional agroecosystems. Allelopathy is directly involved in plant breeding, soil fertility, tillage, plant protection and agricultural systems and can be important in modelling crop productivity and maintaining genetic diversity and ecosystem stability (Weston and Duke 2003). It is very important to know and recognise the allelopathic mechanism, as it is related to the action of allelochemical compounds of plants possessing allelopathic properties between crops and weeds, crops and crops and the toxicity of exudates released by crop and weed residues or roots. Allelopathy is closely related to environmental stress, including the effects of insects and diseases, temperature spikes, variations in nutrient and moisture contents, soil acidity, heavy metal concentrations, radiation and pesticides. These biotic and abiotic factors often increase production of allelochemicals and strengthen the allelopathic potential of plants, which determines their self-defence and survival (Anaya 1999). These factors not only affect the concentration of allelochemicals but also the duration of their allelopathic effect and their breakdown in the environment (Inderjit and Kreating 1999).

Inderjit et al. (2011) proposed that allelopathy should be investigated starting with plant evolution and biogeography and move towards interactions among plants in the plant community, interactions among soil communities and responses to environmental factors as well as processes in the rhizosphere, such as various chemical transformations (Fig. 7.8).

Agricultural crops, weeds and microorganisms are sources of allelochemical compounds. Allelochemicals are biochemical constituents of all organisms that are released into the environment during decomposition of organic matter (Aldrich 1987). Several hundred organic allelochemical compounds released by plants or microbes can affect growth and survival of other plants. Allelopathic inhibition or stimulation occurs due to allelochemicals that either inhibit or stimulate physiological processes in the receiver plant (Einhelling 1995; Vaughn and Boydston 1997).

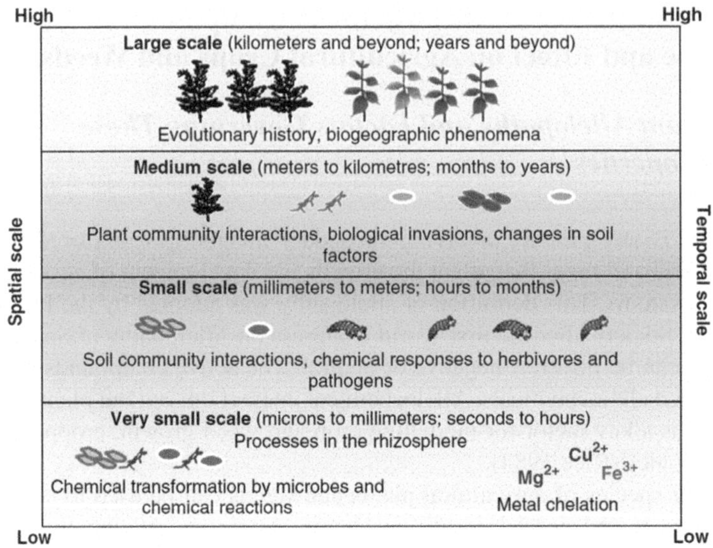

Fig. 7.8 Effects of ecosystem factors, biogeographic variations and coevolutionary relationships on the production, release and activities of allelochemicals along spatial and temporal scales. (Reprinted with permission of [Elsevier, Copyright (2017)]' from Inderjit et al. (2011))

Brassicaceae plants are often distinguished by their allelopathic properties (Malik et al. 2010; Petersen et al. 2001; Haramoto and Gallandt 2005). Laboratory analyses verify the autotoxic effect of oilseed rape residues and their aqueous extracts and root exudates on seed germination and plant growth (Yasumoto et al. 2011). Sunflowers grow poorly in fields where oilseed rape had been cultivated previously. Jafariehyazdi and Javidfar (2011) established that aqueous extracts of oilseed rape residues reduce the germination capacity of sunflower seeds and inhibit their growth. Marcinkeviciene et al. (2013) assessed the effects of aqueous extracts from the areal parts of *Sinapis alba* and *Brassica napus* on spring barley germination and establishment. The highest concentrations of aqueous extracts of the areal parts of the tested plants decreased spring barley germination and inhibited establishment. Hoagland et al. (2008) investigated the effect of oilseed rape meal on germination and growth of plants. They established that oilseed rape meal incorporated in soil inhibits germination of *Triticum aestivum*, *Vicia villosa* and *Echinochoa crus-galli*, whereas it stimulated an increase in biomass of the areal parts.

Uremis et al. (2009) performed laboratory and field experiments to investigate the allelopathic effect of six Brassicaceae plants on johnsongrass (*Sorghum halepense* L. Pers.) and established that white round radish (*Raphanus sativus* L.), garden radish (*R. sativus* L.), black radish (*R. sativus* L. var. *niger*), little radish (*R. sativus* L. var. *radicula*), turnip-rape (*Brassica campestris* L. subsp. *rapa*) and oilseed rape (*B. napus* L. *oleifera* DC) inhibit the growth of johnsongrass. The strongest inhibitory effect was demonstrated by the black radish extract, which had the highest content of allelochemical compounds (benzyl isothiocyanate and allyl isothiocyanate). They concluded that Brassicaceae plants have strong allelopathic inhibitory actions.

Experiments were carried out using Brassicaceae plants as cover crops in plastic tunnels where tomatoes (*Solanum lycopersicum*) were cultivated. The effect of turnip (*Brassica rapa*), oriental mustard (*B. juncea*) and a blend of brown mustard (*B. juncea*) and white mustard (*Sinapis alba*) on yellow nutsedge (*Cyperus esculentus*) and johnsongrass (*S. halepense*) growth was investigated. The glucosinolate analysis revealed that the blend of brown mustard, oriental mustard and turnip produced 26,399, 16,798 and 18,847 μmol m^{-2} glucosinolates, respectively. The Brassicaceae plants reduced the amount of papyrus yellow nutsedge by ≤39% and that of johnsongrass by ≤46%; therefore, they would be useful to apply in an integrated weed control system (Bangarwa and Norsworthy 2014).

Walsh et al. (2014) investigated the allelopathic effect of camelina (*Camelina sativa)* and oilseed rape against wild oat (*Avena fatua*), flax (*Linum usitatissimum*) and radish (*R. sativus*). The leaf washings of camelina and oilseed rape increased the weight of radish, whereas flax seedling weight only increased in response to the oilseed rape leaf washings. The aqueous extracts of camelina and oilseed rape green mass (an entire flowering plant) reduced the germination capacity of wild oat, flax and radish seeds. Seedlings and root mass of wild oat and radish decreased as well. The soil-incorporated green mass of camelina and oilseed rape stimulated the growth of radish biomass but only oilseed rape green residues stimulated the growth of wild oat biomass. Such variations between the experiments and plant species require detailed research into the allelopathic interactions between plant species and a detailed analysis of allelopathic compounds.

7.6.2 Allelochemical Compounds Present in Oilseed Rape and Residues

Allelochemical compounds occur in oilseed rape final products, intermediate products or secondary metabolites of biochemical processes. Most allelochemical compounds are classified as secondary metabolites and appear as the main metabolic by-product and accumulate in the roots, stems, leaves, flowers and seeds. The highest concentration of allelochemical compounds during the growing season is found in leaves and at the end of the growing season in seeds (Gill et al. 1993).

Oilseed rape produces diverse and abundant quantities of phenolic compounds which, as allelochemical compounds, are distinguished by their various ecological and physiological purposes (Pichersky and Gang 2000; Noel et al. 2005). These compounds perform structural functions (e.g. intermediate compounds in the lignification process) and are also important in the general plant defence system against diseases and pests (Lattanzio et al. 2012). Phenolic compounds are a very heterogeneous group that is universally and widely distributed in plants and quite frequently their concentrations are surprisingly high. Biosynthesis of different phenolic compounds in plants has evolved in response to changes in the external environment. Besides their many other functions (e.g. cell wall formation and defence functions), phenolic compounds determine plant colours and specific flavours and smells. The accumulation of these compounds in plant tissues is considered a general adaptive plant response to unfavourable environmental conditions to improve survival status. Moreover, these secondary metabolites are physiologically important as a means to store carbon that can be utilised when nitrogen is limited or photosynthesis slows. The model of phenolic compounds in oilseed rape is complex, as these low-molecular weight substances are distributed unevenly in plants, and differences can be observed at different oilseed rape growth stages. The highest concentrations of these secondary metabolites are present in the plant parts responsible for survival and reproduction (Wink 2003; Noel et al. 2005; Singh and Bharate 2006; Yu and Jez 2008). Low molecular weight phenols are growth regulators or allelochemicals responsible for stabilising biological processes in plants in response to environmental change. Understanding the functional mechanisms of phenolic compounds will allow for more sustainable agriculture including weed and pest control with the help of crop rotation and biocontrol (Popaa et al. 2008).

The key allelochemical compounds synthesised by oilseed rape are glucosinolates. Glucosinolates (beta-thioglucoside-N-hydroxysulphates) are isothiocyanate precursors present in 16 families of dicotyledonous angiosperms, including a large number of edible species. At least 120 different glucosinolates have been identified in these plants, although closely related taxa typically contain only a small number of such compounds (Agerbirk and Olsen 2012). The highest content of these compounds is found in Brassicaceae plants in the order Capparales, to which oilseed rape belongs. Glucosinolates are nitrogen- and sulphur-rich anionic secondary metabolites otherwise known as β-thioglucoside-N-hydroxysulfates or cis-N-hydroximinosulphate esters. Glucosinolates are also called mustard oil glycosides. They are naturally synthesised compounds specific to Brassicaceae that have allelopathic activity (Malik et al. 2010; Petersen et al. 2001; Yasumoto et al. 2011). These secondary metabolites are composed of hydrophilic β-D-glucopyranose, a sulphated functional group of thyohydroximate and a hydrophobic aglycone (otherwise called a radical). Natural glucosinolates differ from each other only in the structure of the aglycone. Glucosinolates are distinguished as aliphatic, aromatic and indole based on the amino acids comprising the aglycone (Fig. 7.9).

Several glucosinolates are specific to different plant families. Oilseed rape contains 14 different types of these compounds. Glucosinolates themselves are not toxic

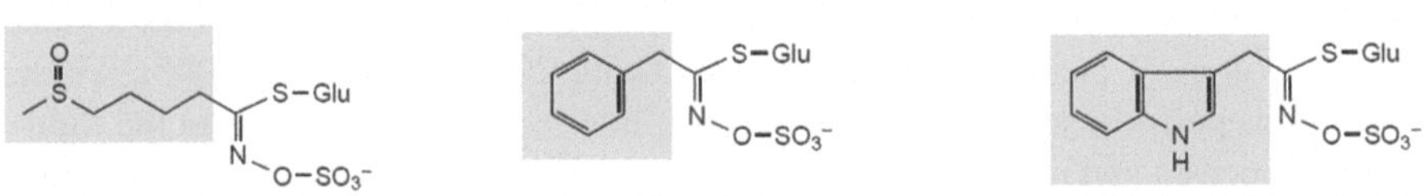

4-Methylsulfinylbutyl-glucosinolate Benzylglucosinolate Indol-3-ylmethylglucosinolate

Fig. 7.9 Examples of aliphatic, aromatic and indole glucosinolate structures. (Reprinted with permission of [Elsevier, Copyright (2017)]' from Wittstock and Halkier (2002))

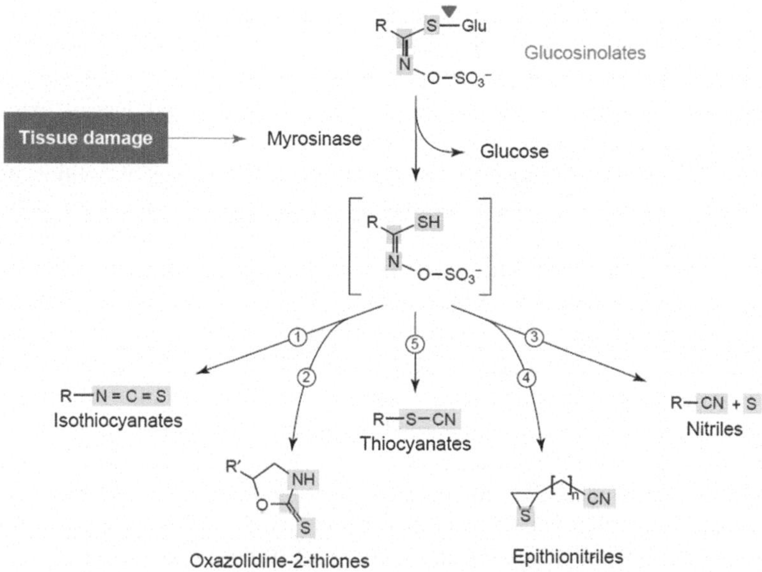

Fig. 7.10 Glucosinolates are hydrolysed by myrosinases upon tissue damage. (Reprinted with permission of [Elsevier, Copyright (2017)]' from [Wittstock and Halkier (2002))

but hydrolysis begins when plant tissue is damaged (Fig. 7.10). Then, myrosin cells release the enzyme myrosinase, which catalyses this process (Borgen et al. 2010). During hydrolysis of glucosinolates, D-glucose and the unstable aglycone thiohydroximate-O-sulphonate uncouple and form several biologically active and most often toxic, volatile compounds (Brown and Morra 1996; Vaughn and Berhow 2005).

The composition of the products of glucosinolate hydrolysis depends mostly on the side chain structure and the reaction conditions. If pH is neutral, an unstable aglycone is transformed into an isotihiocyanate. Oxazolidin-2-thiones are formed if the glucosinolate side chains hydroxylate carbon. An environment containing iron ions (Fe^{+2}) or epitiospecific protein (e.g. rapeseed) favours formation of nitriles. The formation of nitriles requires a pH < 3. Otherwise, sulphur atoms detach after the side chain double bond breaks (nitrile formation is determined by a double bound) and epionitriles are formed. Some glucosinolates are hydrolysed to thiocyanates.

Glucosinolates and myrosinase are the main constituents of the system allowing plants to defend themselves against herbivores, pathogens and weeds. Researchers have called the activity of this system volatile products of glucosinolates hydrolysis, a mustard oil bomb or simply mustard oil (Borgen et al. 2010; Kissen et al. 2009). These studies have pointed out the importance of mustard oil production as well as its ecological and nutritional relevance. The biological availability of glucosinolate degradation products in the soil has been investigated after glucosinolates have been completely hydrolysed. The degradation products are bound to hydrophobic compounds absorbed by humic substances, microbial biomass or form stable complex compounds with soil minerals (Poulsen et al. 2008). Thus, hydrolysis of glucosinolates occurs rapidly and isothiocyanates and other toxic compounds are formed and involved in the matrix of less biologically available compounds in the soil. The complete breakdown of these compounds is a much slower and insufficiently explored process (Gimsing et al. 2006).

Residues of Brassicaceae plants and rapeseed meal, a by-product of extracting rapeseed oil, are characterised by an inhibitory effect on weed seed germination (Hoagland et al. 2008). Petersen et al. (2001) investigated the effect of isothiocyanates, glucosinolate hydrolysis products, present in turnip rape 'on the germination of *Chenopodium album*, *Tripleurospermum perforatum*, *E. crus-galli*, *Vicia villosa*, *Sonchus arvensis* and winter wheat. The five isothiocyanates isolated, such as allyl, *n*-butyl, 3-butenyl, benzyl and 2-phenylethyl, inhibited germination of the tested plants. Brown and Morra (1997) reported that other volatile compounds are released during hydrolysis of glucosinolates present in rape residues and the products of microbial residue decomposition also reduce seed germination. Volatile substances and water-soluble allelochemicals inhibit seed germination of lettuce (*Lactuca sativa*) (Brown and Morra 1996). Thiocyanate ions (SCN^-) have herbicidal characteristics and are used as an active ingredient in commercial herbicides (Borek and Morra 2005). Isothiocyanates and other volatile compounds can be used in crop protection systems against diseases and pests (Gimsing and Kirkegaard 2006; Bjorkman et al. 2011).

Glucosinolates and/or their metabolic products have long been known for their fungicidal, bacteriocidal, nematocidal and allelopathic properties and have recently attracted intense research interest because of their chemoprotective, therapeutic and prophylactic attributes (Matusheski et al. 2006; Halkier and Gershenzon 2006; Hayes et al. 2008; Traka and Mithen 2009). The major focus of research on glucosinolates has been on the negative aspects of these compounds because of the prevalence of certain 'antinutritional' or goitrogenic glucosinolates in oil or meal produced from *Brassicaceae* seeds (Mawson et al. 1993; Taraz et al. 2006). Many of these biologically active and chemically diverse compounds have been identified during the past three decades. Besides Brassicaeae vegetables, these glucosinolates have been found in hundreds of species, many of which are edible or could provide substantial quantities of glucosinolates for isolation, biological evaluation, and potential application as dietary or pharmacological agents. Significant progress has been achieved in research on glucosinolates; however, it is important to clarify the essentials of their biosynthesis, identify the reason for the formation of specific glucosinolates and their role in the metabolic process to understand how and why plants synthesise glucosinolates. These findings will facilitate better use of the potential of glucosinolates in agriculture and medicine (Halkier and Gershenzon 2006).

7.6.3 *Effect of Oilseed Rape on Weeds, Soil Properties and Post-Crop Environment*

An agroecosystem is a man-made environment in which a natural ecosystem is often replaced by a single crop species, leading to reduced plant genetic diversity and often leaving the crop susceptible to a variety of weeds, pathogenic fungi, bacteria, viruses and certain herbivores (Batish et al. 2006).

Because agricultural land is rich in nutrients and there is a period when it is unoccupied by crops, it becomes prime territory for weed invasion, which leads to the high costs associated with weed control in crop fields (Zhang et al. 2011). A major task for contemporary scientific research on plant breeding, soil fertility and tillage, plant protection and cropping systems is to stabilise and sustain agroecosystems (Weston and Duke 2003). A stable functioning agrocenoses is determined by soil organic matter content. The largest quantities of organic matter are returned to the soil in plant residues. They improve soil microbiological activity and the intensity of the biochemical processes occurring in the soil, which has a considerable effect on the vital activity of microorganisms. Some studies indicate that soil enzyme activities increase when plant residues are incorporated into the soil (Wang et al. 2011). The improvements in soil resulting from oilseed rape cultivation are determined by the amount of organic matter remaining in the soil, its chemical composition and the intensity of mineralisation and humification, which depend on the C:N ratio and lignin content. Growing oilseed rape plants develop a matted root system, which entwines the plough layer, aerates it and penetrates deeply (to 3 m) into the subsoil; thus, improving soil structure, aeration, moisture turnover and microorganism activities. Kong et al. (2009) suggested that root surface area and activity have a marked effect on soil enzyme activities. Oilseed rape roots release exudates into the soil, which results in the abundance of various organic compounds in the root rhizosphere (Velicka 2002). Winter oilseed rape, grown for seed, leaves about 5.02 t ha^{-1} residues, i.e. 1.46-times more than cereal. Such a quantity of organic matter has a positive impact on the soil and crop yield for the next 2 years (Velicka 2002). The C:N ratio of rape residues is favourable for decomposition. Due to the higher lignin content than many other crops, rape residues decompose at a slower rate, so their effect on soil and the post-crop environment lasts longer. Various types of oilseed rape residues decompose at different rates. Rape threshing remains decompose at a higher rate (70% within 3 months), therefore the nutrients can be utilised post-crop shortly after their incorporation into the soil. Stubble and roots decompose at a slower rate (50% within 14.5 months); therefore, the nutrients are utilised during later post-crop growth stages. Winter rape roots decompose at the slowest rate (60% within 26.5 months) because they contain a high lignin content, resulting in a delay in the formation of potential nutrients and humic substances (Kriauciuniene 2008). Oilseed rape residues contain many micromycetes, which together with other microorganisms, decompose cellulose, lignin and recalcitrant nitrogen and phosphorus compounds and accumulate nutrients to protect them from being leached. The population of micromycetes varies with the decomposition of rape residues in the soil; some genera disappear and others emerge. The best substrate for these microorganisms is decomposing rape threshing remains. The largest population of these microorganisms is found on threshing remains that decomposed 26 months in the soil. The highest incidence of cellulose decomposers on rape roots was established in month 20 of decomposition (Velicka et al. 2009).

Use of the allelopathic properties of oilseed rape for weed control is potentially important considering crop productivity, genetic diversity conservation and monitoring of ecosystem stability (Singh et al. 2004). Oilseed rape plants release chemical compounds into the environment that can suppress the growth and establishment of other plants growing in their vicinity. However, allelopathic chemicals have other ecological roles, such as plant defence, nutrient chelation and regulation of soil biota in ways that affect decomposition and soil fertility. These ecosystem-scale roles of allelochemicals augment, attenuate or modify their community-scale functions; how plant communities form is a key question in ecology (Inderjit et al. 2011). Plant-released chemicals have a marked impact on organic matter decomposition (Hättenschwiler et al. 2011), pests (Karban et al. 2006), trophic interactions (Hättenschwiler and Jorgensen 2010) and the nitrogen cycle (Northup et al. 1998). Over the past 20 years, only a few of the numerous studies that have been conducted in the field of allelopathy have attempted to reveal allelopathic interactions in a wider context (Metlen et al. 2009; Inderjit et al. 2011; Lankau and Strauss 2007; Tharayil et al. 2009).

Brown and Morra (2009) established that incorporating *Brassicaceae* residues into the soil increases plant-available nitrogen content. The soil where rape residues grew contained a high glucosinolate concentration and higher concentrations of NH_4^+ and NO_2^- ions, compared with the soil in which *Poaceae* residues were incorporated. The positive correlation between these ions and glucosinolate concentration in the residues suggests that the products of glucosinolate hydrolysis are involved in inhibiting soil nitrification. When phenolic compounds present in rape residues are introduced into the soil, they change the dominant direction of the nitrogen turnover cycle from a mineral into an organic form, which reduces likely nitrogen losses and increases the regeneration rate of nitrogen present in rape residues by microbial symbiosis. These substances form complex compounds with aluminium and manganese by reducing their toxicity potential and by fixation with phosphorus in the soil. Phenolic compounds regulate changes in the organic matter by determining their accumulation in the soil and by preventing leaching of cationic forms of nutrients (Northup et al. 1998).

Oilseed rape is a good pre-crop for many crops. The yield of winter wheat grown after winter rape is significantly higher compared with that of monocropped wheat or after barley and does not significantly differ from that of winter wheat grown after a vetch and oat mixture for green mass, clover and early potatoes. Wheat performs better when sown after *Brassicaceae* compared with other pre-crops. The yield of barley grown after winter rape was 17% higher than that of continuously grown barley or after winter wheat. Thus, winter rape has greater value as a pre-crop for barley whereas spring rape is a better pre-crop for potatoes. As a post-crop, barley yields are better after winter rape which was sown after a vetch and oats mixture for fresh mass or after early potatoes. Barley yields are better after spring rape sown after clover (Velicka 2002). Spring rape incorporated into the soil as green manure stimulates the germination of winter wheat sown after 1 month (Zacharovich 2005). Pot experiments showed that oilseed rape residues are an important source of phosphorus for post-crops. A direct relationship has been established between the phosphorus content incorporated into the soil with rape residues and phosphorus content uptake by plants (Pellerin et al. 2007).

Oilseed rape, which increases soil fertility; thus, it is a good pre-crop for cereals, particualry with their specialised crop rotations (Velicka 2002). Oilseed rape is particularly suitable for inclusion in the crop rotation between cereals. Notably, the negative effects of continuous wheat cropping are eliminated by using oilseed rape as the pre-crop. Winter and spring rape are more valuable pre-crops than vetch and oat mixture when growing barley, whereas winter wheat and barley are less valuable as pre-crops than potatoes and first-year clover (Velicka 2002).

It is better to sow winter wheat after oilseed rape than continuously crop or sow after barley (Velicka 1995). Winter rape is a better pre-crop to enhance barley yield, whereas spring rape is best for potato yield. Post-crop barley yields best after winter rape, which was sown after a vetch and oats mixture for green mass or after early potatoes (5.0% yield increase). Barley also yields best (4.17 t ha^{-1}) when spring rape is grown after clover and barley is grown after spring rape, compared with other pre-crops (Velicka 2002).

Some researchers have pointed out the negative effects of oilseed rape as a pre-crop. Spring rape as a pre-crop may favour the spread of *Cercospora* leaf spot in sugar beet crops. Spring rape grown as a pre-crop for sugar beet does not reduce sugar beet root yield, compared with winter wheat; however, it slightly reduces root sugar content (Zemeckis 1993).

The organic matter left after oilseed rape has a positive impact on soil and increases yield of other crops for two subsequent years (Velicka 2002). This depends on the rape post-harvest residue mineralisation and humification processes. German researchers have shown that cereal yield is 0.1–0.2 t ha^{-1} higher after 2 years of oilseed rape, than after other pre-crops (Spielhaus 1980). Swedish researchers have also noted the positive effect of oilseed rape on grain crops 2 years after its cultivation (5–10% yield increase) (Ebbersten 1981). Winter rape warrants a high yield for post-crop barley grown after winter wheat. A particularly pronounced positive effect of winter rape on soil fertility persisted for post-crop barley in the sequences of winter rape–potatoes–barley and winter rape–barley–barley (Velicka 2002). This can be explained by a better nutrient balance and the phytosanitary properties of winter rape. Spring rape improves soil fertility when barley is cropped continuously or when it is grown after poorer pre-crops (winter wheat) (Velicka 2002).

Winter rape is harvested early and after harvesting the soil and is generally weed-free and abundantly entwined by rape roots which improves soil structure and nutrient accumulation and creates favourable conditions for aeration, moisture turnover and microorganism activity. Microbiological processes in the soil after oilseed rape is cultivated are much more intensive than those after cereals because they occur during a warmer period of the year (Makovski 1990). Oilseed rape roots and the above-ground residues in soil decompose to replenish soil humus stocks. There is a close connection between soil humus content and soil structure. The higher the soil humus content, the more water-stable soil particles are bound by humus colloids (Zemeckis 1993).

Although the best soil structure is generally achieved after clover, soil structure is better after winter and spring rape than after winter wheat, a vetch and oat mixture and potatoes (Velicka 2002). This is determined by the amount of rape roots and above-ground residues as well as their chemical composition (Magyla et al. 1997). Soil structure also depends on the area of winter rape in the crop rotation. Water stable soil particle content decreases (due to the mineralisation activity) and soil structure deteriorates when winter rape occupies 60% of the total area in a cereal crop rotation (Velicka 2002).

Winter rape has a more abundant soil root structure than spring rape; therefore, it leaves more organic matter in the soil. Organic matter is decomposed by the saprophytic microflora and is further processed by oligotrophs, and some parts of it (mostly the cyclic compounds) are converted to humus. NH_3, H_2S, H_2 and other compounds are released when microorganisms mineralise humus (Aleksandrova 1980).

The greatest effect on humus content and its composition is exerted by the following factors that govern the course of humification and mineralisation and their intensity: soil texture and mineral composition, physical and chemical properties, particularly the acidity of the soil solution, hydrothermal regime, aeration, plant residue content, their biological and chemical properties, and the composition of microorganisms (Aleksandrova 1980). Humic and fulvic acids are produced during humification. Soil physical and chemical properties deteriorate and the soil becomes more acidic when fulvates are prevalent (Beyer 1996; Arlauskas and Slepetiene 1997; Velicka 2002).

Mineralisation differs from humification. The best conditions for mineralisation are created when soil moisture, temperature and aeration conditions are optimal. During mineralisation, soil organic matter is decomposed to plant-available nutrients, which are rapidly leached from the soil (Aleksandrova 1980). The key objective of modern agriculture is to maintain the balance between these processes and to preserve optimal humus content, as the main accumulator of solar energy, in the soil. This objective is being pursued through crop and soil management practices directed towards soil improvement. Humus composition and the ratio of humic to fulvic acids represents the level of soil amelioration (Velicka 2002).

The amount of nutrients in the soil is largely determined by biological properties of plants, harvesting time and method and meteorological conditions (Wolf and Snyder 2003). Oilseed rape is distinguished by a high demand for nutrients (Velicka 2002), particularly for nitrogen and sulphur. Oilseed rape contains higher concentrations of these nutrients compared with most cereals (Singh et al. 2004). Total nitrogen content decreases by 14–30%, P_2O_5 by 23–44%, and K_2O by 14–47% in crop rotations in which winter and spring rape account for 25, 50 and 75% of the total area after the first rotation (4 years). Changes in the contents of other elements (CaO, MgO, B, Mo, Zn, Co) in the soil are insignificant (Velicka 2002).

7.7 Conclusion

Improvements in soil with winter and spring oilseed rape growth are determined by the peculiarities in the chemical composition and decomposition of their residues. The decomposition of winter and spring oilseed rape residues in the soil depends on their chemical composition, the C:N ratio and the amount of lignin. The influence of agrometeorological conditions on decomposition of different crops residues depends on the residue type and chemical composition. The groups of microorganisms participating in the decomposition of winter and spring oilseed rape residues, their amounts and micromycetes cenosis and distribution depend on the environmental conditions and crop residue type.

The accumulation of phenolic compounds in different parts of oilseed rape depends on the growth stage, whereas the accumulation of phenolic compounds in decomposing residues depends on the duration of decomposition. Quantitative and qualitative composition of glucosinolates and volatile organic compounds depends on the rape residue and the duration of their decomposition in soil. Rape residues and the duration of their decomposition have different effects on the germination and early growth of agricultural crops and weeds.

Research findings suggest that a biological preparation weakens mechanical cutting and breaking properties of winter and spring oilseed rape residues. The force required for cutting and breaking of these plant residues decreases much faster than that needed for plant residues untreated with a biological preparation. The mechanical characteristics of breaking spring rape residue at the beginning of the experiment differed from those of breaking winter rape residue. The force required to break of spring rape residues under natural climatic conditions decreased significantly within the first 2 weeks, while in week 3 the decrease was insignificant. In treatments in which the biological preparation was applied, the force required for to break the residue decreased significantly during the entire experimental period. The best results cutting winter rape and spring rape residue were achieved when cutting was performed with an angled knife, and slide cutting was ensured.

References

Adams MB (2003) Ecological issues related to N deposition to natural ecosystems: research needs. Environ Int 29(2–3):189–199. https://doi.org/10.1016/S0160-4120(02)00179-4

Agerbirk N, Olsen CE (2012) Glucosinolate structures in evolution. Phytochemistry 77:16–45. https://doi.org/10.1016/j.phytochem.2012.02.005

Ågren GI, Bosatta E, Magill AH (2001) Combining theory and experiment to understand effects of inorganic nitrogen on litter decomposition. Oecologia 128(1):94–98. https://doi.org/10.1007/s004420100646

Ahmadi M (2010) Effect of zinc and nitrogen fertilizer rates on yield and yield components of oilseed rape (*Brassica napus* L.). Am Eurasian J Agric Environ Sci 7(3):259–264

Aldrich RJ (1987) Interference between crops and weeds. Allelochemicals: role in agriculture and forestry. ACS Symposium 330:300–311. https://doi.org/10.1021/bk-1987-0330.ch028

Aleksandrova LO (1980) Soil organic matter and process its transformation. Nauka, Leningrad, 286 p (in Russian)

Anaya AL (1999) Allelopathy as a tool in the management of biotic resources in agroecosystems. Crit Rev Plant Sci 18(6):697–739. https://doi.org/10.1080/07352689991309450

Arlauskas M, Slepetiene A (1997) Effects on conventional and minimum soil tilage systems. Diverse fertilization and various crop rotations on the humus composition of loamy soils. The role of humic substances in teh ecosystems and in environmental protection. Wroclaw, pp 513–516

Arlauskiene EA (2001) The prevalence of microorganisms in soil of agroecosystems. Lithuanian soils. Lietuvos mokslas, Vilnius, pp 178–182 (in Lithuanian)

Arvidsson J (2010) Energy use efficiency in different tillage systems for winter wheat on a clay and silt loam in Sweden. Eur J Agron 33(3):250–256. https://doi.org/10.1016/j.eja.2010.06.003

Balota EL, Colozzi-Filho A, Andrade DS, Dick RP (2003) Microbial biomas in soils under different tillage and crop rotation systems. Biol Fertil Soils 38(1):15–20. https://doi.org/10.1007/s00374-003-0590-9

Bangarwa SK, Norsworthy JK (2014) Brassicaceae cover-crop effects on weed management in plasticulture tomato. J Crop Improve 28:145–158

Batish DR, Singh HP, Kohli RK, Dawra GP (2006) Potential of allelopathy and allelochemicals for weed management. Handbook of sustainable weed management. Food Products Press, Binghamton, pp 209–256

Beyer L (1996) Chemical composition of soil organic matter in classical humic compound fractions and bulk samples review. J Plant Nutr Soil Sci 159(6):527–539. https://doi.org/10.1002/jpln.1996.3581590603

Bianchini A, Magalhaes PSG (2008) Evaluation of coulters for cutting suggar cane residue in a soil bin. Biosyst Eng 100(3):370–375. https://doi.org/10.1016/j.biosystemseng.2008.04.012

Bjorkman M, Klingen I, Birch AN, Bones AM, Bruce TJ, Johansen TJ, Meadow R, Molmann J, Seljasen R, Smart LE, Stewart D (2011) Phytochemicals of *Brassicaceae* in plant protection and human health-influences of climate, environment and agronomic practice. Phytochemistry 72(7):538–556. https://doi.org/10.1016/j.phytochem.2011.01.014

Blanchette RA (2000) A review of microbiological deterioration found in archeological wood from different environments. Int Biodeterior Biodegrad 46:189–204. https://doi.org/10.1016/S0964-8305(00)00077-9

Boerjan W, Ralph J, Baucher M (2003) Lignin biosynthesis. Annu Rev Plant Biol 54:519–546. https://doi.org/10.1146/annurev.arplant.54.031902.134938

Borek V, Morra MJ (2005) Ionic tiocyanate (SCN-) production from 4-hydroxybenzyl glucosinolate contained in Sinapis alba seed meal. J Agric Food Chem 53(22):8650–8654. https://doi.org/10.1021/jf051570r

Borgen BH, Thangstad OP, Ahuja I, Rossiter JT, Bones AM (2010) Removing the mustard oil bomb from seeds: transgenic aplation of myrosin cells in oilseed rape (*Brassica napus*) produces MINELES seeds. J Exp Biol 61(6):1683–1697. https://doi.org/10.1093/jxb/erq039

Brown PD, Morra MJ (1996) Hydrolysis products of glucosinolates in *Brassica napus* tissues as inhibitors of seed germination. Plant Soil 181(2):307–316. https://doi.org/10.1016/S0065-2113(08)60664-1

Brown PD, Morra MJ (1997) Control of soil-borne plant pests using glucosinolate-containing plants. Adv Agron 61:167–231. https://doi.org/10.1016/S0065-2113(08)60664-1

Brown PD, Morra MJ (2009) Brassicaceae tissues as inhibitors of nitrification in soil. J Agric Food Chem 57(17):7706–7711. https://doi.org/10.1021/jf901516h

Cakir E, Johnson CE, Raper RL, Schafer RL (1994) The mechanics of cutting plant residue on a rigid surface. ASAE – paper no. 94–1003. American Society of Agricultural Engineers. St. Joseph, Michigan, pp 38–43

Cesevicius G (2007) The effects of soil tillage systems of various intensity and plant residues on soil properties and productivity of agrophytocenosis; summary of doctoral dissertation, Akademija, 24 p

Chen Y, Gratton JL, Liu J (2004) Power requirements of hemp cutting and conditioning. Biosyst Eng 87(4):417–424. https://doi.org/10.1016/j.biosystemseng.2003.12.012

Cramer N (1990) Raps. Anbau und Verwertung. Ulmer, Stuttgart, 148 p (in German)

Donaldson LA (2001) Lignification and lignin topochemistry – an ultrastructural view. Phytochemistry 57(6):859–873. https://doi.org/10.1016/S0031-9422(01)00049-8

Ebbersten S (1981) Växtfoljadsfrador med anknytning till oljeväx todlingen. Sveriges Utsädesforening Tidskrih 91(4):187–197

Eidukeviciene M (2001) The concept of soil formation factors. Eidukeviciene M, Vasiliauskiene V (eds.) Lithuanian soils. Spauda, Vilnius, pp 114–120 (in Lithuanian)

Einhelling FA (1995) Allelopathy: current status and future goals. Allelopathy: organisms, processes and applications. ACS Symposium 582:1–24. https://doi.org/10.1021/bk-1995-0582.ch001

Eitminaviciute I (1994) Soil biota. VU leidykla, Vilnius, 122 p (in Lithuanian)

Eriksen J (2005) Gross Sulphur mineralization-immobilisation turnover in soil amended with plant residues. Soil Biol Biochem 37(12):2216–2224. https://doi.org/10.1016/j.soilbio.2005.04.003

Evansa BW, Snapeb CE, Jarvis MC (2003) Lignification in relation to the biennial growth habitat in brassicas. Phytochemistry 63(7):765–769. https://doi.org/10.1016/S0031-9422(03)00327-3

Farnsworth RL, Giles E, Franzes RW, Peterson D (1993) The residue dimension. Land and Water 9:1–19

Flaig W (1969) Forteschritte der Huminstoff – Chemic. Landbauforschung Völkenrode 19(2):53–66 (in German)

Freytag HE, Luttich M (1988) Nachweis typischer Reaktions – Geschwindigkeitkonstante bei der Mineralisierung der organischen Bodensubstanz (reaktionskinetische Betrachtungsweise). Arch Acker und Pflanzenbau und Bodenkd 32(9):569–575 (in German)

Gill LS, Anoliefo GO, Iduoze UV (1993) Allelopathic effects of aqueous exstract from Siam weed on the growth of cowpea. Chromoleana Newsletters 8:1–11

Gimsing AL, Kirkegaard JA (2006) Gliucosinolate and isothiocyanate concentration in soil following incorporation of *Brassica* biofumigants. Soil Biol Biochem 38(8):2255–2264

Gimsing AL, Sorensen JC, Tovgaard L, Jorgensen AMF, Hansen HCB (2006) Degradation kinetics of glucosinolates in soil. Environ Toxicol Chem 25(8):2038–2044. https://doi.org/10.1897/05-610R.1

Gisi U (1990) Bodenökologie. Georg Thience Verlag, Stuttgart, 304 p (in German)

Golchin A, Oades JM, Skjemstad JO, Clarke P (1994) Soil stucture and carbon cycling. Aust J Soil Res 32(5):1043–1068. https://doi.org/10.1071/SR9941043

Grabber JH (2005) How do lignin composition, structure and cross-linking affect degradability? A review of cell wall model studies. Crop Sci 45:820–831. https://doi.org/10.2135/cropsci2004.0191

Grabber JH, Ralph J, Lapierre C, Barrière Y (2004) Genetic and molecular basis of grass cell-wall degradability. I. Lignin–cell wall matrix interactions. Comptes Rendus Bioogies 327:455–465. https://doi.org/10.1016/j.crvi.2004.02.009

Hadas A, Kautsky L, Goek M, Kara EE (2004) Rates of decomposition of plant residues and available nitrogen in soil, related to residue composition through simulation of carbon and nitrogen turnover. Soil Biol Biochem 36(2):255–266. https://doi.org/10.1016/j.soilbio.2003.09.012

Halkier BA, Gershenzon J (2006) Biology and biochemistry of glucosinolates. Annu Rev Plant Biol 57:303–333. https://doi.org/10.1146/annurev.arplant.57.032905.105228

Hanschmann A (1983) Einfluß von Temperatur und Feuchtigkeit auf die Mineralisierung von Bodenstickstoff. Archiv für Acker- und Pflanzenbau und Bodenkunde 27(5):297–305

Haramoto ER, Gallandt ER (2005) Brassica cover cropping: I. Effects on weed and crop establishment. Weed Sci 53(5):695–701. https://doi.org/10.1614/WS-04-162R.1

Hatfield JL, Stewart BA (1994) Crops residue management. Lewis Publishers, Boca Raton, 219 p

Hättenschwiler S, Jorgensen HB (2010) Carbon quality rather than stoichiometry controls litter decomposition in a tropical rain forest. J Ecol 98(4):754–763. https://doi.org/10.1111/j.1365-2745.2010.01671.x

Hättenschwiler S, Coq S, Barantal S, Handa IT (2011) Leaf traits and decomposition in tropical rainforests: revisiting some commonly held views and towards a new hypothesis. New Phytol 189:950–965. https://doi.org/10.1111/j.1469-8137.2010.03483.x

Hayes JD, Kelleher MO, Eggleston IM (2008) The cancer chemopreventive actions of phytochemicals derived from glucosinolates. Eur J Nutr 47(2):73–88. https://doi.org/10.1007/s00394-008-2009-8

Hemmatian R, Najafi G, Hosseinzadeh B, Tavakoli Hashjin T, Khoshtaghaza MH (2012) Experimental and theoretical investigation of the effects of moisture content and internodes position on shearing characteristics of sugar cane stems. J Agric Sci Technol 14(5):963–974

Henriksen TM, Breland TA (1999) Nitrogen availability effects on carbon mineralization, fungal and bacterial growth and enzyme activities during decomposition of wheat straw in soil. Soil Biol Biochem 31(8):1121–1134. https://doi.org/10.1016/S0038-0717(99)00030-9

Hoagland L, Carpenter-Boggs L, Reganold JP, Mazzola M (2008) Role of native soil biology in *Brassicaceous* seed meal-induced weed suppression. Soil Biol Biochem 40(7):1689–1697. https://doi.org/10.1016/j.soilbio.2008.02.003

Holtze MS, Sorensen SR, Sorensen J, Aamad J (2008) Microbial degradation of the benzonitrile herbicides dichlobenil, bromoxynil and ioxynil in soil and subsurface environments—insights into degradation pathways, persistent metabolites and involved degrader organisms. Environ Pollut 154(2):155–168. https://doi.org/10.1016/j.envpol.2007.09.020

Ince HA, Ugurluay S, Guzel E, Ozcan TM (2005) Bending and shearing characteristics of sunflower stalk residue. Biosyst Eng 92:175–181

Inderjit, Kreating KI (1999) Allelophaty: principles, procedures, processes and promises for biological control. Adv Agron 67:141–231. https://doi.org/10.1016/S0065-2113(08)60515-5

Inderjit, Wardle DA, Karban R, Callaway RM (2011) The ecosystem and evolutionary contexts of allelopathy. Trends Ecol Evol 26:655–662. https://doi.org/10.1016/j.tree.2011.08.003

Jafariehyazdi E, Javidfar F (2011) Comparison of allelopathic effects of some brassica species in two growth stages on germination and growth of sunflower. Plant Soil Environ 57(2):52–56 http://www.agriculturejournals.cz/publicFiles/35211.pdf

Jakiene E (2011) Effect of biological products on sugar-beet crop. Agric Sci 18(2):64–71 (in Lithuanian with Summary in English). https://doi.org/10.6001/zemesukiomokslai.v18i2.2195

Janusiene V (1994) Biological materials circulation, its effects on soil formation and soil fertility. Agric Sci 1:14–18 (in Lithuanian)

Janusiene V (2002) Intensity of decomposition and humification of crop residues and manure in sandy loam soil. Zemdirbyste-Agriculture 77:102–111 (in Lithuanian with summary in English) http://www.lzi.lt/tomai/77TomasINTERN.htm

Jenkinson DS (1965) Studies on the decomposition of plant material in soil. I. Losses of carbon from ^{14}C labelled ryegrass incubated with soil in the field. J Soil Sci 16(1):104–115. https://doi.org/10.1111/j.1365-2389.1965.tb01424.x

Jenkinson DS (1966) Studies on the decomposition of plant material in soil. II Partial sterilization of soil and soil biomass. Eur J Soil Sci 17(2):280–302. https://doi.org/10.1111/j.1365-2389.1966.tb01474.x

Jenkinson DS (1968a) Studies on the decomposition of plant material in soil. III. The distribution of labbeled and unlabelled carbon in soil incubated with 14C-labelled ryegrass. Eur J Soil Sci 19(1):25–39. https://doi.org/10.1111/j.1365-2389.1968.tb01517.x

Jenkinson DS (1968b) The turnover of organic matter in soil. Biochem J 109(2):2 https://www.ncbi.nlm.nih.gov/pmc/articles/PMC1186797/

Jenkinson DS (1977a) Studies on the decomposition of plant material in soil. IV The effect of rate of addition. Eur J Soil Sci 28(3):417–423. https://doi.org/10.1111/j.1365-2389.1977.tb02249.x

Jenkinson DS (1977b) Studies on the decomposition of plant material in soil. V. The effects of plant cover and soil type on the loos of carbon from carbon-14-labeled ryegrass decomposing under field conditions. Eur J Soil Sci 28(3):424–434. https://doi.org/10.1111/j.1365-2389.1977.tb02250.x

Jenkinson DS, Rayner JH (1977) The turnover of soil organic matter in some of the Rothamsted classical experiments. Soil Sci 123:298–305 http://journals.lww.com/soilsci/Abstract/1977/05000/THE_TURNOVER_OF_SOIL_ORGANIC_MATTER_IN_SOME_OF_THE.5.aspx

Jenkinson DS, Rayner JH (2006) The turnover of soil organic matter in some of the Rothamsted classical experiments. Soil Sci 171(6):130–137 http://journals.lww.com/soilsci/Citation/2006/06001/The_Turnover_of_Soil_Organic_Matter_in_Some_of_the.18.aspx

Jenkinson DS, Hart PBS, Rayner JH, Parry LC (1987) Modelling the turnover of soil organic matter in long-term experiments at Rothamsted. Intecol Bull 15:1–8

Jenkinson DS, Andrew SPS, Lynch JM, Goss MJ, Tinker PB (1990) The turnover of carbon and nitrogen in soil. Philos Trans Biol Sci 392(1255):361–367 http://www.jstor.org/stable/76840?seq=1#page_scan_tab_contents

Johnston AE (1986) Soil organic matter effects on soils and crops. Soil Use Manag 2(3):97–105. https://doi.org/10.1111/j.1475-2743.1986.tb00690.x

Kaboneka S, Nivyza JC, Sibomana L (2004) Effects of nitrogen and phosphorus fertilizer addition on wheat streaw darbon decomposition in Burundi acidic soil. In: Bationo A (ed) Managing nutrient cycles to sustain soil fertility in Sub-Saharan Africa. Academy Science Publishers, Nairobi, pp 151–162

Kalbitz K, Solinger S, Park JH, Michalzik B, Matzner E (2000) Controls on the dynamics of dissolved organic matter in soils: a review. Soil Sci 165(4):277–304 http://journals.lww.com/soilsci/Abstract/2000/04000/Controls_on_the_Dynamics_of_Dissolved_Organic.1.aspx

Karban R, Shiojiri K, Huntzinger M, Mccall A (2006) Damage-induced resistance in sagebrush: volatiles are key to intra- and interplant communication. Ecology 87(4):922–930. https://doi.org/10.1890/0012-9658(2006)87%5B922:DRISVA%5D2.0.CO;2

Kazuki S (2004) Sulphur assimilatory metabolism. The long and smelling road. Plant Physiology 136:2443–2450 https://doi.org/10.1104/pp.104.046755

Kissen R, Rossiter JT, Bones AM (2009) The „mustard oil bomb“: not so easy to assemble?! Localization, expression and distribution of the components of the myrosinase enzyme system. Phytochem Rev 8(1):69–86. https://doi.org/10.1007/s11101-008-9109-1

Klimanek EM (1988) Qualität und Umsetzungsverzhalten von Ernte und Wurzelrückständen land – wirtschaftlich genutzter Pflanzenarten: dissertation, FBZ. Müncheberg Bad Lauchstadt. DAL. Akademie, Berlin, 177 p (in German)

Kögel-Knaber J (2002) The macromolecular organic composition of plant and microbial residues as inputs to soil organic matter. Soil Biol Biochem 34(2):139–162. https://doi.org/10.1016/S0038-0717(01)00158-4

Kong L, Wang YB, Zhao LN, Chen ZH (2009) Enzyme and root activities in surface-flow constructed wetland. Chemosphere 76(5):601–608. https://doi.org/10.1016/j.chemosphere.2009.04.056

Kononova MM (1963) Soil organic matter and proceses its transformation. Academy of Sciences Press, Moscow, 314 p (in Russian)

Kriauciuniene Z (2008) Peculiarities of oilseed rape residues decomposition in the loam of *Hypogleyic Cambisol*: summary of doctoral dissertation. Akademija, 26 p

Kriauciuniene Z, Velicka R, Raudonius S (2012) The influence of crop residues type on their decomposition rate in the soil: a litterbag study. Zemdirbyste-Agriculture 99(3):227–236 http://www.lzi.lt/tomai/99(3)tomas/99_3_tomas_str1.pdf

Kushwaha RL, Vaishnav AS, Zoerb GC (1983) Shear strength of wheat straw. Can Agric Eng 25:163–167. http://www.csbe-scgab.ca/docs/journal/25/25_2_163_raw.pdf

Kushwaha RL, Vaishnav AS, Zoberg GC (1986) Soil bin evaluation of disc coutlers under no – till crop residue conditions. Trans ASAE 29(1):40–44. https://doi.org/10.13031/2013.30098

Lal R (1999) Soil quality and soil erosiosn. CRC Press, Boca Raton, 329 p

Lankau RA, Strauss SY (2007) Community complexity drives patterns of natural selection on a chemical defense of *Brassica nigra*. Am Nat 171(2):150–161. https://doi.org/10.1086/524959

Larson WE, Eynard A, Hadas A, Lipiec J (1994) Control and avoidance of soil compaction in practice. In: Soane BD, Ouwerkerk C (eds) Soil compaction in crop production. Elsevier Science, Amsterdam, pp 597–601

Lattanzio V, Cardinali A, Linsalata V (2012) Plant phenolics: a biochemical and physiological perspective. Recent Adv Polyphenol Res 3:1–29

Lejon DPH, Sebastia J, Lamy I, Nowak V, Chaussod R, Ranjard L (2007) Relationships between soil organic status and microbial community density and genetic structure in two agricultural soils submitted to various types of organic management. Microb Ecol 53(4):650–663. https://doi.org/10.1007/s00248-006-9145

Linke C (1998) Direktsaat – eine Bestandsaufnahme unter besonderer Berücksichtigung technischer, agronomischer und ökonomischer Aspekte. Dissertation. – Hohenheim, 482 p (in German)

Liu Q, Ou Y, Qing S, Song C (2007) Cutting force test of sugarcane stalk. Trans Chinese Soc Agric Eng 23(7):90–94

Liu J, Chen Y, Kushwaha RL (2010) Effect of tillage speed and straw length on soil and straw movement by a sweep. Soil Tillage Res 109(1):9–17. https://doi.org/10.1016/j.still.2010.03.014

Lugauskas A (1997) Microbiological material damage. Valstiečių laikrastis, Vilnius, pp 110–122 (in Lithuanian)

Luxhøi J, Magid J, Tscherko D, Kandeler E (2002) Dynamics in invertase, xylanase and coupled quality indices of decomposing green and brown plant rersidues. Soil Biol Biochem 34(4):501–508. https://doi.org/10.1016/S0038-0717(01)00208-5

Magalhaes PSG, Bianchini A, Braunbeck OA (2007) Simulated and experimental analyses of a toothed rolling coulter for cutting crop residues. Biosyst Eng 96(2):193–200. https://doi.org/10.1016/j.biosystemseng.2006.10.014

Magid J, Luxhøi J, Lyshede OB (2004) Decomposition of plant residues at low temperatures seperates turnover of nitrogen and energy rich tissue components in time. Plant Soil 258(1):351–365. https://doi.org/10.1023/B:PLSO.0000016565.14718.4b

Magyla A, Sateikiene D, Slepetiene A (1997) Amount of crop residues, its chemical composition and soil humus in various specialised crop rotations. Zemdirbyste-Agriculture 58:56–75 (in Lithuanian)

Makovski N (1990) Winter rape. Minsk, 50 p (in Russian)

Malik MS, Riley MB, Norsworthy JK, Bridges W (2010) Glucosinolate profile variation of growth stages of wild radish (*Raphanus raphanistrum*). J Agric Food Chem 58(6):3309–3315. https://doi.org/10.1021/jf100258c

Marcinkeviciene A (2003) Influence of cover crops for green manure on barley agrocenoses in sustainable and organic agriculture: summary of doctoral dissertation. Kaunas, 39 p

Marcinkeviciene A, Kriauciuniene Z, Boguzas V, Velicka R (2013) Allelopathic effects of cover crops on spring barley germination and establishment. Int J Food Agric Environ 11(3–4):684–688

Matusheski NV, Swarup R, Juvik JA, Mithen R, Bennett M, Jeffery EH (2006) Epithiospeciffer protein from brroccoli (*Brassica oleracea* L. ssp italica) inhibitit formation of the anticancer agent sulforaphane. J Agric Food Chem 54:2069–2076. https://doi.org/10.1021/jf0525277

Mawson R, Heaney RK, Zdunczyk Z, Kozlowska H (1993) Rapeseed meal-glucosinolates and their antinutritional affects. Flavor and palatability. Nahrung/Food 37(4):336–344. https://doi.org/10.1002/food.19930370405

Metlen KL, Aschehoug ET, Callaway RM (2009) Plant behavioural ecology: dynamic plasticity in secondary metabolites. Plant Cell Environ 32(6):641–653. https://doi.org/10.1111/j.1365-3040.2008.01910.x

Moran KK, Six J, Horwath WR, Kessel C (2005) Role of mineral-nitrogen in residue decomposition and stabile soil organic matter formation. Soil Sci Soc Am J 69(6):1730–1736. https://doi.org/10.2136/sssaj2004.0301

Nadelhoffer K, Giblin A, Shaver G, Laundre J (1991) Effects of temperature and substrate quality on element mineralization in six arctic soils. Ecology 72(1):242–253. https://doi.org/10.2307/1938918

Nazari GM, Jafari A, Mohtasebi SS, Tabatabaeefar A, Sharifi A, O'Dogherty MJ, Rafiee S, Richard G (2008) Effects of moisture content and level in the crop on the engineering properties of alfalfa stems. Biosyst Eng 101(2):199–208. https://doi.org/10.1016/j.biosystemseng.2008.07.006

Noel JP, Austin MB, Bomati EK (2005) Structure-function relationships in plant phenylpropanoid biosynthesis. Curr Opin Plant Biol 8(3):249–253. https://doi.org/10.1016/j.pbi.2005.03.013

Northup RR, Dahlgren RA, Mccoll JG (1998) Polyphenols as regulators of plant-litter-soil interactions in northern California's pygmy forest: a positive feedback? Biogeochemistry 42(1–2):189–220. https://doi.org/10.1023/A:1005991908504

Orlova NE, Bakina LG, Dmitrijeva LE (2002) Features of seasonal transformation of sod-podzolic loamy soil organic matter. Humus and soil formation. Collection of scientific papers. SPbGAU, St. Petersburg, pp 13–21 (in Russinan)

Paul EA, Clark FE (1989) Soil microbiology and biochemistry. Academic, New York, 273 p

Pellerin S, Mollier A, Moler C, Plenchette C (2007) Effect of incorporation of *Brassica napus* L. residues in soil on mycorrhizal „fungus" colonization of roots and phosphorus uptake by maize (*Zea mays* L.). Eur J Agron 26(2):113–120. https://doi.org/10.1016/j.eja.2006.07.007

Petersen J, Belz R, Walker F, Hurle K (2001) Weed suppression by release of isothiocyanates from turnip-rape mulch. Agron J 93(1):37–43. https://doi.org/10.2134/agronj2001.93137x

Pichersky E, Gang DR (2000) Genetics and biochemistry of secondary metabolites in plants: an evolutionary perspective. Trends Plant Sci 5(10):439–445. https://doi.org/10.1016/S1360-1385(00)01741-6

Popaa VI, Dumitrua M, Volf I, Anghel N (2008) Lignin and polyphenols as allelochemicals. Ind Crop Prod 27(2):144–149. https://doi.org/10.1016/j.indcrop.2007.07.019

Poulsen JL, Gimsing AL, Halkier BA, Bjarnholt N, Hansen HCB (2008) Mineralization of benzyl glucosinolate and its hydrolysis product the biofumigant benzyl isothiocyanate in soil. Soil Biol Biochem 40(1):135–141. https://doi.org/10.1016/j.soilbio.2007.07.015

Pupaliene R (2004) The residual effect of different agricultural systems on spring barley agrocoenosis: summary of doctoral dissertation. Kaunas, 21 p

Raudoniene V. (2003) Micromycetes – producers of phenoloxidases and their significance in bioconversion of plant waste: summary of doctoral dissertation, Vilnius, 36 p

Rice EL (1984) Allelopathy, 2nd edn. Academic, London

Robles MD, Burke JC (1997) Legume, grass and conversation reservic program effects on soil organic matter recovery. Ecol Appl 7(2):345–357. https://doi.org/10.1890/1051-0761(1997)007[0345:LGACRP]2.0.CO;2

Ryden L, Migula P, Anderson M (2003) Environmental science. The Baltic University Press, Uppsala, 824 p

Salas AM, Elliot ET, Westfall DG, Colc CV, Six J (2003) The role of particulate organic matter in phosphorus cycling. Soil Sci Soc Am J 67:181–198

Sarauskis E, Köller K, Butkus V (2005) Research on technological parameters to determine the design factors of direct drilling coulters for sugar beets. Lanbauforschung Volkenrode 55(3):171–180

Sarauskis E, Buragiene S, Romaneckas K, Sakalauskas A, Jasinskas A, Vaiciukevičius E, Karayel D (2012) Working time, fuel consumption and economic analysis of different tillage and sowing systems in Lithuania. Proceedings of 11th International scientific conference „Engineering for Rural Development", May 24-25, Jelgava, Latvia, pp 52–59 http://www.tf.llu.lv/conference/proceedings2012/Papers/009_Sarauskis_E.pdf

Sarauskis E, Masilionyte L, Romaneckas K, Kriauciuniene Z, Jasinskas A (2013) The effect of the disc coulters forms and speed ratios on cutting of crop residues in no-tillage system. Bulgarian J Agr Sci 19(3):620–624

Sarauskis E, Masilionyte L, Andriusis A, Jakstas A (2013a) The force needed for breaking and cutting of winter wheat and spring barley straw. Zemdirbyste-Agriculture 100(3):269–276. https://doi.org/10.13080/z-a.2013.100.034

Shkarda M (1985) Production and use of organic fertilizers. Moscow, 364 p (in Russian)

Sidlauskas G (2002) Winter and spring rape (*Brassica napus* L.) yield formation and development relations with environmental factors: habilitation thesis. Lithuaian Institute of Agriculture, Akademija, 142 p (in Lithuanian)

Singh IP, Bharate SB (2006) Phloroglucinol compounds of natural origin. Nat Prod Rep 23:558–591. https://doi.org/10.1039/B600518G

Singh B, Rengel Z, Bowden JW (2004) Canola residues decomposition: the effect of particle size on microbial respiration and cycling of sulphur in a sandy soil. SuperSoil 2004: Proceedings of 3rd Australian New Zeland Soils Conference, University of Sydney, Australia, pp 1–7 http://www.regional.org.au/au/asssi/supersoil2004/s12/oral/1581_singhb.htm

Sjöberg G (2003) Lignin degradation: doctoral thesis. Upsala, 46 p. http://pub.epsilon.slu.se/341/

Smukalski M (1988) Fruchtfolge und Bodenfruchtbarkeit. Fruchtfolgeforschung und Fruchtfolgestaltung Tagungsbericht Berlin 261:81–89 (in German)

Sollins P, Homann P, Caldwell BA (1996) Stabilization and destabilization of soil organic matter: mechanisms and controls. Geoderma 74(1–2):65–105. https://doi.org/10.1016/S0016-7061(96)00036-5

Spielhaus G (1980) Raps möbelt einseitige fruchtfolgen auf. DLG-Mitteilungen 12:674–675 (in German)

Stancevicius A, Boguzas V (1995) Effect of catch crops and fall tillage on the crop yield in the rotation of cereals. In: Problems of field crop husbandry and soil management in Baltic States, Transactions of the Estonian Agricultural University Tartu, vol 182. Estonian Agricultural University, Tartu, pp 118–127

Stanceviius A, Pranaitiene E (1982) Tillage and catch crops influence on plant residue decomposition intensity. Collection of scientific works of Lithuanian Academy of Agriculture, Vilnius, pp 98–100 (in Lithuanian)

Stott DE, Martin JP (1990) Synthesis and degradation of natural and synthetic humic material in soils. MacCarthy P, Clapp CE, Malcolm RL, Bloom PR (eds) Humic substances in soils and crop sciences. Madison, pp 37–63. https://doi.org/10.2136/1990.humicsubstances.c3

Taraz Z, Jalali SMA, Rafeie F (2006) Effects of replacement of soybean meal with rapeseed meal on organs weight, some blood biochemical parameters and performance of broiler chicks. Int J Poult Sci 5(12):1110–1115

Tavakoli H, Mohtasebi SS, Jafari A (2009) Effects of moisture content, internode position and loading rate on the bending characteristics of barley straw. Res Agric Eng 55(2):45–51

Teit R (1991) Soil organic matter: biological and ecological aspects. Mir, Moscow, 399 p (in Russian)

Tharayil N, Bhowmik P, Alpert P, Walker E, Amarasiriwardena D, Xing B (2009) Dual purpose secondary compounds: phytotoxin of Centaurea diffusa also facilitates nutrient uptake. New Phytol 181(2):424–434. https://doi.org/10.1111/j.1469-8137.2008.02647.x

Titova NA, Kogum BM (1991) Transformation of soil organic matter in soil used for agriculture. The results of science and technology. Soil Science and Agricultural Chemistry. Moscow, 8:156 (in Russian)

Traka M, Mithen R (2009) Glucosinolates, isothiocyanates and human helth. Phytochem Rev 8(1):269–282. https://doi.org/10.1007/s11101-008-9103-7

Trinsoutrot I, Nicolardot B, Justes E, Recous S (2000) Decomposition in the filed of residues of oilseed rape grown at two levels of nitrogen fertilisation. Effects on the dynamics of soil mineral nitrogen between succesive crops. Nutr Cycl Agoecosyst 56(2):125–137. https://doi.org/10.1023/A:1009838618133

Tripolskaja L (2005) Organic fertilisers and their effects on the environment. Akademija-Dotnuva, 205 p (in Lithuanian)

Uremis I, Arslan M, Uludag A, Sangun M (2009) Allelopathic potentials of residues of 6 brassica species on johnsongrass (*Sorghum halepense* (L.) Pers.). Afr J Biotechnol 8(15):3497–3501

Vaitauskiene K, Sarauskis E, Naujokiene V, Liakas V (2015) The influence of free-living nitrogen-fixing bacteria on the mechanical characteristics of different plant residues under no-till and strip-till conditions. Soil Tillage Res 154:91–102. https://doi.org/10.1016/j.still.2015.06.007

Van Scholl L, Van Dam AM, Leffelaar PA (1997) Mineralisation of nitrogen from an incorporated catch crop at low temperatures: experiment and simulation. Plant Soil 188(2):211–219. https://doi.org/10.1023/A:1004255102840

Varnaite R (2004) Lignin degradation in rye straw during primary fungi cultivation stages. Ekologija 3:34–37 (in Lithuanian)

Vaughn SF, Berhow MA (2005) Glucosinolate hydrolysis products from various plant sources: pH effects, isolation and purification. Ind Crop Prod 21(2):193–202. https://doi.org/10.1016/j.indcrop.2004.03.004

Vaughn SF, Boydston RA (1997) Volatile allelochemicals realeased by crucifer green manures. J Chem Ecol 23(9):2107–2116. https://doi.org/10.1023/B:JOEC.0000006432.28041.82

Velicka R (1995) Rape for seeds in crop rotation. Problems of filed crop husbandry and soil management in Baltic States. Tartu, pp 128–136

Velicka R (2002) Oilseed rape. Lutute, Kaunas, 320 p (in Lithuanian with Summary in English)

Velicka R, Rimkeviciene M, Kriauciuniene Z, Pupaliene R, Salina O (2009) The effect of cellulose-degrading micro-organisms on the biodestruction of crop residues in the soil. Zemdirbyste-Agriculture 96(1):113–126 http://www.lzi.lt/tomai/96(1)tomas/96_1_tomas_113_126.pdf

Wager MG, Cabrera MI, Ranells NN (1998) Nitrogen and carbon cycling in relation to cover crop residue quality. J Soil Water Conserv 53(3):214–218 http://www.jswconline.org/content/53/3/214.short

Walsh KD, Sanderson D, Hall LM, Mugo S, Hills MJ (2014) Allelopathic effects of camelina (*Camelina sativa*) and canola (*Brasssica napus*) on wild oat, flax and radish. Allelopath J 33(1):83–95

Wang JB, Chen ZH, Chen LJ, Zhu AN, Wu ZJ (2011) Surface soil phosphorus and phosphatase activities affected by tillage and crop residue input amounts. Plant Soil Environ 57:251–257 http://www.agriculturejournals.cz/web/pse.htm?volume=57&firstPage=251&type=publishedArticle

Watkins N, Barraclough D (1996) Gross of N mineralization associated with the decomposition of plant residues. Soil Biol Biochem 28(2):169–175. https://doi.org/10.1016/0038-0717(95)00123-9

Weston LA, Duke SO (2003) Weed and crop allelopathy. Crit Rev Plant Sci 22:367–389. https://doi.org/10.1080/713610861

Wink M (2003) Evolution of secondary metabolites from an ecological and molecular phylogenetic perspective. Phytochemistry 64(1):3–19. https://doi.org/10.1016/S0031-9422(03)00300-5

Wittstock U, Halkier BA (2002) Glucosinolate research in the Arabidopsis era. Trends Plant Sci 7(6):263–270. https://doi.org/10.1016/S1360-1385(02)02273-2

Wolf B, Snyder GH (2003) Sustainable soils: the place of organic matter in sustaining soils and their productivity. Food Products Press, New York, 352 p

Yasumoto S, Suzuki K, Matsuzaki M, Hiradate S, Oose K, Hirokane H, Okada K (2011) Effects of plant residue, root exudate and juvenile plants of rapeseed (*Brassica napus* L.) on the germination, growth, yield, and quality of subsequent crops in successive and rotational cropping systems. Plant Prod Sci 14(4):339–348. https://doi.org/10.1626/pps.14.339

Yu O, Jez JM (2008) Natures assembly line: biosynthesis of simple phenylpropanoids and polyketides. Plant J 54(4):750–762. https://doi.org/10.1111/j.1365-313X.2008.03436.x

Zacharovich VV (2005) Winter wheat cultivation technology based on the use alleopathic properties of cruciferous crops and chemical protection from pests: doctoral dissertation. 120 p (in Russian)

Zekoniene V, Janusiene V (1997) The effect of specialised crop rotations on soil humus content and qualitative composition. Zemdirbyste-Agriculture 59: 37–48 (in Lithuanian with summary in English)

Zemeckis R (1993) Investigations of sugar beet growing in sod-gleyic soils after different predecessors in Lithuanian: summary of doctoral dissertation, Lithuanian Academy of Agriulture, 17 p

Zhang W, Jiang F, Ou J (2011) Global pesticide consumption and pollution: with China as a focus. Proc Int Acad Ecol Environ Sci 1(2):125–144

Chapter 8
Biochar Amendment to Soil for Sustainable Agriculture

Vipin Kumar Singh, Ajay Kumar, and Rishikesh Singh

Abstract The management of surplus lignocellulosic wastes, deteriorating soil quality, food insecurity and changing climate are major issues. The overuse of pesticides and synthetic fertilizers has deteriorated soil natural properties. Therefore, sustainable agriculture based on improvement in soil characteristics is a major requirement to enhance crop productivity. Various soil ameliorant have been proposed for restoring soil health and viability. For instance, biochar produced from plant and animal products improves soil properties. Properties of biochar are influenced by parameters such as type of feedstock material, temperature and residence time.

We review biochar application to soil. First, we evaluate the role of various production conditions such as pyrolysis temperature and feedstock type on biochar characteristics. Then we discuss the following aspects of biochar: habitat modification by biochar application, carbon sequestration, environmental contaminant management, crop productivity, disease management, slow release fertilizers, and risks associated with biochar amendment. Biochar as soil ameliorant leads to reduced greenhouse gas emission, enhanced carbon sequestration, ecological restoration and increased crop productivity. Biochar production for soil amendment is also a good option for waste management and bioenergy production. Another advantage of biochar application to soil is the decreased availability of soil contaminants and improved efficacy of applied agrochemicals.

Keywords Biochar · Carbon sequestration · Slow release fertilizer · Crop productivity · Greenhouse gas · Soil amendment

V. K. Singh (✉) · A. Kumar
Department of Botany, Institute of Science, Banaras Hindu University, Varanasi, India

R. Singh
Institute of Environment & Sustainable Development (IESD), Banaras Hindu University, Varanasi, India

E. Lichtfouse (ed.), *Sustainable Agriculture Reviews 32*, Sustainable Agriculture Reviews 32, https://doi.org/10.1007/978-3-319-98914-3_8

8.1 Introduction

The grain production has enhanced after the introduction of green revolution worldwide. In addition to grain production, surplus amount of crop residue and related wastes are generated which is of prime concern for their management (Lal 2008). Most of these residues are being openly burnt leading to various tangible and intangible consequences on the environment and human health (Crutzen and Andreae 1990). Moreover, from very long time, use of compost and manure derived from animal/plant sources has been replaced by synthetic agrochemicals. Agricultural practices using very large quantities of herbicides, pesticides and fertilizers are known to disturb the natural properties of soil viz. soil organic carbon, soil physicochemical properties and microbial population (Srivastava et al. 2016). Reports are available indicating the reduction in population of soil beneficial microbes after application of chemical fertilizers. Overutilization of agrochemicals in soil has detrimental consequences similar to overuse of antibiotics for human health. Application of excess fertilizers may be effective for few years but after a certain time it leads to soil quality degradation resulting from reduced availability of soil inhabiting organisms (Savonen 1997). For example, growth and viability of nitrogen fixing bacteria is reduced by glyphosate (Santos and Flores 1995) and 2,4-D (Arias and Fabra 1993; Fabra et al. 1997) application in soil. Reduction in soil productivity is a big problem and drawing attention of soil scientist worldwide. In contrast to the physicochemical methods used for soil management practices biological methods are less costly and environment friendly. One of such biological approach involves the application of biochar.

Biochar is a biologically derived material obtained from the thermo-chemical decomposition of organic biomass under reduced oxygen environment in a closed chamber conditions and used for soil amelioration (Lehmann and Joseph 2009; Singh et al. 2015). It has been defined from time to time based on its physicochemical properties and various environmental applications. It is defined as the carbon rich material produced after heating (pyrolysis) of organic material such as wood, leaves or manures in the absence or very little presence of oxygen (Lehmann and Joseph 2009). Shackley et al. (2012) defined biochar as carbon enriched porous material generated by thermochemical decomposition in oxygen depleted environment that bears the characteristics of large pool carbon storage. Verheijen et al. (2010) described biochar as pyrolyzed biomass generated under oxygen deficient or limited condition for application in soil with the purpose of soil quality improvement and carbon sequestration but simultaneously neutralizing the detrimental impact on plant, human and animal system. It differs from other carbonaceous species based on its oxygen to carbon (O:C) ratio ranging in between 0.4 and 0.6 (Schimmelpfennig and Glaser 2012). However, the exact chemical nature of the biochar prodcut depends upon the pyrolysis condition and type of biomass used as shown in Table 8.1 (Lehmann and Joseph 2009).

Biochar is getting higher attention recently due to its nutrient retention capacity as soil nutrient loss is one of the important limitations for the sustainable agriculture

Table 8.1 Impact of pyrolysis temperature and feedstock type on various physico-chemical characteristics of biochar PT = Pyrolysis temperature and SA = Surface area

				% dry weight basis								
Source	PT (°C)	pH	SA (m^2/gm)	C	N	O	H	P	C/N	H_2O	Ash	References
Corn Stover	500	8.90	4.20	25.00	0.60	5.00	1.10	n/a	41.67	9.1	69.00	Spokas et al. (2011)
Corn Stover	515	9.50	4.40	45.00	0.50	1.00	1.70	n/a	90.00	11.5	55.00	Spokas et al. (2011)
Coconut shell	550	8.90	15.10	80.10	0.50	2.50	n/a	n/a	160.20	12.4	n/a	Spokas et al. (2011)
Peanut hulls	481	8.00	1.00	59.00	2.70	12.00	2.30	n/a	21.85	7.2	18.00	Spokas et al. (2011)
Corn cob	400	9.00	< 0.1	80.10	0.60	8.80	3.70	n/a	133.50	3.1	3.70	Spokas et al. (2011)
Sugarcane bagasse	350	5.00	n/a	75.20	0.66	15.80	4.60	n/a	113.94	3.42	3.60	Spokas et al. (2011)
Poultry litter	400	10.30	n/a	42.30	4.20	n/a	n/a	n/a	10.07	n/a	n/a	Spokas et al. (2011)
Cottonseed hull	500	8.50	< 0.1	78.70	2.50	6.90	2.50	n/a	31.48	6.5	7.90	Spokas et al. (2011)
Cottonseed hull	800	7.70	322.00	84.30	0.60	6.60	0.60	n/a	140.50	5.9	9.20	Spokas et al. (2011)
Rape residue	550	n/a	n/a	72.20	1.30	25.60	0.90	n/a	55.54	3.2	21.80	Sánchez et al. (2009)
Sunflower residue	550	n/a	n/a	63.40	1.60	34.30	0.70	n/a	39.63	4.73	28.90	Sánchez et al. (2009)
Wheat straw	n/a	n/a	n/a	43.20	0.61	39.40	5.00	n/a	70.82	n/a	n/a	Spokas et al. (2010)
Rice hulls	n/a	n/a	n/a	38.30	0.83	35.45	4.36	n/a	46.14	n/a	n/a	Spokas et al. (2010)
Olive kernel	800	n/a	n/a	75.68	1.35	12.18	0.79	n/a	56.06	n/a	n/a	Zabaniotou et al. (2008)
Canola straw	500	9.39	n/a	63.40	0.04	n/a	n/a	0.30	1585.00	n/a	18.40	Yuan et al. (2011)
Canola straw	700	10.76	n/a	54.90	0.04	n/a	n/a	0.50	1372.50	n/a	28.55	Yuan et al. (2011)
Soyabean straw	500	10.92	n/a	62.60	0.40	n/a	n/a	0.40	156.50	n/a	17.85	Yuan et al. (2011)
Soyabean straw	700	11.10	n/a	57.90	0.10	n/a	n/a	0.60	579.00	n/a	23.70	Yuan et al. (2011)
Corn straw	500	10.77	n/a	41.90	0.90	n/a	n/a	0.40	46.56	n/a	50.70	Yuan et al. (2011)
Corn straw	700	11.32	n/a	24.50	0.80	n/a	n/a	0.70	30.63	n/a	73.30	Yuan et al. (2011)
Peanut straw	500	10.86	n/a	48.50	1.50	n/a	n/a	0.10	32.33	n/a	32.50	Yuan et al. (2011)
Peanut straw	700	11.15	n/a	47.00	1.50	n/a	n/a	0.12	31.33	n/a	38.50	Yuan et al. (2011)

(Renner 2007). The enhanced nutrient retention property under biochar applied system is due to its high adsorption capacity and inherent stable nutrient source as compared to other organic materials (Chan et al. 2007). However, a proper understanding of soil-crop-climate is imperative for beneficial biochar properties (Singh et al. 2015). Singh et al. (2015) have recently reviewed the crop residue based biochar and suggested it as an effective ameliorant for sustainable agriculture. Due to long term soil residence and resistance against chemical oxidation, biochar serves the function of carbon sink. Impact of biochar application on mycrorrhizal system was investigated by Warnock et al. (2007). They reported the role of biochar as habitat for mycorrhizal association. Such interaction of biochar and mycorrhizae were described to play significant roles in crop productivity, restoration of disturbed ecosystem, carbon sequestration and mitigation of climate change. Biochar soil amendment has been suggested as an effective tool to counter the effect of climate change (Barrow 2012). Furthermore, biochar improves soil productivity by mechanism of carbon sequestration and restoration of degraded lands (Lehmann et al. 2011). Among various advantages conferred by biochar addition to soil ecosystem includes reduced plant disease (Elad et al. 2010), management of soil acidity (Yuan et al. 2011), enhanced plant productivity (Lehmann et al. 2011), nutrient management (Lehmann et al. 2003), minimized greenhouse gas emission (Taghizadeh-Toosi 2011), slow release fertilizers (Manikandan and Subramanian 2013), and soil microbiological activities (Liang et al. 2010). The objective of present chapter is to give a summarized account of current knowledge about the role of biochar in sustainable agriculture. In the present chapter, we would evaluate the role of various pyrolysis conditions and feedstock type on biochar characteristics, followed by its candidature as a sustainable agricultural tool. In terms of sustainable agriculture tool, we emphasized on the role of biochar on: (1) habitat modification, (2) environmental contaminant management, (3) crop productivity, (4) disease management, (5) as slow release fertilizers, (6) in carbon sequestration, and (7) risks associated with biochar amendment to soil.

8.2 Biochar Production Technology

Biochar is produced mainly by three different processes: slow pyrolysis, fast pyrolysis and gasification (Inyang and Dickenson 2015). In fast pyrolysis technique biomass combustion is performed at 425–550 °C in absence of air for very limited time (2 s) while slow pyrolysis uses air deficient combustion at 350–800 °C for minutes to hours. Gasification process involves biomass heating from seconds to hours under presence of oxygen at temperature equivalent to or greater than 800 °C. The properties of biochar in terms of biochar yield, pH, particle size and surface area are dependent on the methods by which they are produced, thus varying under different conditions of generation. Partially pyrolyzed biochar produced under conditions of fast pyrolysis possess the higher content of carbon available to microbial growth thus less suitable for carbon sequestration (Bruun et al. 2011). Thus, application of

Table 8.2 Some commonly used feedstock material for biochar production

Material type	Temperature (°C)	Purpose/application	References
Wheat straw	300, 600	Carbon mineralization	Junna et al. (2014)
Wheat straw	450	Greenhouse gas emission study	Wu et al. (2013)
Wheat straw	450	Carbon sequestration	Cheng et al. (2012)
Wheat straw	450	Nitrate and phosphate removal	Li et al. (2016)
Wheat straw	300	Sorption and leaching of herbicide	Tatarkova et al. (2013)
Wheat straw	300	Remediation of petroleum contaminant	Ying and Chun (2014)
Wheat straw	250–750	Chloropyrifos removal from water	Wang et al. (2015)
Wastewater sludge	300–700	Sludge management	Hossain et al. (2011)
Sugarcane bagasse	600	Agriculture waste management	Inyang et al. (2010)
Orange peel	150–700	Sorption of naphthalene and 1-naphthol	Chen and Chen (2009)
Eucalyptus grandis	300–450	Poultry manure composting	Dias et al. (2010)
Orange peel	250–700	Contaminant removal from water	Chen et al. (2011)
Wood	550	Use as electron donor and acceptor	Saquing et al. (2016)
Rice husk	Not specified	Soil health and crop productivity	Pratiwi and Shinogi (2016)
Rice and bamboo	500, 750	Heavy metal immobilization in soil	Lu et al. (2017)
Bamboo	600	Nitrogen retention in soil	Ding et al. (2010)

partially pyrolyzed biomass promotes the immobilization of soil nitrogen (Brewer et al. 2009). Contrary to fast pyrolysis, slow pyrolysis generates completely pyrolyzed biochar and minimizing the risks of nitrogen immobilization (Bruun et al. 2012).

High temperature pyrolysis enhances fixed carbon content for low ash bearing biochar while same pyrolysis temeperature lowers fixed carbon for more than 20% ash bearing biochar (Enders et al. 2012). Increase in pyrolysis temperature in the range of 450–700 °C favors the biochar production with low particle size (Downie et al. 2009). In contrast, Khanmohammadi et al. (2015) reported increase in particle density and porosity for sewage sludge with increase in pyrolysis temperature. Detailed effect of various pyrolysis conditions on biochar physico-chemical properties is outlined in Table 8.1. Further, a list of commonly used feedstock material for biochar preparation with multiple purposes has been provided in Table 8.2. The following Fig. 8.1 describes the general process of biomass pyrolysis and possible application of various by-products.

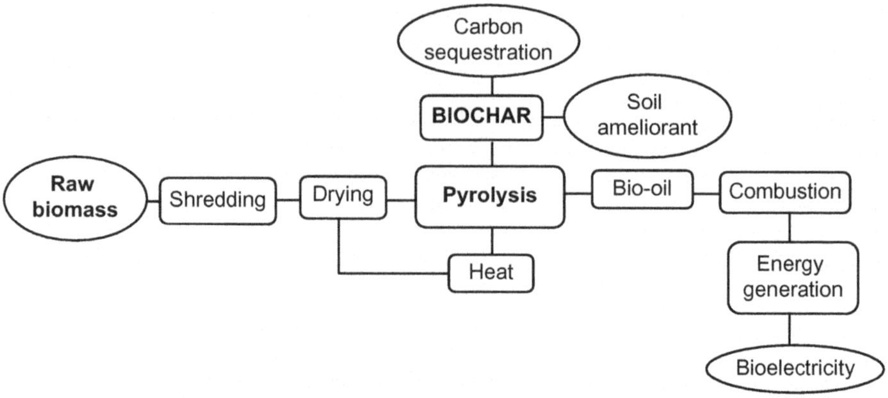

Fig. 8.1 General process of biomass pyrolysis and possible application of various by-products. Source: International Biochar Initiative. Available at www.biochar-international.org/biochar/carbon

8.3 Biochar Application and Alterations in Soil Habitat

The properties of biochar are very much distinct from uncharred soil organic material and are expected to change with time due to natural weathering, soil mineral interaction and microbiological process (Lehmann et al. 2005; Cheng et al. 2008; Cheng and Lehmann 2009; Nguyen et al. 2010).

Biochar can be roughly categorized in three fractions i.e. non labile, labile (leachable) and ash portion. Biochar is known to possess higher content of aromatic carbon especially the fused aromatic carbons as compared to other generally available organic matter. The structure and content of fused aromatic carbon is dependent on temperature; lower temperature favors the occurrence of amorphous carbon while higher one promotes the abundance of turbostratic carbon (Keiluweit et al. 2010; Nguyen et al. 2010). The presence of aromatic carbon structures is the chief reason which facilitate the higher stability of biochar (Nguyen et al. 2010). Despite the chemical stability of biochar, some fractions may be microbially mineralized (Lehmann et al. 2009) and reports are available indicating the increase in microbial activity and abundance (Steiner et al. 2008). The ash portion of biochar bears the mineral elements such as Fe, S, P, K, Mg, and Ca which can be effectively used by the plants and microorganisms.

Differences in physical properties of biochar and soil is responsible for altered soil behavior such as soil tensile strength, hydrodynamics and gaseous exchange which ultimately leads to major impact on soil biotic community. Application of low tensile strength biochar may minimize the soil tensile strength. Lower tensile strength is responsible for easy root elongation, seed germination, efficient nutrient uptake and invertebrate movement through soil thus affecting the prey-predator interactions (Lehmann et al. 2011). Very scarce reports are available suggesting the impact of biochar on soil bulk density (Major et al. 2010). Changes in soil bulk density after biochar application may affect soil water interaction, rooting pattern and soil fauna. The changes may be due to lower biochar density as compared to minerals and presence of micro and macropores thus responsible for holding air and water and reductions in overall bulk density of biochar particles.

Contrary to the organic material present in soil, biochar may remain in particulate phase for very long duration; however, there may be substantial decrease in their size over decades. Biochar also do possess the internal pore spaces which may have important effect on biological processes. Thus, they can be considered as equivalent to soil aggregate. Biochar aggregates are supposed to serve the variety of functions such as organic material protection, habitat provision and availability of proper moisture and nutrient storage (Tisdall and Oades 1982).

8.4 Role of Biochar in Environmental Management

Biochar has the efficiency to adsorb the toxic agrochemicals present in water and soil, thus reducing the health risk of contaminant and paying the way for ecofriendly cost-effective management of polluted soil (Ahmad et al. 2014; Mohan et al. 2014). Biochar application to soil alleviates the effect of toxic organic compounds and reduces phytoaccumulation (Hunter et al. 2010). McLeod et al. (2004, 2007) have reported reduced accumulation of pentachlorobenzenes and benzo(a)pyrene in *Macoma balthica* plants. Reduced accumulation upto 70–87% of polychlorinated biphenyls in invertebrates was reported by Millward et al. (2005) when marine sand was amended with biochar. Yang and Sheng (2003) reported that biochar produced after pyrolysis of rice and wheat feedstocks are 400–2500 more efficient in sorption as compared to soil itself. Sufficient evidences are available which suggest that soil application of biochar can immobilize soil pollutants (Smernik 2009) and thus reducing the risks of phytotoxicity (Beesley et al. 2010; Kim et al. 2015). Yao et al. (2011) applied biochar derived from anaerobically digested sugarbeet for efficient phosphate acquisition. The results suggested their use as slow release fertilizers to manage nutrient deficient soils (Karunanithi et al. 2015; Park et al. 2015a, b).

Higher concentrations of heavy metals in soil impose hazardous risks to plants and human health. Biochar application to soil has advantageous effects in terms of hormones and heavy metal adsorption (Paz-Ferreiro et al. 2014). Biochar soil amendment enhances soil pH, cation exchange capacity and immobilization of toxic heavy metals (Rajapaksha et al. 2015). Biochar possess some important characteristics such as large surface area, presence of micropores, variety of associated functional groups and high pH (Chen and Lin 2001) which make them suitable candidate for their use as efficient biosorbent to immobilize the heavy metals (Ahmad et al. 2014). Uchimiya et al. (2011a, b) has reported the efficient stabilization of copper and lead in acidic soil by surface functional groups assisted ion exchange present in biochar. Contrary to this, reductions in exchangeable aluminium metals after poultry manure based biochar soil application have also been reported by Chan et al. (2008). Both electrostatic and non-electrostatic interactions are reported to be involved in lead sorption, however, non-electrostatic forces were more commonly involved one. Biochar possesses the capability of enhancing the nonelectrostatic interactions with metals and thus restricting their bioavailability to plants (Jiang et al. 2012). Another mechanism of metal immobilization in soil was ascribed by the complexation of metals with phosphate present in biochar material to produce metal-phosphate precipitate (Cao et al. 2009).

Zhang et al. (2017) tested the affinity of sludge derived biochar towards the metals like lead and zinc. They found higher lead adsorption as compared to zinc but the co-presence of lead minimized the zinc adsorption. Process of metal adsorption onto sludge derived biochar was temperature dependent. The mechanism of sorption as revealed by photoelectron spectroscopy was metal-phosphate complex formation. Recently Qian et al. (2016) reported the mechanism of zinc sorption on pine needle and wheat straw derived biochar prepared at two different temperature. Higher temperature produced biochar were more efficient than low temperature pyrolyzed biomass. They observed the metal precipitation on the surface of biochar which was mainly due to the surface functional groups such as hydroxyl, carbonate and silicate ions. The zinc complexes formed on the biochar surfaces were zinc hydroxide, hydrozinchite and hemimorphite. Studies are also available regarding the role of faecal matter and cow dung derived biochar in cadmium phytoavailability (Woldetsadik et al. 2016). Faecal matter derived biochar proved superior to cow dung derived biochar and lime for effective immobilization of cadmium in soil.

8.5 Role of Biochar on Crop Productivity

Biochar application to soil has potential role in crop improvement, soil fertility and thus crop productivity. Liu et al. (2013) conducted the experiment for testifying the biochar effect on crop productivity under different agricultural conditions. Good responses were recorded for pot experiments, acidic soil and sandy textured soil as compared to field experiment, neutral and loam and silt soil, respectively. Dry land plants were more responsive as compared to paddy. Biochar produced at high pyrolyzing temperature were proved more effective in crop productivity due to liming effect. Uzoma et al. (2011) carried out experiments regarding the effect of cow dung derived manure to evaluate the effect on maize yield. Biochar mixing was performed with sandy soil at different rates to see the effect on test crop in the study. Marked improvement in maize productivity as well as nutrient uptake was noticed after increase in application rate of biochar. Increase in net water use efficiency of maize was also noticed due to rise in hydraulic conductivity. Soil quality was also improved after addition of biochar. Deb et al. (2016) investigated their study under field conditions with varying nutrient status. They found that soil nutrient status is more important determining factor than the parent originating material of biochar and crop type. The effect of biochar was found more effective for nutrient (phosphorus) deficient soil resulting into the higher crop productivity. Effect of biochar together with the phosphate solubilizing microbes indicated that there was significant increase in crop productivity for phosphate deficient soils as compared to phosphate enriched soil. Jeffery et al. (2011) conducted a meta-analysis to see the impact of biochar on productivity; significant but small increase in productivity was found due to liming effect of biochar application and enhancement in soil water holding capacity. Poultry litter based biochar formulation was more effective as compared to biosolids which had negative impact on crop productivity. Khan et al. (2013) suggested the sewage sludge derived biochar for effective management of rice cultivation in acidic soil. Sewage sludge based biochar application is attractive in the sense of wider availability but the main limitation comes out in the form of higher background concentration of heavy metal/metalloids. Using sewage sludge as a potential source of biochar may reduce the bioavailability of toxic metals and metalloids to soil and plant ecosystem. There was increase in parameters such as soil pH, total nitrogen, and soil organic carbon; significant reductions in availability of metals like arsenic, nickel, chromium, cobalt and lead but not zinc, copper and cadmium. Effect of plant salt stress can be alleviated by application of higher doses of biochar (Akhtar et al. 2015; Kim et al. 2016). Reductions in salt-induced mortality were recorded for plants. Rice hull originated biochar was applied to reclaimed tidal land soil containing higher level of soluble salts and exchangeable sodium ions to see the responses on maize crop. Biochar application caused increase in soil organic carbon and water stable aggregate while decrease was observed for exchangeable percent of sodium ions. Biochar added soil was found to contain higher content of phosphate which promoted the maize yield. The salt effect alleviated after biochar application to soil was mainly due to the reduced uptake of sodium ions and higher accumulation of potassium ion which competed with former for inside cellular transport. Akhtar et al. (2015) conducted pot experiment in controlled greenhouse to see the responses of potato crop after biochar amendment to soil. Soil irrigated with three different salt (NaCl) concentrations under biochar addition at the rate of 0% and 5% were studied for potato crop responses. Biochar soil amendment had ameliorating effect on salt stress which was mainly due to sodium adsorption by biochar. Under different soil salinity conditions, increment were recorded in important parameters like photosynthetic rate, stomatal conductance, tuber yield, and shoot biomass after biochar soil application as compared to control. Ameliorating effect of biochar was also described as a function of high content of cellular potassium ion storage, decreased sodium ions and sodium potassium ratio. Thus biochar application may be suggested as an effective practice for management of usar saline soil for sustainable agriculture.

8.6 Role of Biochar in Disease Management

Biochar application to soil has emerged as an emerging technology for the effective management of disease. Elmer and Pignatello (2011) conducted experiments to understand the effect of biochar on *Fusarium* mediated crown and root rot disease in *Asparagus*. Biochar addition to soil caused reduction in root lesions induced by *Fusarium* as compared to control. The reduction in disease may be associated with reduced iron availability to pathogen. Reduced nutrient availability to pathogen after biochar addition has opened the new ways to develop disease resistance in crop plants. Elad et al. (2010) have demonstrated the beneficial effect of biochar in terms of disease resistance. Induction in disease resistance against fungal pathogen *Botrytis cinerea* (gray mold), *Leveillula taurica* (powdery mildew) attacking on pepper and tomato and pest (*Polyphagotarsonemus latus*) infesting on pepper were seen after biochar addition to soil. The effective biochar dose for disease resistance was found in the range of 1–5%. Long term study has revealed the significant reductions in severity of disease after biochar addition to soil, however, the final 25 days results indicate no differences in disease severity for biochar treated and untreated soil. The suppressed activity of pathogen may arise by the variety of mechanisms like (i) activation of beneficial microbial communities exhibiting the property of competition, antibiosis, parasitism (ii) plant growth promotion by enhancing solubility and availability of nutrients to plants and increased soil water holding capacity; and (iii) biochar induced activation elicitors involved plant defense mechanisms. Harel et al. (2012) explored the possibility of using biochar for disease management practice. They elaborated the impact of biochar derived from citrus wood and greenhouse waste against fungal pathogen *Botrytis cinerea, Colletotrichum acutatum* and *Podosphaera apahanis*. Tests were performed with 1–3% biochar amended to potting mixture. Biochar addition suppressed the growth of fungi employing different infection strategy which was also supported by expression of defense related genes. Plants grown with 1% greenhouse waste for 25 days had significant reduction in disease severity as compared to no reduction observed in those grown under same condition for 10 days only. Increase in dose of greenhouse waste biochar (3%) resulted into similar effects on mature and young plants grown for 15 and 10 days only. Reductions in disease severity were also supported by expression of defense related genes such as FaPR1, Faolp2, Fra a3, Falox and FaWRKY1 in disease free strawberry plants providing direct evidence of disease resistance mechanism by biochar addition. Zwart and Kim (2012) evaluated the effect of biochar on *Phytophthora* induced disease in tree species *Quercus rubra* (L.) and *Acer rubrum* (L.). Biochar amendment to potting media caused alleviation in horizontal expansion of lesions in tree seedlings of *Quercus rubra* and *Acer rubrum*, while under similar conditions reductions in vertical expansion of disease were recorded for *Acer rubrum* only. Application of 5% biochar produced enhancement effect in stem water potential of *Quercus rubra* and higher biomass in *Acer rubrum* as compared to control plants. George et al. (2016) have investigated the effect of biochar on resistance enhancement in carrot against the nematode mediated root lesion in carrot. They hypothesized the biochar soil amendment as potent nematicide for enhancing resistance in host plant. Biochar of different types and zeolites (5%) were applied to analyze the effect on carrot biomass under root infection caused by nematode. Approximately 80% reductions in nematode mediated root lesions were found after application of all biochar except pine wood derived biochar, however, infection reduction upto 96% was recorded after spelt husk biochar application. Biochar application to infected plants produced two to four times higher biomass as compared to infected plants not supplemented with biochar. It was ascribed that biochar produced from different sources exhibited differential effect on pathogen infection rate and host biomass. Reports are available which presented the withdrawal of many nematicides due to risks associated to human health and environment. If biochar application to soil system could develop as an effective control measurement technique toward disease management, it may substitute/minimize the large scale application of synthetic chemicals to safeguard our soil natural environment. Lu et al. (2016) investigated and verified the positive effect of biochar on tomato bacterial wilt disease. Biochar derived from peanut shell and wheat straw was applied to soil system to investigate the consequences on soil microbial characteristics and tomato bacterial wilt disease. Both biochar amendments proved effective in wilt disease suppression from 28.6% to 65.7%. Disease development in pathogen infested plants was delayed after biochar addition to soil. Significant reductions in soil population of *Ralstonia solanacearum* were recorded after biochar application to soil. Bacterial population of *Ralstonia solanaceraum* negatively impacts the soil system by altering the beneficial bacterial, and actinomycetes community. Improvement in soil population of bacteria and actinomycetes were observed after biochar addition to soil harboring the wilt disease pathogen *Ralstonia solanaceraum*. Biochar application tended to increase the activity of important soil enzymes like urease and neutral phosphatase.

8.7 Role of Biochar in Development of Slow Release Fertilizer

Very large amount of applied fertilizers are washed out or made unavailable to plants. Surface adsorption-complexation of nutrients to biochar has emerged as a novel technology for development of slow release fertilizer so that loss of fertilizers during field application can be minimized. Manikandan and Subramaniam (2013) has tested the urea intermixed biochar (pyrolysis at 350 °C) derived from *Prosopis juliflora* wood for development of slow release nitrogen fertilizer. Different forms of nitrogen were applied in the range of 20–200 mM to biochar and tested their release pattern. Among the nitrogenous substrate, 1:1 ratio biochar mixed with urea depicted the best suited option to be formulated and developed as slow release fertilizer. Cai et al. (2016) studied the ammonium ion sorption and desorption behavior of agriculturally derived waste material. Agriculture crop (maize, pomelo peel, and banana) waste derived biochar had the ability of long term retention of more than 90% nutrient material (ammonium ion) so that it can be proposed for the future generation slow release nitrogen fertilizer. The main functional groups of crop residue waste derived biochar associated with ammonium ion sorption were carboxyl and keto groups. Low temperature generated biochar possessed higher ammonium ion adsorption efficiency as compared to high temperature pyrolyzed biomass due to abundance of oxygen bearing functional groups. Esfandboda et al. (2017)

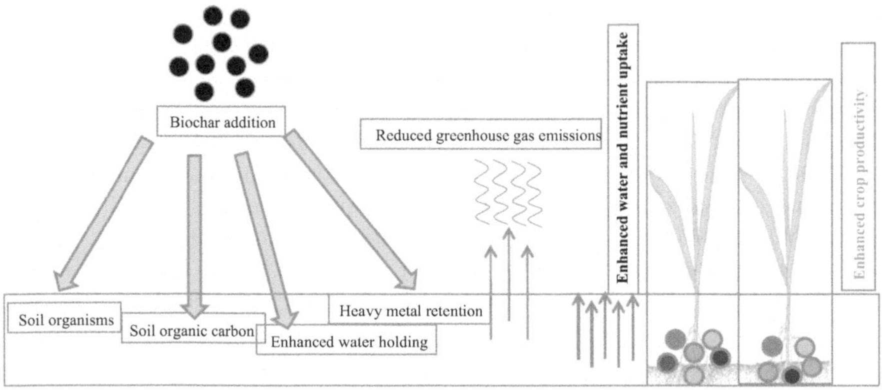

Fig. 8.2 Multifaceted advantages of biochar for sustainable agriculture

evaluated the effect of acidic and alkaline biochar amendment on ammonium ion retention ability. The acidic biochar treated alkaline soil had low leaching via volatilization for nitrogen. The higher retention ability could be attributed by the good presence of oxygen containing functional groups. The treatment technology bears innovation in the sense of restoration and remediation of alkaline soil which generally does not favour the growth of plants.

The multiple benefits of biochar in soil ecosystem can be represented by following Fig. 8.2.

8.8 Role of Biochar in Carbon Sequestration

Increase in atmospheric greenhouse gas concentration is one of the important factors responsible for global warming. Human interferences in carbon cycling by excessively using fossil fuel are one of the main reasons of global warming. The basic strategy for controlling the greenhouse gas relies on preventing atmospheric emission and storing it in suitable geological environments. However, the first option for controlling greenhouse gas remained impossible due to very high energy demand in developing countries. Keeping in view for controlling the greenhouse gas emission, biochar application to soil system may serve as a good option in near future for sustainable agricultural practices.

Carbon sequestration is the phenomena of long term carbon pool storage to mitigate the effects of global warming. One of the most important effects of biochar application to soil agroecosystem is enhancement of soil carbon content. The biochar application to soil together with bioenergy crop production efficiently stores more carbon than is emitted in the atmosphere (Roberts et al. 2010). According to Lehmann (2007), soil amendment with biochar increases more than 20% atmospheric carbon sequestration.

Biochar is considered to be stable in the soil for more than 1000 years (Zimmerman 2010; Ahmad et al. 2014) and this is the basic criteria for selecting the biochar for carbon sequestration purpose. Biochar is highly resistant to chemical decomposition even when subjected to strong weathering conditions prevailing in tropical climate (Schneider et al. 2011).

There are several reports suggesting the reductions in greenhouse gas emission after biochar application to soil (Castaldi et al. 2011; Mao et al. 2012). But, the effect of biochar on different greenhouse gases such as carbon-di-oxide, methane and nitrous oxide varies significantly. Feng et al. (2012) reported decreased rate of methane production from paddy field after application of biochar to soil. The reductions were noticed due to significant changes in ratio of methanogenic to methanotrophic community abundance (Feng et al. 2012). Results are also available indicating no changes in methane flux (Castaldi et al. 2011). On contrary, the increase in total methane emission after biochar application is also reported (Zhang et al. 2010), which could be ascribed by increased activity of methanotrophs (Spokas et al. 2010). The another possibility of increased methane emission may be expected by decomposition of biochar-sorbed hydrocarbon compounds which may act as a substrate and thus inducing methane generation by reducing methane oxidation process (Spokas and Reicosky 2009).

Biochar application has sometimes reported to result into the decreased nitrous oxide emission under laboratory incubation experiments (Case et al. 2012, 2014; Smith et al. 2010). This could be possible by effect of biochar on microbial community involved in nitrogen fixation. In the experiment conducted by Castaldi et al. (2011), biochar treated plots were found to exhibit significant reductions in nitrous oxide flux ranging from 26% to 79% as compared to control plots. In an experimental study conducted by Rondon et al. (2007), fixed nitrogen content was enhanced from 50% without biochar application to 72% with added biochar simultaneously with reduced N_2O flux. High nitrogen containing biochar were reported to induce nitrous oxide emission (Spokas and Reicosky 2009; Van Zwieten et al. 2010). Biochar application to soil improves the nitrogen incorporation into soil system very efficiently (Clough and Condron 2010) by affecting the process of nitrogen cycling, ammonia adsorption and storage (Spokas et al. 2012).

8.9 Risks Associated with Biochar Application

Despite of certain advantages linked with biochar application certain disadvantages may not be avoided. The rising negative points associated with biochar application may be described in terms of toxic heavy metal source, suppression of activity of applied herbicides and pesticides, and toxic impact on soil microbial population.

Biochar may possess higher amount of polycyclic aromatic hydrocarbons, volatile organic compounds, some toxic metals like cadmium, copper, chromium, zinc and nickel which can strongly affect the vital cellular physiological processes. Toxic polycyclic aromatic hydrocarbons generated after incomplete combustion of feedstock materials are often recalcitrant in nature and their toxicity vary according to pyrolysis temperature. Reports are available which demonstrated the higher toxicity effect in maize after application of biochar generated at higher pyrolysis temperature as compared to low temperature derived biochar (Busch et al. 2012). The toxicity was not found after application of biochar derived at low temperature pyrolysis. High temperature pyrolysis gave rise to polycyclic aromatic hydrocarbons, naphthalene which was responsible for suppressed root and shoot length.

Few studies have reported the enhanced (more than 30 times) bioavailability of copper and arsenic simultaneously with increases in soil organic carbon and pH after biochar soil application. The bioavailability is not always increased for every toxic metal after biochar application to soil. Biochar amended soil may also possess the property of very much reduced microbial biodegradation. Due to surface binding property biochar addition to soil may adsorb applied agrochemicals such as pesticides which may make them unavailable or minimally available to plants. Studies have also been conducted elucidating the effect of biochar variability, dose rate, particle size distribution and time after application on the degradation, leaching and adsorption of herbicide simazine (Jones et al. 2011). Differences were also observed for freshly prepared and aged biochars on the degradation, solubility, transportation, bioavailability, soil distribution of herbicide with the resultant effect of reduced herbicide biodegradation. However, the reduced herbicide transport, degradation and availability to plants can be managed by using large particle size biochar.

Few compounds present in biochar such as polycyclic aromatic hydrocarbons, formaldehyde, cresols or xylenols may pose negative effect to soil microbial system by acting as bactericide or fungicide. The content of these toxic substances vary according to pyrolysis temperature of feedstock. Both positive and negative reports are available regarding the impact of biochar on soil earthworm population. According to a study carried out by Wen et al. (2009) there was reduction in the bioavailability of pentachlorphenol to earthworm in biochar amended soil as compared to control soil. Contrary to results of Tammeorg et al. (2014), biochar application caused the fall in soil water potential which was responsible for biochar avoidance by earthworm. The application of biochar in plant soil system must be implemented in such a way so that enhanced crop productivity can be attained. In a number of studies, higher dose of biochar have been proved to impose negative effect on soil nutrient status due to reduction in mineralizable soil organic carbon pool. Another factor responsible for reduced crop productivity may be linked with higher ash content of biochar providing higher salt to soil.

Sufficient numbers of documents are available supporting the reduced emission of greenhouse gases such as methane, nitrous oxide and carbon dioxide. However, contradictory reports of increased greenhouse gas emissions after biochar soil application are also published. Liu et al. (2014) demonstrated the increased emission of nitrous oxide after biochar amendment to rice crop system probably due to enhancement in soil soluble organic carbon and ammonium ion. Song et al. (2016) suggested that greenhouse gas emission from soil system after biochar application

Table 8.3 Summary of the benefits and risks associated with biochar

Benefits associated with biochar application	Probable risks associated with biochar application
1. Increase in overall soil quality	1. May act as source of heavy metals
2. Reduced bioavailability of toxic metals	2. Suppressed activity of applied herbicides
3. Reduced greenhouse gas emission	3. Changes in natural microbial community
4. Increase in overall crop productivity	4. Differential behavior under field condition
5. Efficient and enhanced carbon sequestration	5. Low availability of applied nutrients
6. Slow release fertilizer after nutrient addition	6. Reduced degradation of soil contaminant
7. Increased resistance against diseases	7. Sometimes phytotoxic nature

depends upon (i) field and laboratory conditions, (ii) upland and lowland cultivation practice, (iii) longevity of experimental set-up, (iv) biochar feedstock parent material, and (v) temperature used for biochar production.

Both advantages and disadvantages are associated with the biochar application to soil. Few of the important points are presented in Table 8.3.

8.10 Conclusion

Biochar amendment to soil has attracted the scientists from worldwide. Biochar application to soil has promising future in soil management and thus agricultural sustainability for human livelihood. Biochar may serve as a very good option for managing crop residue in ecofriendly way. Most of the studied experiments have supported the soil application of biochar but the negative reports on soil quality, soil microbial population, greenhouse gas emission, and crop productivity cannot be ignored. The effect of biochar also varies according to its dose. Before biochar application to soil crop system, thorough study of its probable impact should be carried out. Most of the reports are in the view of its application throughout the world but few negative impacts have also been reported thus limiting its wide applicability. Biochar may act as good substitute for synthetic chemicals due to its soil nutrient enhancement ability, soil water holding capacity, activation of disease resistance, induction of beneficial microbes responsible for enhanced crop productivity. Biochar act as efficient adsorbant for various toxic metals, metalloids and agrochemicals such as herbicides and pesticides thus reducing their bioavailability to plants and minimizing the toxicity to human and soil system. Further, their ability to sequester larger concentrations of carbon dioxide from the atmosphere has opened the new way for controlling global warming effect. However, still more field research is needed before its full application in soil-crop system because many times field and laboratory experiments vary very much. Future research on development of slow release fertilizer by using biochar is also one of the most challenging research areas for sustainable agriculture.

Acknowledgement The author is very much thankful to Head, Department of Botany. Thanks and welcomes are also to Eric Lichtfouse for allowing us to write the chapter.

References

Ahmad M, Lee SS, Lim JE, Lee SE, Cho JS, Moon DH, Hashimoto Y, Ok YS (2014) Speciation and phytoavailability of lead and antimony in a small arms range soil amended with mussel shell, cow bone and biochar: EXAFS spectroscopy and chemical extractions. Chemosphere 95:433–441

Akhtar SS, Andersen MN, Liu F (2015) Biochar mitigates salinity stress in potato. J Agron Crop Sci 201:368–378

Arias RN, Fabra PA (1993) Effects of 2,4-dichlorophenoxyacetic acid on *Rhizobium* sp. growth and characterization of its transport. Toxicol Lett 68:267–273

Barrow CJ (2012) Biochar: potential for countering land degradation and for improving agriculture. Appl Geogr 34:21–28

Beesley L, Moreno-Jimenez E, Gomez-Eyles JL (2010) Effects of biochar and greenwaste compost amendments on mobility, bioavailability and toxicity of inorganic and organic contaminants in a multi-element polluted soil. Environ Pollut 158:2282–2287

Brewer CE, Schmidt-Rohr K, Satrio JA, Brown RC (2009) Characterization of biochar from fast pyrolysis and gasification systems. Environ Prog Sust Ener 28:386–396

Bruun EW, Hauggaard-Nielsen H, Norazana I, Egsgaard H, Ambus P, Jensen PA, Johansen KD (2011) Influence of fast pyrolysis temperature on biochar labile fraction and shortterm carbon loss in a loamy soil. Biomass Bioenergy 35:1182–1189

Bruun EW, Ambus P, Egsgaard H, Hauggaard-Nielsen H (2012) Effects of slow and fast pyrolysis biochar on soil C and N turnover dynamics. Soil Biol Biochem 46:73–79

Busch D, Kammann C, Grünhage L, Müller C (2012) Simple biotoxicity tests for evaluation of carbonaceous soil additives: establishment and reproducibility of four test procedures. J Environ Qual 41:1023–1032

Cai Y, Qi H, Liu Y, He X (2016) Sorption/desorption behavior and mechanism of NH_4^+ by biochar as a nitrogen fertilizer sustained-release material. J Agric Food Chem 64:4958–4964

Cao XD, Ma L, Gao B, Harris W (2009) Dairy-manure derived biochar effectively sorbs lead and atrazine. Environ Sci Technol 43:3285–3291

Case SD, McNamara NP, Reay DS, Whitaker J (2012) The effect of biochar addition on N2O and CO2 emissions from a sandy loam soil-the role of soil aeration. Soil Biol Biochem 51:125–134

Case SDC, McNamara NP, Reay DS, Whitaker J (2014) Can biochar reduce soil greenhouse gas emissions from a Miscanthus bioenergy crop. GCB Bioenergy 6:76–89

Castaldi S, Riondino M, Baronti S, Esposito FR, Marzaioli R, Rutigliano FA, Vaccari FP, Miglietta F (2011) Impact of biochar application to a Mediterranean wheat crop on soil microbial activity and greenhouse gas fluxes. Chemosphere 85:1464–1471

Chan KY, Van ZL, Meszaros I, Downie A, Joseph S (2007) Agronomic values of greenwaste biochar as asoil amendment. Aust J Soil Res 45:629–634

Chan KY, Van Zwieten L, Meszaros I, Downie A, Joseph S (2008) Using poultry litter biochars as soil amendments. Aust J Soil Res 46:437–444

Chen B, Chen Z (2009) Sorption of naphthalene and 1-naphthol by biochars of orange peels with different pyrolytic temperatures. Chemosphere 76:127–133

Chen JP, Lin M (2001) Equilibrium and kinetics of metal ion adsorption onto a commercial H-type granular activated carbon: experimental and modeling studies. Water Res 35:2385–2394

Chen B, Chen Z, Shaofang LV (2011) A novel magnetic biochar efficiently sorbs organic pollutants and phosphate. Bioresour Technol 102:716–723

Cheng CH, Lehmann J (2009) Ageing of black carbon along a temperature gradient. Chemosphere 75:1021–1027

Cheng CH, Lehmann J, Engelhard M (2008) Natural oxidation of black carbon in soils: changes in molecular form and surface charge along a climosequence. Geochim Cosmochim Acta 72:1598–1610

Cheng Y, Cai ZC, Chang SX, Wang J, Zhang J (2012) Wheat straw and its biochar have contrasting effects on inorganic N retention and N_2O production in a cultivated black Chernozem. Biol Fertil Soils 48:941–946

Clough TJ, Condron LM (2010) Biochar and the nitrogen cycle: introduction. J Environ Qual 39:1218–1223

Crutzen PJ, Andreae MO (1990) Biomass burning in the tropics: impact on atmospheric chemical and biogeochemical cycles. Science 250:1669–1678

Deb D, Kloft M, Lässig J, Walsh S (2016) Variable effects of biochar and P solubilizing microbes on crop productivity in different soil conditions. Agroecol Sustain Food Syst 40:145–168

Dias BO, Silva CA, Higashikawa FS, Roig A, Sánchez-Monedero MA (2010) Use of biochar as bulking agent for the composting of poultry manure: effect on organic matter degradation and humification. Bioresour Technol 101:1239–1246

Ding Y, Liu YX, Wu WX, Shi DZ, Yang M, Zhong ZK (2010) Evaluation of biochar effects on nitrogen retention and leaching in multi-layered soil columns. Water Air Soil Pollut 213:47–55

Downie A, Crosky A, Munroe P (2009) Physical properties of biochar. In: Lehmann J, Joseph S (eds) Biochar for environmental management: science and technology. Earthscan, London, pp 13–32

Elad Y, David DR, Harel YM, Borenshtein M, Ben Kalifa H, Silber A, Graber ER (2010) Induction of systemic resistance in plants by biochar, a soil applied carbon sequestering agent. Phytopathology 100:913–921

Elme WH, Pignatello JJ (2011) Effect of biochar amendments on mycorrhizal associations and *Fusarium* crown and root rot of *Asparagus* in replant soils. Plant Dis 95:960–966

Enders A, Hanley K, Whitman T, Joseph S, Lehmann J (2012) Characterization of biochars to evaluate recalcitrance and agronomic performance. Bioresour Technol 114:644–653

Esfandboda M, Phillips IR, Miller B, Rashti MR, Lana ZM, Srivastava P, Singh B, Chen CR (2017) Aged acidic biochar increases nitrogen retention and decreases ammonia volatilization in alkaline bauxite residue sand. Ecol Eng 98:157–165

Fabra A, Duffard R, Evangelista DDA (1997) Toxicity of 2,4-dichlorophenoxyacetic acid in pure culture. Bull Environ Contam Toxicol 59:645–652

Feng Y, Xu Y, Yu Y, Xie Z, Lin X (2012) Mechanisms of biochar decreasing methane emission from Chinese paddy soils. Soil Biol Biochem 46:80–88

George C, Kohler J, Rillig MC (2016) Biochars reduce infection rates of the root-lesion nematode Pratylenchus penetrans and associated biomass loss in carrot. Soil Biol Biochem 95:11–18

Harel YM, Elad Y, Rav-David D, Borenstein M, Shulchani R, Lew B, Graber ER (2012) Biochar mediates systemic response of strawberry to foliar fungal pathogens. Plant Soil 357:245–257

Hossain MK, Strezov V, Chan KY, Ziolkowski A, Nelson PF (2011) Influence of pyrolysis temperature on production and nutrient properties of wastewater sludge biochar. J Environ Manag 92:223–228

Hunter WH, Gan J, Kookana RS (2010) Bioavailability of hydrophobic organic contaminants in soils and sediments. In: Huang PM, Senesi N (eds) Biophysico-chemical processes of anthropogenic organic compounds in environmental systems. Wiley, Hoboken

Inyang M, Dickenson E (2015) The potential role of biochar in the removal of organic and microbial contaminants from potable and reuse water: a review. Chemosphere 134:232–240

Inyang M, Gao B, Pullammanappallil P, Ding W, Zimmerman AR (2010) Biochar from anaerobically digested sugarcane bagasse. Bioresour Technol 101:8868–8872

Jiang TY, Jiang J, Xu RK, Li Z (2012) Adsorption of Pb (II) on variable charge soils amended with rice-straw derived biochar. Chemosphere 89:249–256

Jeffery S, Verheijen FGA, van der Velde M, Bastos AC (2011) A quantitative review of the effects of biochar application to soils on crop productivity using meta-analysis. Agric Ecosyst Environ 144:175–187

Jones DL, Edwards-Jones G, Murphy DV (2011) Biochar mediated alterations in herbicide breakdown and leaching in soil. Soil Biol Biochem 43:804–813

Junna S, Bingchen W, Gang X, Hongbo S (2014) Effects of wheat straw biochar on carbon mineralization and guidance for large-scale soil quality improvement in the coastal wetland. Ecol Eng 62:43–47

Karunanithi R, Szogi A, Bolan N, Naidu R, Loganathan P, Hunt PG, Vanotti MB, Saint CP, Ok YS, Krishnamoorthy S (2015) Phosphorus recovery and reuse from waste streams. Adv Agron 131:173–250

Keiluweit M, Nico PS, Johnson MG, Kleber (2010) Dynamic molecular structure of plant-derived black carbon (biochar). Environ Sci Technol 44:1247–1253

Khan S, Wang N, Reid BJ, Freddo A, Cai C (2013) Reduced bioaccumulation of PAHs by Lactuca satuva L. grown in contaminated soil amended with sewage sludge and sewage sludge derived biochar. Environ Pollut 175:64–68

Khanmohammadi Z, Afyuni M, Mosaddeghi MR (2015) Effect of pyrolysis temperature on chemical and physical properties of sewage sludge biochar. Waste Manage Res 33:275–283

Kim HS, Kim KR, Kim HJ, Yoon JH, Yang J, Ok Y, Owens G, Kim KH (2015) Effect of biochar on heavy metal immobilization and uptake by lettuce (*Lactuca sativa* L.) in agricultural soil. Environ Earth Sci 74:1249–1259

Kim HS, Kim KR, Yang JE, Ok YS, Owens G, Nehls T, Wessolek G, Kim KH (2016) Effect of biochar on reclaimed tidal land soil properties and maize (*Zea mays* L.) response. Chemosphere 142:153–159

Lal R (2008) Carbon sequestration. Philos Trans Royal Soc B 363:815–830

Lehmann J (2007) A handful of carbon. Nature 447:143–144

Lehmann J, Joseph S (2009) Biochar for environmental management: an introduction. In: Lehmann J, Joseph S (eds) Biochar for environmental management science and technology. Earthscans, London, pp 1–12

Lehmann J, da Silva JP, Steiner C, Nehl T, Zech W, Glaser B (2003) Nutrient availability and leaching in an archaeological Anthrosol and a Ferralsol of the Central Amazon basin: fertilizer, manure and charcoal amendments. Plant Soil 249:343–357

Lehmann J, Liang B, Solomon D, Lerotic M, Luizão F, Kinyangi F, Schäfer T, Wirick S, Jacobsen C (2005) Near-edge X-ray absorption fine structure (NEXAFS) spectroscopy for mapping nano-scale distribution of organic carbon forms in soil: application to black carbon particles. Global Biogeochem Cycles 19:GB1013

Lehmann J, Czimczik C, Laird D, Sohi S (2009) Stability of biochar in soil. In: Lehmann J, Joseph S (eds) Biochar for environmental management: Science and technology. Earthscan, London, pp 183–205

Lehmann J, Rillig M, Thies J, Masiello C, Hockaday WC, Crowle D (2011) Biochar effects on soil biota–a review. Soil Biol Biochem 43:1812–1836

Li J, Lv G, Bai W, Liu Q, Zhang Y, Song J (2016) Modification and use of biochar from wheat straw (Triticum aestivum L.) for nitrate and phosphate removal from water. Desalin Water Treat 57:4681–4693

Liang B, Lehmann J, Sohi SP, Thies JE, O'Neill B, Trujillo L, Gaunt J, Solomon D, Grossman J, Neves EG, Luizão FJ (2010) Black carbon affects the cycling of non-black carbon in soil. Org Geochem 41:206–213

Liu X, Zhang A, Ji C, Joseph S, Bian R, Li L, Pan G, Paz-Ferreiro J (2013) Biochar's effect on crop productivity and the dependence on experimental conditions-a meta-analysis of literature data. Plant Soil 373:583–594

Liu X, Ye Y, Liu Y, Zhang A, Zhang X, Li L, Pan G, Kibue GW, Zheng J, Zheng J (2014) Sustainable biochar effects for low carbon crop production: a 5-crop season field experiment on a low fertility soil from Central China. Agric Syst 129:22–29

Lu Y, Rao S, Huang F, Cai Y, Wang G, Cai K (2016) Effects of biochar amendment on tomato bacterial wilt resistance and soil microbial amount and activity. Int J Agron 2016:Article ID 2938282

Lu K, Yang X, Gielen G, Bolan N, Ok YS, Niazi NK, Xu S, Yuan G, Chen X, Zhang X, Liu D, Song Z, Liu X, Wang H (2017) Effect of bamboo and rice straw biochars on the mobility and redistribution of heavy metals (Cd, Cu, Pb and Zn) in contaminated soil. J Environ Manag 186:285–292

Major J, Lehmann J, Rondon M, Goodale C (2010) Fate of soil-applied black carbon: downward migration, leaching and soil respiration. Global Chang Biol 16:1366–1379

Manikandan A, Subramanian KS (2013) Urea intercalated biochar–a slow release fertilizer production and characterization. Indian J Sci Technol 6:5579–5584

Mao JD, Johnson RL, Lehmann J, Olk DC, Neves EG, Thompson ML, Schmidt-Rohr K (2012) Abundant and stable char residues in soils: implications for soil fertility and carbon sequestration. Environ Sci Technol 46:9571–9576

McLeod PB, Van Den Heuvel-Greve MJ, Allen-King RM, Luoma SN, Luthy RG (2004) Effects of particulate carbonaceous matter on the bioavailability of benzo[a]pyrene and 2,20,5,50-tetrachlorobiphenyl to the clam, *Macoma balthica*. Environ Sci Technol 38:4549–4556

McLeod PB, Van Den Heuvel-Greve MJ, Luoma SN, Luthy RG (2007) Biological uptake of polychlorinated biphenyls by *Macoma balthica* from sediment amended with activated carbon. Environ Toxicol Chem 26:980–987

Millward RN, Bridges TS, Ghosh U, Zimmerman JR, Luthy RG (2005) Addition of activated carbon to sediments to reduce PCB bioaccumulation by a polychaete (*Neanthes arenaceodentata*) and an amphipod (*Leptocheirus plumulosus*). Environ Sci Technol 39:2880–2887

Mohan D, Sarswat A, Ok YS, Pittman CU (2014) Organic and inorganic contaminants removal from water with biochar, a renewable, low cost and sustainable adsorbent-a critical review. Bioresour Technol 160:191–202

Nguyen B, Lehmann J, Hockaday WC, Joseph S, Masiello CA (2010) Temperature sensitivity of black carbon decomposition and oxidation. Environ Sci Technol 44:3324–3331

Park JH, Cho JS, Ok YS, Kim SH, Kang SW, Choi IW, Heo JS, Delaunee RD, Seo DC (2015a) Competitive adsorption and selectivity sequence of heavy metals by chicken bone-derived biochar: batch and column experiment. J Environ Sci Health 50:1194–1204

Park JH, Ok YS, Kim SH, Cho JS, Heo JS, Delaune RD, Seo DC (2015b) Evaluation of phosphorus adsorption capacity of sesame straw biochar on aqueous solution: influence of activation methods and pyrolysis temperatures. Environ Geochem Health 37:969–983

Paz-Ferreiro J, Lu H, Fu S, Méndez A, Gascó G (2014) Use of phytoremediation and biochar to remediate heavy metal polluted soils: a review. Solid Earth 5:65–75

Pratiwi EPA, Shinogi Y (2016) Rice husk biochar application to paddy soil and its effects on soil physical properties, plant growth, and methane emission. Paddy Water Environ 14:521–532

Qian T, Wang Y, Fan T, Fang G, Zhou D (2016) A new insight into the immobilization mechanism of Zn on biochar: the role of anions dissolved from ash. Sci Rep 6:33630. https://doi.org/10.1038/srep33630

Rajapaksha AU, Ahmad M, Vithanage M, Kim KR, Chang JY, Lee SS, Ok YS (2015) The role of biochar, natural iron oxides, and nanomaterials as soil amendments for immobilizing metals in shooting range soil. Environ Geochem Health 37:931–942

Renner R (2007) Rethinking biochar. Environ Sci Technol 41:5932–5933

Roberts KG, Gloy BA, Joseph S, Scott NR, Lehmann J (2010) Life cycle assessment of biochar systems: estimating the energetic, economic, and climate change potential. Environ Sci Technol 44:827–833

Rondon MA, Lehmann J, Ramírez J, Hurtado M (2007) Biological nitrogen fixation by common beans (*Phaseolus vulgaris* L.) increases with biochar additions. Biol Fertil Soils 43:699–708

Santos A, Flores M (1995) Effects of glyphosate on nitrogen fixation of free-living heterotrophic bacteria. Lett Appl Microbiol 20:349–352

Sánchez ME, Lindao E, Margaleff D, Martínez O, Morán A (2009) Pyrolysis of agricultural residues from rape and sunflowers: production and characterization of bio-fuels and biochar soil management. J Analyt Appl Pyrolysis 85:142–144

Saquing JM, Yu YH, Chiu PC (2016) Wood-derived black carbon (biochar) as a microbial electron donor and acceptor. Environ Sci Technol Lett 3:62–66

Savonen C (1997) Soil microorganisms object of new OSU service, Good Fruit Grower

Schimmelpfennig S, Glaser B (2012) One step forward toward characterization: some important material properties to distinguish biochars. J Environ Qual 41:1001–1013. https://doi.org/10.2134/jeq2011.0146

Schneider MPW, Lehmann J, Schmidt MWI (2011) Charcoal quality does not change over a century in a tropical agro-ecosystem. Soil Biol Biochem 43:1992–1994

Shackley S, Carter S, Knowles T, Middelink E, Haefele S, Sohi S, Cross A, Haszeldine S (2012) Sustainable gasification-biochar systems? A case-study of rice-husk gasification in Cambodia, part 1: context, chemical properties, environmental and health and safety issues. Energy Policy 42:49–58

Singh R, Babu JN, Kumar R, Srivastava P, Singh P, Raghubanshi AS (2015) Multifaceted application of crop residue biochar as a tool for sustainable agriculture: an ecological perspective. Ecol Eng 77:324–347

Smernik RJ (2009) Biochar and sorption of organic compounds. In: Lehmann J, Joseph S (eds) Biochar for environmental management: science and technology. Earthscan, London

Smith JL, Collins HP, Bailey VL (2010) The effect of young biochar on soil respiration. Soil Biol Biochem 42:2345–2347

Song X, Pan G, Zhang C, Zhang L, Wang H (2016) Effects of biochar application on fluxes of three biogenic greenhouse gases: a meta-analysis. Ecosyst Health Sustain 2:1–13

Spokas KA, Reicosky DC (2009) Impacts of sixteen different biochars on soil greenhouse gas production. Ann Environ Sci 3:179–193

Spokas KA, Baker JM, Reicosky DC (2010) Ethylene: potential key for biochar amendment impacts. Plant Soil 333:443–452

Spokas KA, Cantrell KB, Novak JM, Archer DW, Ippolito JA, Collins HP, Boateng AA, Lima IM, Lamb MC, McAloon AJ, Lentz RD, Nichols KA (2011) Biochar: a synthesis of its agronomic impact beyond carbon sequestration. J Environ Qual 41:973–989

Spokas KA, Cantrell KB, Novak JM, Archer DW, Ippolito JA, Collins HP, Boateng AA, Lima IM, Lamb MC, McAloon AJ, Lentz RD (2012) Biochar: a synthesis of its agronomic impact beyond carbon sequestration. J Environ Qual 41:973–989

Srivastava P, Singh R, Tripathi S, Raghubanshi AS (2016) An urgent need for sustainable thinking in agriculture – an Indian scenario. Ecol Indic 67:611–622

Steiner C, Das KC, Garcia M, Förster B, Zech W (2008) Charcoal and smoke extract stimulate the soil microbial community in a highly weathered xanthic ferralsol. Pedobiologia 51:359–366

Taghizadeh-Toosi A (2011) Biochar incorporation into pasture soil suppresses in situ nitrous oxide emissions from ruminant urine patches. J Environ Qual 40:468–476

Tammeorg P, Parviainen T, Nuutinen V, Simojoki A, Vaara E, Helenius J (2014) Effects of biochar on earthworms in arable soil: avoidance test and field trial in boreal loamy sand. Agric Ecosyst Environ 191:150–157

Tatarkova V, Hiller E, Vaculi M (2013) Impact of wheat straw biochar addition to soil on the sorption, leaching, dissipation of the herbicide (4-chloro-2-methylphenoxy)acetic acid and the growth of sunflower (*Helianthus annuus* L.). Ecotoxicol Environ Saf 92:215–221

Tisdall JM, Oades JM (1982) Organic matter and water-stable aggregates in soils. J Soil Sci 33:141–163

Uchimiya M, Wartelle LH, Klasson KT, Fortier CA, Lima IM (2011a) Influence of pyrolysis temperature on biochar property and function as a heavy metal sorbent in soil. J Agric Food Chem 59:2501–2510

Uchimiya M, Chang S, Klasson KT (2011b) Screening biochars for heavy metal retention in soil: role of oxygen functional groups. J Hazard Mater 190:432–441

Uzoma KC, Inoue M, Andry H, Fujimaki H, Zahoor A, Nishihara E (2011) Effect of cow manure biochar on maize productivity under sandy soil condition. Soil Use Manag 27:205–212

Van Zwieten L, Kimber S, Morris S, Chan KY, Downie A, Rust J, Joseph S, Cowie A (2010) Effects of biochar from slow pyrolysis of papermill waste on agronomic performance and soil fertility. Plant Soil 327:235–246

Verheijen F, Jeffery S, Bastos AC, van der Velde M, Diafas I (2010) Biochar application to soils. A critical scientific review of effects on soil properties, processes and functions. European Commission, Italy

Wang P, Yin Y, Guo Y, Wang C (2015) Removal of chlorpyrifos from waste water by wheat straw-derived biochar synthesized through oxygen-limited method. RSC Adv 5:72572–72578

Warnock DD, Lehmann J, Kuyper TW, Rillig MC (2007) Mycorrhizal responses to biochar in soil – concepts and mechanisms. Plant Soil 300:9–20

Wen B, Li RJ, Zhang S, Shan XQ, Fang J, Xiao K, Khan SU (2009) Immobilization of pentachlorophenol in soil using carbonaceous material amendments. Environ Pollut 157:968–974

Woldetsadik D, Drechsel P, Keraita B, Marschner B, Itanna F, Gebrekidan H (2016) Effects of biochar and alkaline amendments on cadmium immobilization, selected nutrient and cadmium concentrations of lettuce (*Lactuca sativa*) in two contrasting soils. Springer Plus 5:397

Wu F, Jia Z, Wang S, Chang SX, Startsev A (2013) Contrasting effects of wheat straw and its biochar on greenhouse gas emissions and enzyme activities in a Chernozemic soil. Biol Fertil Soils 49:555–565

Yang YN, Sheng GY (2003) Enhanced pesticide sorption by soils containing particulate matter from crop residue burns. Environ Sci Technol 37:3635–3639

Yao Y, Gao B, Inyang M, Zimmerman AR, Cao X, Pullammanappallil P, Yang L (2011) Biochar derived from anaerobically digested sugar beet tailings: characterization and phosphate removal potential. Bioresour Technol 102:6273–6278

Ying ZW, Chun TJ (2014) Remediation of wheat-straw-biochar on petroleum-polluted soil. J Agric Res Environ 31:259–264

Yuan JH, Xu RK, Zhang H (2011) The forms of alkalis in the biochar produced from crop residues at different temperatures. Bioresour Technol 102:3488–3497

Zabaniotou A, Stavropoulos G, Skoulou V (2008) Activated carbon from olive kernels in a two-stage process: industrial improvement. Bioresour Technol 99:320–326

Zhang A, Cui L, Pan G, Li L, Hussain Q, Zhang X, Zheng J, Crowie D (2010) Effect of biochar amendment on yield and methane and nitrous oxide emissions from a rice paddy from tai Lake plain, China. Agric Ecosyst Environ 139:469–475

Zhang W, Huang X, Jia Y, Rees F, Daniel TCW, Qiu R, Wang H (2017) Metal immobilization by sludge-derived biochar: roles of mineral oxides and carbonized organic compartment. Environ Geochem Health 39:379–389. https://doi.org/10.1007/s10653-016-9851-z

Zimmerman AR (2010) Abiotic and microbial oxidation of laboratoryproduced black carbon (biochar). Environ Sci Technol 44:1295–1301

Zwart DC, Kim SH (2012) Biochar amendment increases resistance to stem lesions caused by *Phytophthora* spp. in tree seedlings. Hort Sci 47:1736–1740

Chapter 9
Soil Quality and Agricultural Sustainability in Semi-arid Areas

Msafiri Yusuph Mkonda and Xinhua He

Abstract Soil quality and agricultural sustainability are required to feed about nine billion people by the year 2050. To feed such a population, the planet ought to increase food production by 60%. To attain agricultural sustainability, there should be a balance among biophysical, economic and social dimensions under which soil quality is a core aspect. It is worthwhile to explore soil quality versus agricultural sustainability in sub-Saharan countries because the population is expected to increase by 80%. This chapter reviews the current agronomic practices in countries characterized by semiarid agro-ecological zones and their implications to soil quality and agricultural sustainability, using Tanzania as a case study.

We found that agro-pastoralism based on maize, sorghum, millet, sheep, cattle and cow is a current dominant agricultural system but with low yields. Monoculture has contributed to the degradation of soil quality. Drought has raised issues to already stressed ecosystems and made rain-fed agriculture a vulnerable and unsustainable livelihood for smallholder farmers. This situation has reduced the per capita grain harvested area from 0.6 to less than 0.4 ha and thus, affected for more than 70% the smallholder farmers' livelihoods. Fortunately, areas using fertilizations of animal manure and other organic soil management practices have increased soil fertility and crop yields from 0.82 tn ha^{-1} under no-fertilization to 1.8 tn ha^{-1} under organic fertilization.

Keywords Agricultural sustainability · Climate change · Ecology · Food security · Nutrient use efficiency · Organic fertilizations · Semiarid · Smallholder farmers · Soils quality · Tanzania

M. Y. Mkonda (✉)
Centre of Excellence for Soil Biology, College of Resources and Environment, Southwest University, Chongqing, China

Department of Geography and Environmental Studies, Solomon Mahlangu College of Science and Education, Sokoine University of Agriculture, Morogoro, Tanzania

X. He
Centre of Excellence for Soil Biology, College of Resources and Environment, Southwest University, Chongqing, China

School of Biological Sciences, University of Western Australia, Crawley, Australia
e-mail: xinhua.he@uwa.edu.au

E. Lichtfouse (ed.), *Sustainable Agriculture Reviews 32*, Sustainable Agriculture Reviews 32, https://doi.org/10.1007/978-3-319-98914-3_9

9.1 Introduction

Increasing needs for food due to global population has forced agricultural systems to be the main concern to addressing the problem (Vermeulen et al. 2012; and IPCC 2014). Global food demand has increased rapidly due to the fact that, the growth of cereal grain is 1% while that of population is 3% (FAO 2006, 2008; and 2012). This demands has been more rampant from the last two decade of twentieth century and the first decade of twenty-first century (Monfreda et al. 2008; and Branca et al. 2013). During this time, the per capita cereal produced has decreased from 150 to 130 kg per person in Africa while increasing in Asia and South America from 200 to 250 kg per person (FAO 2008, 2012; and Sieber et al. 2015). This has led to the growing demand of sustainable agriculture to attain optimal food security (Monfreda et al. 2008; and FAO 2012).

The pressure from increasing population, limited arable land and increasing scenarios of climate change have compelled experts to assess agricultural sustainability (Branca et al. 2013; Ahmed et al. 2011; and Rowhani et al. 2011a). The decline of agricultural productivity is evidenced by a number of scenarios such as frequent food insecurity, hunger and malnutrition cases (URT 2007, 2012; and FAO 2012). While the decline of per capita grain harvested areas at global level is the most alarming indicator of agricultural unsustainability in sub-Saharan Africa (UNEP 2011; and Poppy et al. 2014). The per capita harvested area declined from 0.23 ha in 1980s to 0.12 ha in early 2000s in the region (FAO 2006; UNEP 2012; and Branca et al. 2013). In addition, irrigated land has declined from 0.047 ha in 1980s to 0.044 in 2000s (URT 2007; FAO 2013; and Duru et al. 2015). Unfortunately, the Sub-Saharan Africa and more especially in the arid, semiarid tropical climates experience the most consequences of climate stress (Giller et al. 2009; Branca et al. 2013; Chai et al. 2015; and Pauline et al. 2016).

In Tanzanian semiarid tropics, long-term monoculture practices have significantly decreased total carbon, nitrogen contents and other important minerals (Bockstaller et al. 1997; Hartemink 1997; Medeiros et al. 1997; Sosovele et al. 1999; Monfreda et al. 2008; and Msongaleli et al. 2015). This situation had later declined the level of soil quality and crops production and therefore affecting the livelihoods of over 70% of the smallholders in the area (Duru et al. 2015; and URT 2014; 2012). The continued stresses from climate change have exacerbated the vulnerability of these farmers and later on elevated food insecurity and abject poverty (Paavola 2008; Lema and Majule 2009; Yanda 2015; Kangalawe 2016; and Kangalawe et al. 2016).

The annual food deficit was approximated to 50% in the area because the little obtained yields were consumed within 3–6 months leaving the people under severe starvation for the rest of the year (URT 2007, 2012 and 2014). More bad years still happen in the area and has skyrocketed food insecurity and abject poverty among the smallholder farmers (Ahmed et al. 2011; Rowhani et al. 2011b; Kangalawe and

Lyimo 2013; and URT 2014). From that point, improvement of soil quality is very important. According to the context, good agronomic practices is an immediate resolution (Andrews 1998; Lal 1998; Andrews and Carroll 2001).

The improvement of soil ecology is a primary factor to elevate crop yields and food security (Doran and Zeiss 2000; Lichtfouse et al. 2009; and FAO 2013). Organic fertilizations increases the accumulation of soil organic matter that raise cations ions in the soil and facilitate the uptake of important nutrients especially nitrogen, phosphorus and potassium for crops growth. A number of studies have recommended soils organic management as good proposition to achieving agricultural sustainability (Lal 1998; Bationo et al. 2006; and Kimaro et al. 2015). Organic manure has increased soil organic matter replenishment where applied. Practically, soils replenishments has been significantly higher in the homestead areas than far farmland. This is because most farmers have no reliable means of transporting these manure to far distance (McDonagh et al. 2001; Thierfelder and Wall 2009; Partey and Thevathasan 2013; and Msongaleli et al. 2015). Therefore, crops yields were significantly higher in the homestead than distant farms.

Organic soil managements have significantly increased crops yields by more than 40% and they act as climate smart practices in the midst of changing climate (Vanlauwe 2004; and Giller et al. 2009). Similarly, they serve as a sustainable measure for environmental services (URT 2007). Among other things, they create favorable condition that catalyzes biological functions of mycorrhizas and other soil microorganisms (McDonagh et al. 2001). In this environment, the influence of mycorrhiza fungi on plant nutrient uptake and growth, resistances to pathogens is significantly higher than in no-fertilization (Birch-Thomsen et al. 2007; and Wall et al. 2013). Therefore, agricultural sustainability will be attainable only when soil quality is optimal. Under such a situation, conservation of environment is particular important to yield optimal agricultural outputs. The mutual environment and agricultural synergies can ensure increased crops yield, food security, and ecological conservation (UNEP 2011, 2012; and Poppy et al. 2014).

However, in semiarid areas of Tanzania, food production is among the main driver of environmental degradation, and if not well addressed, it may completely destroy the ecosystems. Therefore, food policies should be respectful to environment and more especially on the fragile ecosystems in order to achieve the win-win situation between the two aspects. This study explored the role of good agronomic practices (e.g. organic fertilization and other soil organic management) in soil quality improvement and agricultural sustainability in the semiarid tropics agro-ecological zone of Tanzania, and other countries where these organic managements are practiced.

9.2 Tanzania

The present study focuses on Tanzania, an Eastern African country with rich biodiversity. Based on altitude, precipitation pattern, dependable growing seasons and average water holding capacity of the soils and physiographic features, Tanzania has seven agro-ecological zones although there are numerous micro ones. Agro-ecological zones refers to the geographical areas exhibiting similar climatic conditions that determine their ability to support rain-fed agriculture (Bockstaller et al. 1997; and URT 2007). Therefore, climate varies over different agro-ecological zones while agricultural systems and crops produced depend on agricultural knowledge of the farmers.

Tanzania semiarid zones covers a number of regions and it has been grouped into two major parts. These are central and southern. The central region includes Dodoma, Singida, Northern Iringa, some parts of Arusha, Shinyanga (URT 2007). These regions are located at 1000–1500 m altitudes and receive unreliable unimodal rainfall ranging from 500 to 800 mm per annum. In general, the topography of the area is characterized by undulating plains with rocky hills and low scarps with a well-drained low fertile soils. It has an alluvial hardpan and saline soils in Eastern Rift Valley and Lake Eyasi and black cracking soils in Shinyanga (URT 2007).

The southern parts involves the regions of Morogoro (except Kiliombero and Wami Basins, and Uluguru Mountains), Lindi and Southwest of Mtwara. These regions are located at 200–600 m altitudes and receive unreliable unimodal rainfall ranging from 600 to 800 mm per annum. Overall, the topography of the area is characterized flat or undulating plains with rocky hills, moderate fertile loams and clays in South Morogoro, infertile sand soils in center (URT 2007). The growing season in both parts starts from December to March however, it has been changing due change of onset and cessation of rainfall caused by global climate change.

To write this paper, we reviewed more than 50 journal papers conducted within the topic, area or/and nearby area with similar climates. We selected the scientific papers published in authentic journals mostly from the web of science. Mainly from journals with high impact factors and number of citation of a particular paper. The most recent publications were given priority in the selection. Analyses and modification of some data were done to suit the study objectives. We considered all publication ethics including the seeking of permission to journal authors where necessary. The review was done to meet the standards of Sustainable Agricultural Reviews.

9.3 Dominant Agricultural Systems and Crops Produced

Small scale agro-pastoralism is a dominant livelihoods in the area (Sosovele et al. 1999; and Lema and Majule 2009). It produces maize, leguminous, millet and sorghum, sheep, goat, cattle and donkey for both food and sale. The Maasai tribe forms the core of this semi-nomadic system in the area. Agro-pastoralism prevails in Arusha, Singida, Shinyanga and Morogoro regions just to mention some (URT 2007). Extensive semi-nomadic grazing and small scale cultivation of drought tolerant roots and cereals are practiced to cope with climatic conditions (URT 2007; Lema and Majule 2009; and Kangalawe and Lyimo 2013).). However, the level of adoption is based on the biophysical characteristics of the place (see Table 9.1), but mixed farming is not yet effectively adopted to harvest full potentials.

Table 9.1 Farming systems in the study area

Farming systems types	Rank of Farming Systems
Maize/legume system	1
Livestock/Sorghum-millet system	2
Pastoralist system	3
Agro pastoralist system	4
Cassava/cashew/coconut system	5
Agroforestry (the enclosed systems *"ngitiri"* in Shinyanga)	6
Wetland paddy/sugarcane system in water sources	7

Livestock/sorghum-millet system is prevalent in Shinyanga. Sorghum, millet and maize are dominant. Sugarcane are the least adopted farming system due to nature of climate. It can only be adopted in fewer areas with swamps and wetlands. (Source: Modified from Sosovele et al. (1999))

In few areas where manure fertilizations is practiced, the yields from maize, millet and sorghum have increase from 0.82 to 1.8 tn ha^{-1} under no-fertilization and organic fertilization respectively. The same has applied on ecology by increasing the influence of mycorrhiza fungi and other microorganism to perform biological functions (Hartemink 1997; Partey and Thevathasan 2013; and Kimaro et al. 2015). These are benefits that outsmart the areas under no-fertilizations. In this aspects, organic manure raise cation ions that increase the uptake of important soil nutrients by the plants. As well, this acts as a buffer to avoid nutrients loss through leaching.

9.4 Soil Quality Status at Agroecosystem level and Farm Level

Soil quality can be defined in a number of ways depending on the community context and understanding (Doran and Zeiss 2000). It is regarded as a capacity of the soils to perform specific function (Lal 1998). Similarly, Doran and Parkin (1994) asserted that soil quality is a capacity of the soils to functions with the surrounding ecosystem to sustain biological productivity, conserve the quality of the environment as well as safeguarding the health of animals and plants. They also pointed that it can be a measure of the soil condition in relation to the requirements of one or more species and/or to any human needs. Generally, soil quality can be understood differently from various discipline (Karlen et al. 2003). This study assessed the influence of soil quality on agricultural sustainability to estimate the impacts of diverse agricultural systems to peoples' livelihoods in semi-arid and tropical climates. These practices either increase or/and decrease soil fertility and thus, the whole process has implications to the sustainability of agriculture and food security in various countries.

According to Lal (1998) soil degradation often prevails due to monoculture and land-use conversion and thus, reducing soils quality and fertility. This degradation may vary over soils types such as Chromic Luvisols, Cambisols, Histosols etc. (Hartemink 1997; and Glaser et al. 2001). In Tanzania (i.e. with diverse agro-ecological zones) and other tropical countries, this soil quality decline has been affecting crop yields and thus, putting food security in risk. Various soil models predict that this decline may be more pronounced in future if substantial interventions are not taken (Hartemink 1997; and Bationo et al. 2006). Under such a situation, food security and malnutrition will continue affecting the vulnerable societies especially in sub-Saharan Africa.

To improve the agro-ecosystems, good agricultural practices (i.e. organic soil management) seem to increase soil fertility/quality at farm level and entire ecosystems (Andrews and Carroll 2001). Manure fertilizations and other forms of organic soils management serve as adaptation measures and optional livelihoods to the vulnerable communities in most semi-arid areas of Tanzania and Africa in general. Manure offers optimal ingredients of nitrogen, carbon, phosphorus and potassium for plant growth. These nutrients also create favorable condition for mycorrhiza fungi to function well especially in helping the plants to optimize nutrient uptake, growth and resistances to pathogens (Glaser et al. 2001; Vanlauwe et al. 2014; and Kimaro et al. 2015). In Shinyanga, Dodoma and parts of Morogoro regions organic fertilization has significantly increased crops yields under smallholders farming. And animal manure has been a major source of organic fertilizations (see Fig. 9.1).

Various studies show that under organic fertilizations; organic soil nutrients i.e. carbon, nitrogen and phosphorus were significantly higher than under no-fertilizations (Hartemink 1997). It was further realized that these nutrients are always abundant on top soil 0–20 cm than below 30 cm due to continued fertilizations (Thierfelder and Wall 2009; Partey and Thevathasan 2013; and Msongaleli et al. 2015). Ecologically, this implies that crops with shallow roots can trap sufficient nutrients because their roots excel within the nutrient abundant and thus, giving more yields probably than the one with deep roots that seem to trap nutrient beyond the nutrient storage zone. That situation was contrary to areas under no-fertilizations where a bit deep soils had numerous nutrient than top soils (Hartemink 1997). The major reason for this difference is that under no-fertilization the top soil is under severe utilization while the beneath layer is a bit of resting or with little disturbances from anthropogenic activities. To alleviate this, the soils under no-fertilizations need to undergo organic fertilizations to restore its fertility and ecological functions.

Fig. 9.1 The animal manure deposited at farm level before fertilizations. The fertilization can be done through even spreading in the whole farm, i.e. when manure are plenty, or applying in the seeding holes only, i.e. when manure are scarce

9.5 Agricultural Sustainability

Agricultural sustainability is among the most concern in the era of global climate change (Paavola 2008; Yanda 2015; Kangalawe 2016; and Kangalawe et al. 2016). It is approximated that by 2050 there will be an increase in population for two billion and making over 8 billion people all dwelling on the Planet Earth. To feed all these population, we need to increase food production for 60% (UNEP 2012; Poppy et al. 2014). And about 80% of these two billion people will be residing in developing countries especially Sub-Saharan region. Despite of that fact, the region is expected to produce the least of the required food. Overall, the region among the most vulnerable regions to environmental stress especially climate change impacts (Lema and Majule 2009; Ahmed et al. 2011; and Rowhani et al. 2011a) thus, immediate interventions are needed to curb both short and long-term challenges.

Precisely, agricultural industry can be beneficial and sustainable if it operates to meet the food requirements of the present population without compromising the needs of the future generation (Lichtfouse et al. 2009; UNEP 2011, 2012). Since agriculture involves people and environment synergies, there should be a balance among the involved dimensions (see Fig. 9.2). These dimension include biophysical, economic and social (Lal 1998). Biophysical involves the quantity of output (Mg of yield/ha) while economic and social dimensions refers to the value of gross or net, and the capacity of the system to support farming respectively (see Fig. 9.2).

9.6 Integrative Effects of Soil Quality on Agricultural Sustainability and Environment

There is a close link between soil quality, economic progress and environmental quality (Andrews and Carroll 2001). Under such a situation, a decline in soil quality always lead to degradation of environmental and reduction of agricultural productivity (FAO 2006; Monfreda et al. 2008; and UNEP 2012) and thus increasing food insecurity. In Tanzania and other developing countries, the degradation of environmental quality has brought serious problems of hunger, food insecurity and poverty at large. The semiarid tropic regions such as Shinyanga, Singida and Dodoma, monoculture has degraded soils nutrient to the extent that replenishment has been a difficult alternative (Herdt and Steiner 1995; URT 2007; Monfreda et al. 2008; Kimaro et al. 2015).

In the labile of degraded soils, the available nutrients are likely to be found at deep soils than top ones due to permanent cropping (Hartemink 1997). However, our review indicated that organic soil management and some intensification agriculture can improve soil quality and crops yield in the area. Thus, it should be understandable that, to achieve agriculture sustainability with socio-economic and socio-ecological potentials, depend on the conservation of soils quality (see Fig. 9.3).

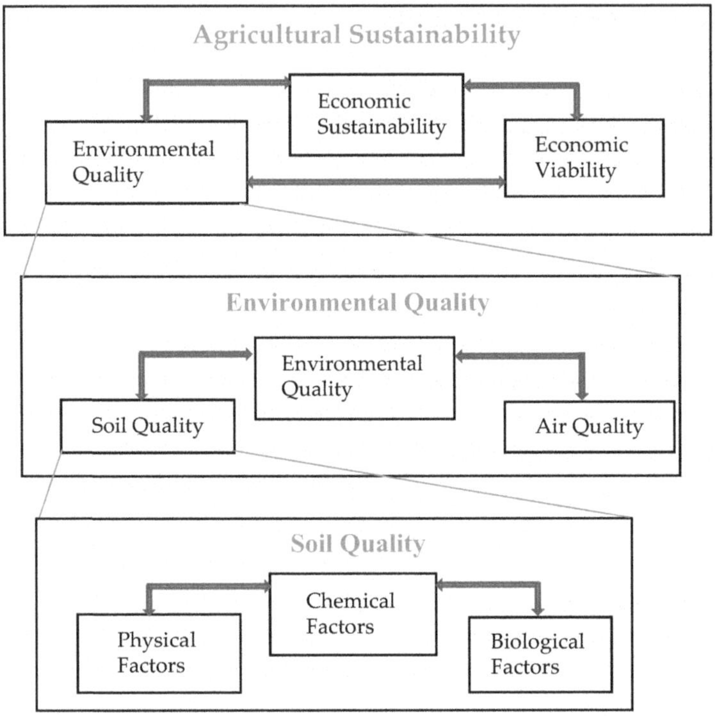

Fig. 9.2 Soil quality, environmental quality, and agriculture sustainability synergies. Soil fertility improvement creates favorable environmental condition for crop production and environmental conservation. Source: Modified from Andrews (1998)

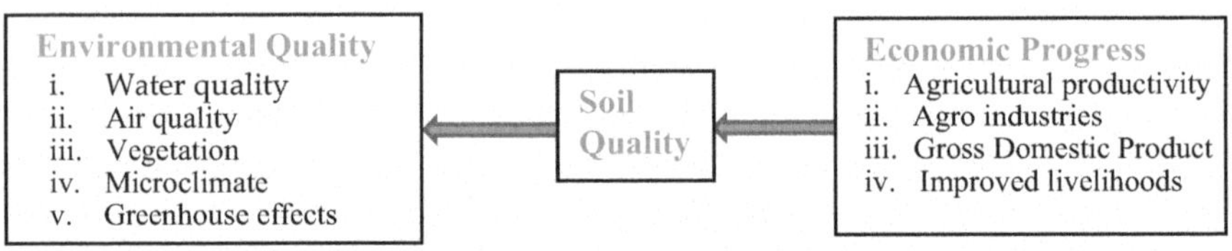

Fig. 9.3 Relationship among soil, environment and economic progress. Economic development is a good tool in ensuring agricultural production through the use of advanced farm instruments. This in turn raise the gross domestic product of the country and its people. Consequently, this ensures the maintenance of soil and environmental management. The whole process enables the provision and sustainability of environmental services. (Source: Modified from Lal (1998))

9.7 Climate Change and Soil Quality

Climate has a significant influence to soil quality (Doran and Parkin 1994; and Doran and Zeiss 2000). By nature, semiarid zone has high climate variability with some extreme stresses to agriculture and biodiversity (Yanda 2015; Mkonda and He 2017c). Rainfall and temperature either increase or decrease soil quality through the processes of degradations or formation and accumulation of important soil minerals.

Table 9.2 Percentage of annual increase of yield due to good agronomic practices in the semiarid areas of Tanzania

Crops	Area %/year	Production % /year	Production%
Maize	9.4	0.8	3.9
Sorghum	0.7	0.1	1.1
Millet	2.8	0.4	3.8
Groundnuts	8.3	0.6	7.4
Sunflower	0.1	2.4	4.2

Source: Extracted from Lema and Majule (2009), and Msongaleli et al. (2015)

The accumulation of soil carbon, total nitrogen, phosphorus and potassium just to mention a few is highly attributed by climate (IPCC 2000; Mkonda and He 2017d).

The temperature influences the decomposition of organic matter, mineralization and immobilization of soil nutrients (Doran and Parkin 1994; Lal 1998; and IPCC 2000).Similarly, high temperature increase the level of carbon offset through decomposition. Therefore, reducing carbon content in the soil, an important nutrient for crop production. On other hand, rainfall can have two side effects as it either increase or decrease soils nutrients. High rainfall facilitate the production of a wide range of plant biomass while downscaling the dominance of minerals which readily dissolve in water and therefore increasing the rate of mineral mobilization while low rainfall can reduce the production of important nutrients such as phosphorus that need abundant water.

Therefore, it is healthy for both temperature and rainfall to be kept at average. Otherwise, their extremes can have ecological repercussions. As intervention to the problem, Kalhapure et al. (2013) suggested that under drought condition, effective soil organic management can increase optimal amount of minerals in soil. Under such a condition, carbon (C) which is potential for C sequestration, seem to be abundant than other soils nutrients.

In general, the abundance of carbon and other important minerals in the soils depend on depth, types of agronomic practices and level of organic fertilization. Similarly, the abundance the soils nutrients the higher the yields. Table 9.2 show how good agronomic practices (e.g. organic fertilization and conservative tillage) has increased crops yields from maize, millet, sorghums, groundnuts and sunflower in the study area.

The statistics in Table 9.2 are based on the average of 10 years i.e. 2000–2010. Therefore, it can be recommended that smallholder farmers should apply good agronomic practices to attain socio-ecological achievements.

9.8 Integrative Adaptation and Mitigation Strategies for Agricultural Sustainability

There is a close link between the adaptation and mitigation practices, and sustainable agriculture (Andrews 1998; and Birch-Thomsen et al. 2007). These practices promotes soil quality and increase carbon sequestration for the betterment of both the

Table 9.3 Adaptation and mitigation practices in semiarid areas of Tanzania

Adaptation	Mitigation
Adoption of drought-tolerant crops and animal breeds	Reduced or more efficient use of chemical fertilizers
Adjustments in irrigation practices and systems	Management of water sources especially wetlands
Changes in timing of planting	Reduced tillage
Conservation of crop and livestock genetic	Planting of biofuels and trees for fuel wood
Crops rotation or production systems	Use of improved feeding practices for livestock
Conservation of agrobiodiversity	Planting of fast-growing tree plantations

Source: Extracted from Lema and Majule (2009), and Kangalawe and Lyimo (2013)

present and future generations (Lichtfouse et al. 2009; and Duru 2015). Farmers in the area have been doing such a combined activities by knowingly or unknowingly.

Adaptation options may include a wide range of approaches designed to reduce the vulnerability and enhance the adaptive capacity of agricultural systems to climate change impacts (Yanda 2015; and Mkonda and He 2017c) while mitigation options involve activities that increase carbon stocks above and below ground, that reduce direct agricultural emissions (carbon dioxide, methane, nitrous oxides) anywhere in the lifecycle of agricultural production; and actions that prevent the deforestation and degradation (Kangalawe and Lyimo 2013) as seen in Table 9.3.

9.9 Experience from Other Countries

Although ecological management for agriculture is important for every parts of the world, the Sub-Saharan African and other tropical parts of the world need it the most (Pretty et al. 2006). Most of these areas are facing food shortage due to the existence of production-limiting constraints faced by resource-poor farmers that include: shrinking farm sizes and inequitable land-distribution patterns, depleted soils and limited use of fertilizer and soil amendments (either organic and inorganic), unreliable rainfall and lack of irrigation capacity, and limited access to improved varieties and seed distribution systems (Hartemink et al. 2008; and Okeyo et al. 2014). Food and Agriculture Organization (2013) pointed that most small-scale farms both in in Africa are less than 2 hectares and they are dependent on household members as a sole source of labor force. To underpin this discussion, we earmark the agricultural systems that are practiced in East, West and Southern Africa.

In East Africa, climate smart agriculture has been under implementation for some couple years aiming at conserving the environment and improving crop yields (Osman-Elasha et al. 2006; and Mkonda and He 2017a). The implementation of climate smart agriculture has been done through projects and programs. These programs (i.e. funded by both local and international organs meant to improve food security and climate resilience among farmers in the region. Food and Agricultural

Organization of the United Nation is a lead institution in this aspect. In this aspect, some areas/zones have benefited from these programs (Solomon et al. 2007). However, high diversity in agro-ecological zone impedes the implementation of these projects. This is amplified by climate change impacts. In Kenya, Rusinamhodzi et al. (2011) pointed out that the adoption of conservation agriculture especially under rain-fed maize production, would improve the yields. This idea was supported by Kimaro et al. (2015) who asserted the same when proposing the optimization of yields along the Uluguru Mountain in Tanzania. In addition, agroforestry systems i.e. woodlots has significant contribution to ecological improvements tenable for agricultural production (Christensen 1988; and Nyadzi et al. 2006). The findings of these studies underpinned that the potential and actual optimization of yields had its base from adequate soil quality improvement in the area. They concluded by endorsing organic soil management against long-term chemical fertilization which appeared to affects the ecosystems (Mkonda and He 2017b).

According to various reports by IPPC (2014), FAO (2013) and other findings from various studies, East Africa is among the worst vulnerable region in Africa. This vulnerability is intensified by climate change impacts which have been hitting the region for couple of years. In addition, poverty, market value of resources, rapid population growth and technology are among the underlying factors affecting agricultural development in the region. In fact the poor performance of Agriculture sector has significant impacts to gross domestic product of Tanzania, Kenya, Uganda, Rwanda, Burundi and Sudan depends on agriculture for about 50–70%). The adoption of various agricultural systems has also been impeded by land conflict especial in countries where land and its implementation is somewhat loose. This conflict has been rising even where formal governance of access to land is in place, government land regulations often conflict with customary laws of land tenure in Africa. In addition, competitive prices have only led to more land acquisition by both domestic and foreign investors with many local farmers being left out because of their weak financial muscles to compete. Therefore, it is essential to improve the ecological condition by undertaking all possible necessary steps ranging from crops genetics, intensive irrigation, fertilization and institutional framework to optimize crop yields.

In the Southern African countries such as Zimbabwe, Zambia, Malawi, Botswana, Mozambique and Angola; agricultural intensification is given high attention to alleviate the predominant food shortage in the region. Here, intensive agriculture ranges from crop production, livestock rearing, forestry and fish farming (Nyong et al. 2007; and Duru 2015). For example, Malawi attempts to improve fishing industry by applying different techniques like animal manure to feed the fish in the ponds (Blythe 2013). This program has significantly increased yields especially "tilapia" that eventually has raised income through selling. In this aspect, sustainability is measured in terms of environmental, economic, social and cultural aspects. Attaining many of these aspect during the production process is regarded as agricultural sustainability. On the other hand, the growing demand of organic products in the world market has risen the desire to adopt organic farming. Principally, this system gives little yields but of high value. Now that, it is worthwhile to ensure food security than safety in order to solve the immediate challenges of food shortage.

West Africa is another important region where agricultural systems need restoration to improve soil quality (Nezomba et al. 2010). Despite the agricultural diversity, the region practices both traditional and modern agricultural systems (Bationo et al. 2006). The majority of the farming systems are traditionally practiced and they range from the extensive (i.e. shifting cultivation and nomadic herding) to more intensive and specialized types of farming (such as compound farms and terrace farming). Shifting cultivation is an extensive agricultural system which mainly involve 'slash-and-burn' cleared land alternates with a fallow period. The system degrades the environment as it involves serious deforestation. The cut materials are burned to allow the plantation of crops like yams, sorghum, millet, maize and cassava depending on the ecological zone (Nyong et al. 2007). On other hand fallowing involve the resting of the cultivated areas for regrowth of natural vegetation and rejuvenation of soil fertility (quality) through nutrient cycling, addition of litter and suppression of weeds. In most cases, the resting period can be 4–5 years however, ideally the longest period can range between 10 and 20 years.

In Liberia, the traditional agriculture of the Loma people involves farmers planting crops in fertile man-made soil known as *'anthropogenic dark earth'*. This man-made highly fertile soil, which is used for growing crops, forms in the same localized areas, building up over generations (Kareemulla et al. 2017). The soil is created inevitably by everyday domestic life, from deposits of charred and fresh organic matter, including manure, bones, ash, charcoal and ceramics. It is evident that this traditional agriculture has twice the energy efficiency of either 'slash and burn' rice production and hunting and gathering. However, the sustainability is this farming systems is at "cross road" because it is limited by 'sacred' forests, which form around current settlements and cover areas of fertile man-made soil which used to be towns in the past. On top of that, customary laws prohibit these forests being cleared for farming, as some trees are believed to have mystical 'medicinal' power, and also because of the presence of graves.

On the other hand, Mali is highly vulnerable to the threat of soil fertility decline and food deficit (Kalra et al. 2013). A series of development organizations have promoted inappropriate "new green revolution" technologies that depend on external inputs rather than local abilities and resources, and food aid has become a fall-back resolution to alleviate food shortages. Women are particularly vulnerable, and face particular challenges in accessing productive resources (land, water, credit) and receiving technical advisory services. Therefore, Mali is in need of long-term solutions for small-scale farmers to optimize crop production and ecosystems services.

Further, a wide range of crop cultivars and species have been introduced to cope with the global and local environmental change. This has been done through different programs funded by both local and international organs e.g. FAO. For example, in Senegal about 14 high-yielding, early maturing and drought resistant dry cereal varieties have been developed thus, have succeeded to optimize productivity by at least 30% (Duru 2015). Alongside, this program has benefitted more than 423,000 farmers in the country whose yields have boomed after adopting the new varieties and thus, they have become more resilient to climate shocks. Likewise, in other countries such as Guinea, Sierra Leone, Liberia and Côte d'Ivoire, the new saline-tolerant rice varieties, climate-smart irrigation techniques and better soil fertility management increased rice yields of more than 100,000 farmers (Nyong et al. 2007).

This change (i.e. adoption of new agricultural technology and crop cultivars) in region have been influenced by a number of reasons such as introduction of Asian and New World Crops, population expansion, the need for spices and agricultural raw materials for industry; expansion of cassava production into marginal areas where other crops often fail, and introduction of mechanization into farming and adoption of new techniques just to mention a few.

India, the second most populous country in the world, its priority has been to elevate agricultural yields, maintain food security and ensure the availability of industrial raw materials (Kalra et al. 2013). However, the country has great diversity in agro-climatic zones with as many as 127 zones under five agro-ecosystems such as rain-fed, arid, irrigated, coastal and hilly systems (Kareemulla et al. 2017). In that respect, there are spatial and temporal differences in agricultural systems tenable to meet the local challenges. However, for agriculture to be sustainable, it needs to walk along with of ecological, economic, cultural and social sustainability. Another aspect that prompts agricultural differences is population density. In India, West Bengal, Bihar, Himachal Pradesh, Punjab, Bihar, Uttar Pradesh, Jharkhand and Kerala are among the major states with high population density of over 800 persons per square kilometer (Kareemulla et al. 2017). Thus far, population necessitates the intensive agriculture rather that organic and extensive farming. Intensive agriculture can give more yields even in a small geographical area. Therefore, intensive agriculture forms the major agricultural system in India.

China, the most populous country in the Planet has significant contribution to global agricultural sustainability (Tilman et al. 2002). With diverse climatic region, China applies different farming systems to meet this spatial biophysical characteristics. Most dry areas such as Northwest and Central China apply intensive irrigation in agriculture while other parts that still receive reliable rainfall depend on rain-fed (Sharma and Minhas 2005). On other hand, the intensive high-yield agriculture is dependent on addition of fertilizers, especially industrially produced NH_4 and NO_3. This is done to accrue high yields for food and industrial raw materials. Unfortunately, only 30–50% of applied nitrogen fertilizer 40, 41 and ~45% of phosphorus fertilizer 42 is taken up by crops. This means, a significant amount of the applied nitrogen and a smaller portion of the applied phosphorus is lost from agricultural fields and thus, polluting the environment.

While fertilization is highly emphasized, the agricultural systems in most dry areas is limited of irrigation. In this respect, the availability of water is essential for agricultural production in these dry areas (Sharma and Minhas 2005). Nevertheless, despite of strongly influencing local agricultural development; excessive utilization of water resources plays a vital role in accelerating environmental degradation (Li et al. 2009). In arid land of northwest China, the water consumption for agriculture accounts for approximately 90% of the total water uses (Li et al. 2010) but the average available water is less than $1635 \times 108\ m^3$ per year, only 5.8% of the China average level. Now that, this tells that agricultural sustainability is promising where there is no shocks or immediate demand of environmental services that can exert more pressure on resources utilization (Huang et al. 2012). Otherwise it is difficult to maintain environmental conservation while optimizing crop yields.

On the other hand, according to European Union (2012) Europe plays great roles in both practicing and funding agricultural sustainability around the globe. Europe strongly believes that agriculture that is environmentally, economically and socially sustainable and can make a vital contribution in our response to the most urgent challenges especially reducing poverty and ensuring food security. The report further elaborates that increasing demand of organic products at global level has raised organic agriculture in Europe. It is envisaged apart from giving quality yields, this farming system ensures constant provision of environmental services. For example, the southwest regions of Spain and southern Portugal the "*Dehesa*" *is* a very specific Mediterranean system of extensively grazed, wooded pasture that shows the multifunctional role of forests. Their intrinsic characteristics and management practices ensure the provision of a wide range of environmental services such as biodiversity, soil conservation, and carbon storage. In these areas farmers rear Iberian pig species known as 'pata negra', which feed on acorns of oak trees.

Besides, Europe has been a main partner and donor of the Global Rinderpest Eradication Campaign in collaboration with the World Organization for Animal Health (OIE) and FAO, contributing 390 million € over the last 50 years (www.oie.int/en/for-the-media/rinderpest/). The European Union is also supporting local communities in building capacities to restore and sustainably manage their dryland ecosystems, improve their marketing "activities" as well as support dialogues among stakeholders to share knowledge, ideas and priorities. A good example of the supported countries includes: Jordan, Mali, Botswana and Sudan which most of their areas are dryland.

9.10 Conclusion

This study assessed the influence of soil quality on agricultural sustainability. We found that, monoculture is the dominant degradation activity in semiarid tropics of Tanzania. Through that, optimal amount of soils nutrients get lost. Unfortunately, climate change impacts have stressed the already affected environment and sterilized all biological functions of the soil especially mycorrhiza fungi (especially primary and ecto-mycorrhizas). Under such a situation, soil quality decreases, crops yields are lowered and the risk of food insecurity increases. However, in few areas good agronomic practices have significantly elevated soil quality through organic fertilizations and other organic soil managements. Animal manure appeared to have significant contribution to organic fertilizations. It provides substantial nutrients such as nitrogen, phosphorus, potassium and carbon to the soils. In those areas, yields of food crops have increased to 1.8 tn ha^{-1} compared to 0.82 tn ha^{-1} under soils with no-fertilizations. This is implies that if serious organic fertilization is done, we can attain agricultural.

Acknowledgement The authors are thankful to the Editor-in-Chief of Sustainable Agricultural Reviews, Professor Eric Lichtfouse for inviting them to write a chapter in "Ecology for agriculture". They are also appreciative to the reviewers for the constructive comments and insights.

References

Ahmed S, Deffenbaugh N, Hertel T, Lobell D, Ramankutty N, Rios A, Rowhani P (2011) Climate volatility and poverty vulnerability in Tanzania. Glob Environ Chang 21(2011):46–55

Andrews SS (1998) Sustainable agriculture alternatives: ecological and managerial implications of poultry litter management alternatives applied to agronomic soils. PhD dissertation. University of Georgia, Athens, GA, USA

Andrews SS, Carroll CR (2001) Designing a soil quality assessment tool for sustainable agroecosystem management. Ecol Appl 11(6):1573–1585

Bationo A, Kihara J, Vanlauwe B, Waswa B, Kimetu J (2006) Soil organic carbon dynamics, functions and management in West African agro-ecosystems. Agric Syst 94(1):13–25. https://doi.org/10.1016/j.agsy.2005.08.011

Birch-Thomsen T, Elberling B, Bjarne F, Magid J (2007) Temporal and spatial trends in soil organic carbon stocks following maize cultivation in semi-arid Tanzania. Springer, East Africa.

Blythe JL (2013) Social-ecological analysis of integrated agriculture aquaculture systems in Dedza, Malawi. Environ Dev Sustain 15:1143–1155. https://doi.org/10.1007/s10668-012-9429-6

Bockstaller C, Girardin P, van der Werf HMG (1997) Use of agro-ecological indicators for the evaluation of farming systems. Eur J Agron 7:261–270

Branca G, Lipper L, McCarthy N, Jolejole M (2013) Food security, climate change, and sustainable land management. Agron Sustain Dev 33:635–650. https://doi.org/10.1007/s13593-013-0133-1

Chai Q, Gan Y, Zhao C, Xu H, Waskom R, Niu Y, Siddique K (2015) Regulated deficit irrigation for crop production under drought stress, Agron Sustain Dev. Diplomacy of a New State. NOK Publishers, New York

Christensen B (1988) Effects of animal manure and mineral fertilizer on the total carbon and nitrogen contents of soil size fractions. Biol Fertil Soils 5:304–307

Doran JW, Parkin TB (1994) Defining and assessing soil quality. In: Doran JW, Coleman DC, Bezdicek DF, Stewart BA (eds) Defining soil quality for a sustainable environment. SSSA, Inc., Madison

Doran J., Zeiss M. (2000). Soil health and sustainability: managing the biotic component of soil quality. Appl Soil Ecol. 15, 1, pg 3–11 https://doi.org/10.1016/S0929-1393(00)00067-6

Duru M (2015) How to implement biodiversity-based agriculture to enhance ecosystem services. Agron Sustain Dev 35:1259–1281. https://doi.org/10.1007/s13593-015-0306-1

Duru M, Therond O, Fares M (2015) Designing agroecological transitions. Agron Sustain Devaluation of farming systems. Eur J Agron 7:261–270

European Union (2012) Agriculture and Rural Development http://ec.europa.eu/agriculture

FAO (2006) Conservation agriculture for SARD and food security in Southern and Eastern Africa. Terminal report, Kenya and Tanzania

FAO (2008) Investing in sustainable agricultural intensification: the role of conservation agriculture. A Framework for Action, Rome

FAO (2012) Conservation agriculture and sustainable crop intensification in Karatu District, Tanzania, Integrated crop management, vol 15. Food and agriculture organization of the United nations (FAO), Rome

FAO (2013) Climate-smart agriculture sourcebook. FAO, Rome

Giller K, Witter E, Corbeels M, Tittonell P (2009) Conservation agriculture and smallholder farming in Africa: The heretics' view. Field Crops Res 114:23–34

Glaser B, Lehmann J, Führböter M, Solomon D, Zech W (2001) Carbon and nitrogen mineralization in cultivated and natural savanna soils of Northern Tanzania. Biol Fertil Soils 33:301–309. https://doi.org/10.1007/s003740000324

Hartemink AE (1997) Soil fertility decline in some Major Soil Groupings under permanent cropping in Tanga Region, Tanzania. Geoderma 75:245–229

Hartemink AE, Veldkamp T, Bai Z (2008) Land cover change and soil fertility decline in tropical regions. Turk J Agric For 32:195–213

Herdt RW, Steiner RA (1995) Agricultural sustainability: concepts and conundrums. In: Barnett in East Africa: the impacts of resource variability on human security Clim Change. https://doi.org/10.1007/s10584-010-9884-8

Huang Y, Li YP, Chen X, Ma YG (2012) Optimization of the irrigation water resources for agricultural sustainability in Tarim River Basin, China. Agric Water Manag 107:74–85. https://doi.org/10.1016/j.agwat.2012.01.012

IPCC (2000) Land use, land use change and forestry, a special report of the IPCC (Watson RT, Noble IR, Bolin B, Ravondranath NH, Verardo DJ, Dokken DJ, eds). Cambridge University Press, Cambridge

IPCC (2014) Climate Change 2014 impacts, adaptation, and vulnerability. Part A: global and sectoral aspects. Contribution of Working Group II to the fifth assessment report of the Intergovernmental Panel on Climate Change (Field CB, Barros VR, Estrada, Genova RC, Girma B, Kissel ES, Levy AN, MacCracken S, Mastrandrea PR, White LL, eds). Cambridge University Press, Cambridge/New York

Kalhapure A, Shete B, Dhonde M (2013) Integrated nutrient management in maize (Zea Mays L.) for increasing production with sustainability. Int J Agric Food Sci Technol 4(3):195–206 ISSN 2249-3050, http://www.ripublication.com/ ijafst.htm

Kalra RKB, Tonts AM, Siddique KHM (2013) Self-help groups in Indian agriculture: a case study of farmer groups in Punjab, Northern India. Agroecol Sustain Food Syst 37:509–530

Kangalawe R (2016) Climate change impacts on water resource management and community livelihoods in the southern highlands of Tanzania. Clim Dev. https://doi.org/10.1080/17565529.2016.1139487

Kangalawe R, Lyimo G (2013) Climate Change, Adaptive Strategies and Rural Livelihoods in Semiarid Tanzania. Nat Res 2013(4):266–278. https://doi.org/10.4236/nr.2013.43034

Kangalawe R, Mung'ong'o C, Mwakaje A, Kalumanga E, Yanda P (2016) Climate change and variability impacts on agricultural production and livelihood systems in Western Tanzania. Clim Dev. https://doi.org/10.1080/17565529.2016.1146119

Kareemulla K, Venkattakumar R, Samuel M (2017) An analysis on agricultural sustainability in India. Curr Sci 112(258):2

Karlen D, Ditzlerb C, Andrewsa S (2003) Soil quality: why and how? Geoderma 114:145–156. https://doi.org/10.1016/S0016-7061(03)00039-9

Kimaro A, Mpanda M, Rioux J, Aynekulu E, Shaba S, Thiong'o M, Mutuo P, Abwanda S, Shepherd K, Neufeldt H, Rosenstock TS (2015) Is conservation agriculture 'climate-smart' for maize farmers in the highlands of Tanzania? Springer-Verlag, Berlin/Heidelberg. Nutr Cycl Agroecosyst. https://doi.org/10.1007/s10705-015-9711-8

Lal R (1998) Basic concepts and global issues: soil quality and agricultural sustainability. In: Lal R (ed) Soil quality and agricultural sustainability. Ann Arbor Science, Chelsea, pp 3–12

Lema MA, Majule AE (2009) Impacts of climate change, variability and adaptation strategies on agriculture in semi-arid areas of Tanzania: the case of Manyoni District in Singida Region. Tanzan J Environ Sci Technol 3:206–218

Li YP, Huang GH, Wang GQ, Huang YF (2009) FSWM: a hybrid fuzzy-stochastic water management model for agricultural sustainability under uncertainty. Agric Water Manag 96:1807–1818. https://doi.org/10.1016/j.agwat.2009.07.019

Li W, Li YP, Li CH, Huang GH (2010) An inexact two-stage water management model for planning cultural irrigation under uncertainty. Agric Water Manag 97:1905–1914. https://doi.org/10.1016/j.agwat.2010.07.005

Lichtfouse E, Navarrete M, Debaeke P et al (2009). Agron Sustain Dev 29:1. https://doi.org/10.1051/agro:2008054

McDonagh JF, Birch-Thomsen T, Magic J (2001) Soil organic matter decline and compositional change associated with cereal cropping in Southern Tanzania. Land Degrad Dev:13–26. https://doi.org/10.1002/ldr.419

Medeiros C, Serrano R, Martos J, Girón V (1997) Effect of various soil tillage systems on structure development in a Haploxeralf of central Spain. Soil Technol 11(2):197–204. https://doi.org/10.1016/S0933-3630(96)00131-6

Mkonda MY, He XH (2017a) Conservation agriculture in Tanzania. In: Lichtfouse (ed) Sustainable agriculture reviews 22. Springer, Switzerland, pp 309–324. https://doi.org/10.1007/978-3-319-48006-0_10

Mkonda MY, He XH (2017b) Long-term chemical fertilization in Tanzania. In: Lichtfouse (ed) Sustainable agriculture reviews 25. Springer, Switzerland, pp 261–276. https://doi.org/10.1007/978-3-319-58679-3_9

Mkonda MY, He XH (2017c) Yields of the major food crops: implications to food security and policy in Tanzania's semi-arid agro-ecological zone. Sustainability 9:1490. https://doi.org/10.3390/su9081490

Mkonda MY, He XH (2017d) Sustainable environmental conservation in East Africa through agroforestry systems: a case of the eastern arc mountains of Tanzania. Int J Sustain Green Energy 6(4):49–56. https://doi.org/10.11648/j.ijrse.20170604.11

Monfreda C, Ramankutty N, Foley JA (2008) Farming the planet. 2. Geographic distribution of crop areas, yields, physiological types, and net primary production in the year 2000. Glob Biogeochem Cycles 22:GB1022. https://doi.org/10.1029/2007GB002947

Msongaleli B, Rwehumbiza F, Tumbo S, Kihupi N (2015) Impacts of climate variability and change on rain-fed sorghum and maize: implications for food security policy in Tanzania. J Agric Sci 7(5). https://doi.org/10.5539/jas.v7n5p124

Nezomba H, Tauro TP, Mtambanengwe F, Mapfumo P (2010) Indigenous legume fallows (indifallows) as an alternative soil fertility resource in smallholder maize cropping systems. Field Crop Res 115:149–157

Nyadzi G, Janssen B, Oenema O (2006) Analysis of the effects of rotational woodlots on the nutrition and yield of maize following trees in western Tanzania. Agric Ecosyst Environ 116:93–103

Nyong A, Francis A, Osman-Elasha B (2007) The value of indigenous knowledge in climate change mitigation and adaptation strategies in the African Sahel. Mitig Adapt Strat Glob Chang 12:787–797

Okeyo A, Muchera-Muna M, Mugwe J, Ngetich K, Mugendi D, Diels J, Shisanya C (2014) Effects of selected soil and water conservation technologies on nutrient losses and maize yields in the central highlands of Kenya. Agric Water Manag 137:52–58

Osman-Elasha B, Goutbi N, Spanger-Siegfried E, Dougherty B, Hanafi H, Zakieldeen S, Sanjak A, Atti H, Elhassan, H (2006) Adaptation strategies to increase human resilience against climate variability and change: lessons from the arid regions of Sudan. AIACC Working Paper No. 42

Paavola J (2008) Livelihoods, vulnerability and adaptation to climate change in Morogoro. Tanzan Environ Sci Policy 11:642–654. https://doi.org/10.1016/j.envsci

Partey ST, Thevathasan NV (2013) Agronomic potentials of rarely-used agroforestry species for smallholder agriculture in Sub-Saharan Africa: an exploratory study. Commun Soil Sci Plant Anal 44(11):1733–1748

Pauline N, Vogel C, Grab S, Liwenga E (2016) Smallholder farmers in the Great Ruaha River sub-Basin of Tanzania: coping or adapting to rainfall variability? Clim Dev:1–14. https://doi.org/10.1080/17565529.2016.1184607

Poppy GM, Chiotha S, Eigenbrod F, Harvey C, Honzák M, Hudson M, Dawson T (2014) Food security in a perfect storm: using the ecosystem services framework to increase understanding. Philos Trans R Soc Lond Ser B Biol Sci 369(1639):20120288. https://doi.org/10.1098/rstb.2012.0288

Pretty J, Noble A, Dixon J, Hine J, Penning de Vries F, Morison J (2006) Resource-conserving agriculture increases yields in developing countries. Environ Sci Technol 40:1114–1119

Rowhani P, Lobell DB, Linderman M, Ramankutty N (2011a) Climate variability and crop production in Tanzania. Agric For Meteorol Sci Direct 151(2011):449–460

Rowhani P, Degomme O, Guha-Sapir D, Lambin E (2011b) Malnutrition and conflict in East Africa: the impacts of resource variability on human security. Clim Change 105:207–222. https://doi.org/10.1007/s10584-010-9884-8

Rusinamhodzi L, Corbeels M, van Wijk MT, Rufino MC, Nyamangara J., Giller K. E. (2011). A meta-analysis of long-term effects of conservation agriculture on maize grain yield under rain-fed conditions. Agron Sustain Dev 31: 657-673.

Sharma BR, Minhas PS (2005) Strategies for managing saline/alkali waters for sustainable agricultural production in South Asia. Agric Water Manag 78:136–151. https://doi.org/10.1016/j.agwat.2005.04.019

Sieber S, Tscherning K, Graef F, Sergio U, Paloma G (2015) Food security in the context of climate change and bioenergy production in Tanzania: methods, tools and applications, vol 15. Springer, Berlin/Heidelberg, pp 1163–1168. https://doi.org/10.1007/s10113-015-0834-x

Solomon T, Snyman H, Smit G (2007) Cattle-rangeland management practices and perceptions of pastoralists towards rangeland degradation in the Borana zone of southern Ethiopia. J Environ Manag 82:481–494

Sosovele H, Schechambo F, Kisanga D (1999) Rethinking natural resource degradation in semi-arid Sub-Saharan Africa: the case of semi-arid Tanzania. ODI, University of Dar Es Salaam, Tanzania

Thierfelder C, Wall PC (2009) Effects of conservation agriculture techniques on infiltration and soil water content in Zambia and Zimbabwe. Soil Tillage Res 28:209–220. https://doi.org/10.1111/j.1475-2743.2012.00406.x

Tilman D, Cassman KG, Matson PA, Naylor R, Polasky S (2002) Agricultural sustainability and intensive production practices. Nature 418:671–677

UNEP (2011) Food and ecological security: identifying synergy and trade-offs. UNEP, Nairobi http://www.unep.org/ecosystemmanagement/Portals/7/Documents/unep_policy_series/Food and Ecological solutionsJS.pdf

UNEP (2012) Avoiding future famines: Ecological foundation sustainable food systems. United Nations Environment Program (UNEP), Nairobi

URT (2007) United Republic of Tanzania, National Adaptation Program of Action (NAPA). Division of Environment, Vice President's Office, Dar es, Salaam

URT (2012) [Tanzania] National Sample Census of Agriculture 2007/2008 Small Holder Agriculture: Regional Report – Singida Region (Volume Vm). United Republic of Tanzania, Ministry of Agriculture, Food Security and Cooperatives, Ministry of Livestock Development and Fisheries, Ministry of Water and Irrigation, Ministry of Agriculture, Livestock and Environment, Zanzibar

URT (2014) Review of food and agricultural policies in the United Republic of Tanzania, MAFAP Country Report Series. FAO, Rome

Vanlauwe B (2004) Integrated soil fertility management research at TSBF: the framework, the principles, and their application. In: Bationo A (ed) Managing nutrient cycles to sustain soil fertility in Sub-Saharan Africa. Academy Science Publishers, Nairobi

Vanlauwe B., Wendt J., Giller K. et al. (2014). A fourth principle is required to define conservation agriculture in sub-Saharan Africa: the appropriate use of fertilizer to enhance crop productivity. Field Crops Res 155. pg 10–13.

Vermeulen S, Campbell B, Ingram J (2012) Climate change and food systems. Annu Rev Environ Resour 37(1):195–222. https://doi.org/10.1146/annurev-environ-020411-130608

Wall PC, Thierfelder C, Ngwira A, Govaerts B, Nyagumbo I, Baudron F (2013) Conservation agriculture in Eastern and Southern Africa. In: Jat RA, Graziano de Silva J (eds) Conservation agriculture: global prospects and challenges. CABI, Cambridge ISBN-13: 9781780642598

Yanda P (2015) Climate change implications for management and use of agricultural biodiversity resources in Africa. Environ Ecol Res 3(2):35–43 http://www.hrpub.org. https://doi.org/10.13189/eer.2015.030201

Chapter 10
Organic Agriculture for Food Security in Pakistan

Amir Raza, Saeed A. Asad, and Wisal Mohammad

Abstract The challenge of the twenty-first century is to ensure food security under prevailing climate change, notably in developing countries such as Pakistan. Current farming practices are far less effective in increasing soil resilience as they tend to decline soil fertility and organic matter contents. Organic farming systems improve organic matter, water holding capacity, porosity, biodiversity and productivity. Agriculture is the dominant user of fresh water resources in Pakistan. Water holding capacity of soils is generally higher under organic farming than conventional farming. This chapter presents the impacts of climate change on food security in Pakistan. We also compare the performance of organic and conventional agriculture in developed countries.

Keywords Climate change · Food security · Mitigation · Pakistan · Sustainable agriculture

10.1 Introduction

Global food production needs to be increased by up to 60% to feed 3 billion additional population by 2050. Climate change may make the task of achieving food security even more challenging, especially in the developing world as adaptation of food production sector to the effects of climate change will be imperative for survival (FAO 2009). In Pakistan, water, food and energy nexus is brutally imbalanced by a globally changing climate. Agriculture is the most vulnerable sector to climate change and it may lead to long term implications for agro-based economy. Agricultural productivity is affected by a number of climate change indicators including rainfall patterns, temperature rise, changes in sowing and harvesting

A. Raza (✉) · W. Mohammad
Nuclear Institute for Food and Agriculture, Peshawar, Pakistan

S. A. Asad
Centre for Climate Research and Development, COMSATS University, Islamabad 45550, Pakistan

E. Lichtfouse (ed.), *Sustainable Agriculture Reviews 32*, Sustainable Agriculture Reviews 32, https://doi.org/10.1007/978-3-319-98914-3_10

schedule, shortening of crop growth period, water availability, evapotranspiration and land suitability, thereby, making Pakistan amongst the leading countries worst hit by climate change (Kreft et al. 2015).

Empirical evidences revealed that climate change may increase the temperature by up to 4 °C by 2080, having serious implications for agriculture through effects on water, crops, soils and livestock (IPCC 2014). Reduced productivity of crops and livestock due to heat stress is identified as a major threat to food security (Devendra 2012). The country's limited capacity to adopt to climate change may result in compromised food and water security at national level. In Pakistan, land area is mostly arid and semi-arid where approximately, 60% of the area receives less than 250 mm of rainfall; its rivers are predominantly fed by glacier melts in the Hindukush- Karakoram-Himalaya region, which are reported to be receding rapidly due to global warming (Rasul et al. 2008).

Pakistan has minor contribution to the global Green House Gas (GHG) emissions but economic losses due to climate change are estimated to be $20 billion. Adaptation and mitigation can help to abridge the negative effects of climate change (UNFCCC 2006). The main mitigation potential lies in the capacity of agricultural soils to sequester CO_2 through building organic matter. This potential can be realized by employing sustainable agricultural practices, such as those within organic farming systems. These practices include the use of organic fertilizers and crop rotations i.e. legumes and cover crops etc. Mitigation is also achieved in organic agriculture through the avoidance of open biomass burning and reducing the use of chemical fertilizers, the production of which causes emissions from fossil fuel use (Niggli et al. 2009; Brodin 2016).

Organic farming practices contribute to adaptation through increasing soil organic matter and water retention capacity, creating more stable and fertile soils and reducing vulnerability to drought, extreme precipitation events, floods and water logging. Adaptation is supplemented by increased agro-ecosystem diversity of organic farms, based on management decisions, reduced nitrogen inputs and the absence of chemical pesticides. The high diversity together with the lower input costs of organic agriculture is key to reducing production risks associated with extreme weather events (de Moraes Sá et al. 2017; Jacobi et al. 2017; Untenecker et al. 2017).

This chapter presents an overview of the expected impacts of climate change on soil, crops, water and livestock. The vulnerability of food security under existing farming practices and potential of organic farming to mitigate the adversaries of climate change are discussed.

10.2 Land and Water Resources of Pakistan

10.2.1 Land Resources

Pakistan has total 79.6 million hectares (Mha) of land, out of which about 25 Mha is under cultivation. 86% of cultivable area is irrigated. 6.69 Mha of the cultivated lands are affected by varying levels of salinity. 8.3 Mha of cultivable wastelands could be used for agriculture (SDPI 2009) if irrigation water is made available.

10.2.2 Water Resources

Pakistan is blessed with rich surface and subsurface water resources. The major source of surface water in Pakistan is river Indus and its tributaries such as Chenab, Jhelum, Ravi, Sutlej and Beas receiving water from melting of glaciers in the Hindukush-Karakoram-Himalaya mountain ranges in the north and the monsoon rainfall. Indus flows contributes 141 out of total 191 maf of water flowing through Pakistan.

10.2.2.1 Irrigation System of Pakistan

The canal irrigation system in Pakistan was developed more than 100 years ago and is recognized as the world's largest integrated irrigation system (Sohag and Mahessar 2004). It originates from glaciers reserves in the Hindukush- Karakoram-Himalaya ranges in the north of Pakistan and consists of two large reservoirs on Indus River System (IRS), namely Tarbela Dam on river Indus in Khyber Pakhtunkhwa province and Mangla Dam on the river Jhelum in Azad Jammu and Kashmir, the allied state of Pakistan. Both dams have combined live storage capacity of 17.34 billion cubic meter (BCM). According to Indus River System Authority (2013), the storage capacity of these reservoirs has reduced by about 20%, to 14.69 billion cubic meter by the year 2011 due to siltation and deforestation whereby latter is highest in Pakistan compared with the remaining world (WWF 2009), and is considered as the major reason behind dam siltation.

Deforestation, siltation of reservoirs and rise in temperature may intensify the impact of climate change. It has projected that if siltation continues at the current rate, the combined storage capacity of these reservoirs is likely to reduce to 12.33 billion cubic meter (10 million acre feet) in the next 10 years. With the increased frequency and intensity of floods and droughts due to climate change, there will be even greater need to store the surplus water during peak flows.

A brief description of irrigation system is provided in Table 10.1.

10.2.2.2 Rainfall

Pakistan has arid to semi-arid climate with annual precipitation of 150–250 mm except in the north of the country that has humid to sub-humid conditions with total rainfall up to 1500 mm (Archer et al. 2010). The rainy season during July–September contributes 60–80% of total annual precipitation (Asif 2013).

10.2.2.3 River Flows

In Pakistan, the Indus Basin covers 65% of the country's total territory flowing through all the four provinces (FAO 2011). On average, 125 cubic kilometers of water is diverted in to the Indus Basin Irrigation System out of approximately 180

Table 10.1 Pakistan's irrigation system

Items	Description
Major reservoirs	2
Barrages / headworks	18
Link canals	12
Canal systems	45
Length of watercourses	107,000 km
Length of canals	56,073 km
Average Canal water diversion	17.34 billion cubic meter
Groundwater abstractions	51.31 billion cubic meter
Tube wells	1,000,000
Irrigated area	44.5 million acres

Source: Indus River System Authority (2013)

cubic kilometer of water entering the basin annually (Yu et al. 2013). 50–80% of the average river flows in the Indus River System are fed by snow and glacier melt in the Hindu Kush-Karakoram part of the Hindukush- Karakoram-Himalaya mountain ranges. The Upper Indus Basin, mostly the Karakoram part, has more than 5000 glaciers which cover a total glaciated area of about 15,000 sq. km. Their importance can be seen from the fact that the stored volume of ice in these glaciers is equivalent to about 14 years of average Indus River System flows. With the rise in sea level caused by climate change, the minimum flow requirements will also go up in future. The Indus River, flowing from Hindu Kush, Karakoram and Himalayan mountains and its tributaries are central to water security in Pakistan and stands as a primary source of fresh water along with replenishing its ground water resources.

10.2.2.4 Ground Water Status

Ground water is being used to supplement canal water for boosting crop yield in Pakistan. With decline in water flow from canals particularly for those farmers having holdings at tail end, the use of tube wells continued to rise for irrigation purposes. The sustainability of irrigated agriculture is facing a new threat of ground water shortage mainly because of over exploitation of water.

10.3 Impact of Climate Change

10.3.1 Land Resources

The potential effects of changing climate and higher atmospheric CO_2 on soils are highly interactive and complex. Empirical evidences suggest that land use for arable crops will decline because of yield reductions (Bouwman 1990; Jaggard et al. 2007).

Climate change may affect soil environment through nutrient depletion, soil erosion, loss of organic matter and soil biodiversity. The negative impact of climate change may be accelerated by unprecedented rainfall, flooding, drought, extensive shifts in land use and deforestation. In Pakistan, the soil resources are prone to all these indicators of climate change. The introduction and use of new high yielding hybrid maize varieties may increase the erosion rates because of wider row spacing and may also reduce soil thickness and yield of crops sown after the maize harvest. 27% more soil erosion was recorded when rainfall increased by up to 15% (Boardman and Favis-Mortlock 1993).

Floods and droughts are key climate change triggered events affecting soils across the country and their impacts may vary with duration and magnitude of event. The impacts may also vary with soil cover, soil management history, technology and knowledge available with farmers to adapt to climate related hazards. Floods may render the soil unfit for cultivation through damage to soil structure by erosion, loss of top productive soil layers, water logging and anoxia (Misra 2013). Different soil types and land use influence the soil vulnerability to climate change. For example, intensively cultivated soils, soils with low organic matter contents, bare soils, soils on slopes, loamy sands and fine sandy loams are more vulnerable to climate triggered extreme events. Similarly, rain-fed areas are at high vulnerability to the adversaries of climate change due to poor economic conditions of farmers with limited capacity to adopt new technologies according to the changing scenarios. 5 Mha of the cultivated area (24% of the total cultivated area) in Pakistan is rain-fed (Aslam 2016).

10.3.2 Water Resources

As already discussed in Sect. 10.2.2.3, water flow in the Indus basin is mainly derived from snow and glacier melts in the Hindukush- Karakoram-Himalaya mountains, so it is influenced by temperature and precipitation changes affecting the accumulation and ablation of snow and glaciers respectively in the Upper Indus basin (Immerzeel et al. 2010; Yu et al. 2013; Shrestha et al. 2015). Glaciers dependent rivers of South Asia have been projected to be the climate change hot spots due to their sensitivity to climate change indicators (De Souza et al. 2015). Climate change will have a relatively low impact on overall discharge in the near future (until 2050), but shifts in peak flows are likely, whereby peak run-offs may be shifting to earlier in the year. These changes in peak flows accompanied by changing monsoon rainfall patterns may reduce agricultural produce (Imran et al. 2014).

In Pakistan, mean temperature has risen by 0.6–1 °C and precipitation has declined by 30–40% but with rising intensity of monsoon. There is an observed 0.5–0.7% increase in solar radiation over southern half of the country. With these effects, the glacial cover of the country is melting rapidly. Gebre and Ludwig (2014) projected that mean monthly maximum temperatures will continue to increase in the future and maximum summer temperatures could increase by 1–4 °C in the period between 1970 and 2100. However, the temperature in the Himalayas and other high elevation regions may increase three times higher that of the global average (Parry et al. 2007; Rasul et al. 2011). The estimated time for glaciers melt is 45 years after which significant decline in water flow would result into 40–50% less water availability for irrigation thereby causing food insecurity in 62% (74) districts of the country (Amir 2008).

Lack of environmental flows to the deltaic area is likely to result in water scarcity, sea level rise and exposing around 2.26 million inhabitants. In the upstream, the areas around the river are under the threat of glacial lake outburst floods, which collectively are the major climate-related threats Pakistan faces presently. Pakistan 's further depreciation from a water-stressed to a water-scarce nation, due to anthropogenic activity and a changing climate, influences the country 's ability to cope with rising food and energy demands.

Many factors including demographic growth, primitive inefficient methods used for irrigating the agricultural crops and over exploitation of ground water and climate change are the major impediments to limit the country's water security. More recently, a report by International Monetary Fund (IMF 2015) demonstrates that Pakistan is physically going to be water scarce by 2025. On the other hand, accompanied by gradual increase in temperature, agricultural crop water requirements are increasing particularly in summer. The increasing gap between supply and demand of water would have serious repercussions for the agriculture sector which withdraws more that 85% of water flowing in the Indus River. According to Farooqi et al. (2005), water and agricultural sectors are likely to be worst affected by changes in climatic scenarios. Fresh water availability is going to be severely impacted by climate change accompanied by severe water shortages in the arid and semi-arid regions and increasing tendency of floods in river deltas.

"As water quantity and quality are interlinked, any shift in water flow would impact its quality accordingly. For example, depleted and lower water levels enhance pollutant concentrations whereas, flooding increases turbidity and flush the contaminants in to the water bodies. In the coming decades, surface water temperatures are going to increase as a result of increasing air temperatures. Warmer air temperatures may also result in lowering the water levels and decreased duration of ice cover. Consequently, these changes may lower the levels of dissolved oxygen, elevated levels of phosphorus accompanied with odor and taste problems" (Farooqi et al. 2005).

Pakistan has limited scope for expanding the supplies of water, so it will have to go for improving the efficiency of water use in all the sectors, particularly in the agriculture sector. A detailed account on how water use efficiency can be improved for sustainable agriculture is already available (see Raza et al. 2012a, b). The major climate change related threats to water security in Pakistan include increased variability of river flows influenced by changing patterns of monsoon and winter rains and loss of natural reservoirs in the form of glaciers. Along with these, severity of extreme weather events, increased demand of irrigation water because of higher evaporation rates increasing sedimentation, changes in the seasonal pattern of river flows due to early start of snow and glacier melting and glaciers retreat are also the major climate change consequences on the water resources of Pakistan (Rasul 2010). The threats to water security are highlighted in Fig. 10.1.

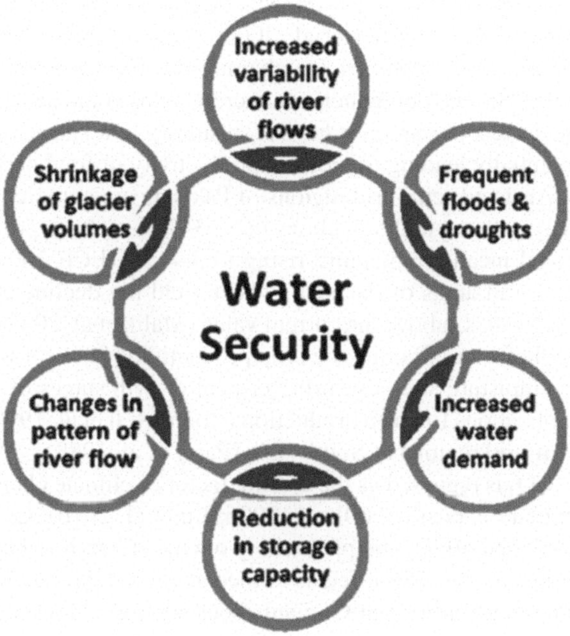

Fig. 10.1 Threats to water security due to climate change

10.3.3 Crop Production

The environment for agriculture is drastically changing since last 10,000 years due to human-induced climate change. Amongst climate change indicators, heat stress is found to be a severe threat for global food security (IPCC 2007). Due to the increasing carbon intensity, global warming is expected to increase gradually if no mitigation efforts are put in place to address it. Agricultural productivity and food security would be directly affected by rise in temperature (Schmidhuber and Tubiello 2007; Ainsworth and Ort 2010). According to Battisti and Naylor (2009), most of the agricultural zones around the world are projected to be most likely exposed to the record average high temperatures by the end of twenty-first century.

Among the most significant impacts of climate change is the potential increase of food insecurity and malnutrition (Zewdie 2014). Like many other countries in south Asia, agriculture in Pakistan is the most vulnerable sector to potential risks triggered by climate change. Under warming trends, glacier melting may accelerate accompanied by early start of melt season thus altering the crop growing seasons in the coming decades.

The temperature being the main thermal agent controlling the plant metabolic processes would ultimately influence the biomass production and thus grains and fruits (Hay and Walker 1989). Increasing trend in global warming may result in limited photosynthesis, reduced light interception accompanied by increased phenological development (Tubiello et al. 2007). Many researches in the past have concluded that negative impacts of global warming would be more pronounced in tropical area (Fischer et al. 2005) compared with temperate regions where these changes may result in lengthening the crop growing season (Olesen and Bindi 2002). Significant reductions in crop yields could also result from elevated temperatures speeding up crop growth, resulting in a shortening of the growth period and potentially leading to reduced productivity of food and fodder crops (GoP 2013).

Arid and semi-arid regions of Pakistan are at more risk regarding wheat productivity, where yield reductions up to 34% have been projected to occur by 2050 with global mean temperature rising to 3–4 °C (ECF 2004). However, under irrigated semi-arid areas of Pakistan, wheat yield my decline up to 30% when temperature rises by 4 °C above the current value (Malik et al. 2005). Temperature is not the only single factor affecting the crop productivity, but heat waves, dry spells and limiting soil moisture stresses during critical growth stages of the crop would have serious implications for crop production (Rounsevell et al. 1999) particularly in the arid and semi-arid regions of South Asia (Parry et al. 1988).

CO_2 is regarded as the driving factor of climate change, however, its direct effect on plant is positive (Olesen 2006; Lu et al. 2016). CO_2 enriches atmosphere positively and affects the plants in two ways. First, it increases the photosynthesis process in plants. This effect is termed as carbon dioxide fertilization effect. This effect is more prominent in C_3 plants because higher level of CO_2 increases rate of fixed carbon and also suppresses photorespiration. Second, increased level of CO_2 in atmosphere decreases the transpiration by partially closing of stomata and hence declines the water loss by plants. Both aspects enhance the water use efficiency of plants causing increased growth (Sayed 2011).

A research study conducted by Global Change Impact Studies Centre in 2009, evaluated the impact of climate change on wheat crop in four different agro-ecological zones namely northern mountainous region, northern sub-mountainous region, southern arid plains, and southern semi-arid plains. The data showed that with the increase in average temperature, the interval of the growth period of wheat will be shorten for all agro-ecological zones in Pakistan. Therefore, with a temperature increase in the range of 1–5 °C, the productivity of wheat will increase in the mountainous regions, however, the wheat yield will decrease drastically in the sub-mountainous, arid and semi –arid areas (GCISC 2009).

Impacts of climate change may vary among crops. Janjua et al. (2010) speculated 8–10% decrease in wheat yield by 2030 noting thereby that dependence on wheat may shift to coarse grains like sorghum, millets, coarse rice and oats. Worth noting that these reductions in cereal production would be observed due to reduced supply of water influenced by climate change. Yield reductions for this staple crop are projected to be more pronounced in mountainous areas of Pakistan where temperature increase of 1.5–3 °C may result in the yield declines of up to 24% in the north western province of Pakistan. Precipitation increases of up to 15% during the cropping season has been projected to have non-significant impact on the yield of wheat (Hussain and Mudasser 2007).

Sugarcane being an important cash crop in Pakistan is very sensitive to extreme weather and climate related events such as atmospheric CO_2, rainfall and temperature. The potential negative impacts of climate indicators on sugarcane production are very much evident in many developing countries including Pakistan, Zimbabwe, and Fiji. A case study from Fiji in 1994 indicated up to 50% yield reductions of sugarcane under drought conditions (Gawander 2007). Increasing temperatures may accelerate the evapotranspiration and reduce the amount of water available in the soil to make the planting of sugarcane very difficult and may eventually increase the use of surface water (de Carvalho et al. 2015). These negative impacts are more striking in Sindh province, the southern Pakistan alongside the coastal areas. Although increasing temperature and CO_2 may enhance the vegetative growth of crop under controlled settings and pot experiments (Vu and Allen Jr 2009), but sugarcane production is highly vulnerable to climate change indicators like heat, frost, flooding and drought (Knox et al. 2010; Chandiposha 2013).

Under the changing climate scenarios, rice production may be reduced by up to 20% in Pakistan (Tariq et al. 2014). Consequences of climate change on maize production could be even more severe where temperatures increase up to 1.8 °C, will reduce the commodity production by 20% beyond 2050 (Khaliq 2008). The negative impacts of rising temperature and changing rainfall patterns would be more striking for southern areas of Pakistan. It has been projected that a 1 °C rise in temperature during the vegetative and flowering stages of cotton growth would reduce yield by 24% and 8%, respectively (Raza and Ahmad 2015). The yield losses of major crops of Pakistan due to climate change are summarized in Table 10.2.

Triggered by these climatic changes, the likely consequences for agricultural sector could be shifts in cropping patterns and crop rotations and significantly reduced production of Pakistan's main cash crops including; wheat, rice, sugarcane, cotton and maize (Abid et al. 2016). The ways by which climate change may impact and hamper the productivity of agriculture sector are depicted in Table 10.3.

10.3.4 Livestock and Fisheries

Like crop production, livestock sector is also highly vulnerable to the climate change particularly in the developing countries like Pakistan (Musemwa et al. 2012; Moreki and Tsopito 2013). This vulnerability exists as adaptation strategies are either lacking at policy level or low education level of livestock farmers. Climate

Table 10.2 Climate change induced yield losses of major crops in Pakistan

Crop	Yield Losses (%)	Reference
Wheat	8–34	Malik et al. (2005)
		Janjua et al. (2010)
Rice	Up to 20	Tariq et al. (2014)
Cotton	8–24	Raza and Ahmad (2015)
Sugarcane	Up to 50	Gawander (2007)
Maize	Up to 20	Khaliq (2008)

Table 10.3 Impacts of climate change on productivity of agriculture

S.No.	Impacts of climate change
1	Increased frequency of climate change events may affect management interventions
2	Shortening of life cycle of crops due to increase in temperatures
3	Reduction in crop yield and decrease in production from dairy and fisheries sector
4	Changes in pattern of river flow may disturb existing farming systems
5	Increase in water demands under rising temperature to sustain crops in the field
6	Land degradation due to extreme weather events liked droughts and floods

change has been reported to have direct or indirect effects on livestock. Direct effects include heat exchange between the animal and its niche. These direct effects are all influenced by climate change indicators like warming temperatures and shifts in rainfall patterns. Indirect effects triggered by climate change include animal feed shortages, reduced feed production due to changes in land use patterns to prioritize food production and exposure to pests and pathogens (Moreki and Tsopito 2013).

Climate change affects physiology of livestock and may hinder growth, milk and meat production. Zewdu et al. (2014) observed reduced fertility and milk production by dairy animals in response to heat stress. Changing climatic conditions may impact livestock sector by having serious ramifications on pastures and forage production, feed availability and price, livestock production, their health and reproduction ability, diseases and pest attack, water quantity and quality, biodiversity and loss of genetic diversity and soil infertility (Silanikove and Koluman 2015). The livestock dependent on grazing are likely to be more vulnerable to globally rising temperatures than the industrialized or managed systems. This may be due to low rainfall, and direct effects of high temperatures and solar radiations on livestock (Nardone et al. 2010). Pastures in the arid and semi-arid regions are more vulnerable to climate change depending upon geographical conditions and magnitude and severity of extreme weather events (Sautier et al. 2013).

Heat sensitivity negatively affects milk production by cows which is directly related to less feed intake under warmer conditions. Reduced milk production is also linked to metabolic adaptations in response to heat stress which are activated by milk-born negative feedback system down regulating the synthesis, secretion of milk and blood flow and glucose uptake by mammary glands. Cows subjected to heat stress had elevated level of insulin (Baumgard and Rhoads 2012, 2013; Rhoads et al. 2013). Warming temperatures inhibit the productivity of livestock and dairy animals by reducing their appetite, slowing growth accompanied by less milk and meat productivity. Reproduction also decreases which is further hampered by diseases and pests. Pests have tendency to expand in the newly climate stressed areas, with ability to survive better in the warming winters, hence their control requires increased use of pesticides (Silanikove and Koluman 2015). Livestock production is anticipated to decline by 30% as rangelands become increasingly stressed by longer droughts and by human and animal migration around riverine areas. Animals' water requirements increase with temperature, but in many places climate change is likely

Table 10.4 Effects of climate change on livestock

Direct effects	Indirect effects
Heat stress	Shortage of feed for animal
Increase in water requirements	Reduction in fertility rates
Reduction in milk and meat production	Extinction of breeds of animals
Reduction in appetite	Reduction in weight gains by animals
Elevated levels of insulin	Incidence of pests and diseases

to mean that water becomes scarce and supplies become more unpredictable. The direct and indirect effects of changing climate on livestock are summarized in Table 10.4.

Climate change also threatens the survival of the strategic reservoir of crop and livestock genetic resources needed to adapt production systems to future challenges. As conditions change, crop varieties and breeds may be abandoned by peasants and livestock farmers, and may become extinct if steps are not taken to ensure their conservation. Catastrophic extreme weather events such as; floods and droughts, which in many parts of the world are expected to become more frequent particularly in Pakistan because of climate change, can pose an immediate threat to the survival of breeds and varieties that are raised only in specific small geographical areas.

Fisheries are key sector to support economies and important social structures in many countries especially in developing economies (Allison et al. 2009). Empirical evidences indicate that climate change is the latest threat to the world's fast declining fish stocks (UNEP 2008; Cochrane et al. 2009). Marine fisheries are and will be exposed to ocean acidification, sea level rise, improper mixing of water column, increasing sea surface temperatures, storm intensity and changing rainfall patterns influenced by climate change. The sensitivity of fish stocks to these changes will determine the distribution and adaptability to these changes and access to marine resources by fisher folk (Johnson and welch 2010).

Global models predict an increase by 5–12% in the wind speed of tropical cyclones as a result of warming temperatures (IPCC 2007), which is likely to affect shallow and coastal fish habitats, limiting access to fish stocks and hence fishing (FAO 2007). Extreme weather events will also disturb the nutrient balance of marine ecosystems affecting productivity and distribution of fish stocks (Garcia and Grainger 2005).

Average air temperatures around the globe have increased by 0.74 °C, over the last century and are projected to increase by 4 °C towards the end of twenty-first century. Oceans have absorbed approximately 80% of the additional heat thereby increasing global ocean temperature by 0.5 °C since 1961. Sea surface temperatures are projected to increase over the current century with tropical oceans experiencing the increases from 1 to 3 °C in many regions (IPCC 2007; Lough 2007). Minor increases in ocean temperatures would have serious implications on the fish physiology, distribution, life cycle events and biodiversity (Munday et al. 2008; Boyce et al. 2008; Brierley and Kingsford 2009). Similar is the case with CO_2 concentration which is being over absorbed by the oceans since 1750. This additional CO_2

reacts with sea water to form carbonic acid and lowering the pH of sea water (McNeil and Matear 2007). Oceanic pH is projected to decrease by 0.35 units (Kleypas et al. 2006; IPCC 2007) rendering the ocean water more acidic than at any time in the past 300 million years (Caldeira et al. 2003).

Sea level is rising due to increased melting of glaciers and ice sheets because of warming air temperatures. According to IPCC (2007), average sea level is rising at 3.1 mm per year, with larger rises projected up to 1.4 m (Hansen 2007) depending on the rapid melting of ice caps and glaciers (Schneider 2009). Resultantly, coastal habitats would be impacted to the maximum extent, destroying the fisheries infrastructure and millions of people living in the coastal areas (UNEP 2007) whose livelihood and subsistence is dependent on coastal fisheries.

Marine ecosystems in Pakistan will be exposed directly to changing climate variables and indirectly to ecosystem responses to climate change. Research conducted on the fisheries provide a clear indication of vulnerability of fish population to climate change, which may affect species richness, reproduction potential, abundance, distribution and community structure (Munday et al. 2008; Brierley and Kingsford 2009). Among others include fishing mortality, pollution, habitat loss, disturbances and introduction of alien species. The two subsectors of agriculture namely livestock and fisheries have been paid least attention which need to be addressed both at research and policy level.

10.4 Organic Farming for Mitigating Climate Change

In previous sections, a detailed picture of climate change impacts on different sectors of agriculture and water resources have been presented. It is pertinent to mention here that contribution of Pakistan towards global greenhouse gases emission is less than 1% but still it ranks among the top ten countries worst influenced by climate change. As Pakistan is predominantly an agricultural country, where farming is mainly for subsistence. Any extreme event may jeopardize the national food security and thus needed to opt for a sustainable agricultural system. Under the prevailing scenarios, the sustainable productivity is the need of the hour and organic farming could help the farming community to live with climate change as will be highlighted in the synthesis below:

IPCC recommends to adopt sustainable farming systems with reduced reliance on external inputs (e.g. rotations which include legume crops). Sustainable agricultural practices such as organic agriculture strongly reduce the reliance on external inputs by recycling wastes as nutrient source, using nitrogen-fixing plants, improving cropping systems and landscapes, avoiding synthetic pesticides, integrating crops and animals into a single farm production sector and including grass clover leys for fodder production, while avoiding purchase of feed concentrates. Intercropping of deep and shallow rooted crops also increases productivity and nutrient efficiency through nitrogen resource management. In order to avoid nutrient losses, especially soils need to be kept covered permanently by crops in an optimized sequence. In organic agriculture, the inclusion of cover and catch crops is both a conventional and modern practice of farming (Thorup-Kristensen et al. 2003).

The fertilizer use efficiency is low in conventional farming and high levels of reactive nitrogen (NH_4, NO_3) in soils may contribute to the emission of nitrous oxides. The fertilizer use efficiency further decreases with increasing fertilization, because a significant part of the fertilizer either leaches down to the water bodies or evaporates in to the atmosphere rather than being up taken by plants. Recycling nitrogen by using manure and nitrogen fixing plants enhances soil fertility under organic and low external input agriculture. Organic and green manures as well as nitrogen from legumes can be managed very precisely due to the design of the crop rotations including cover and catch crops (Thorup-Kristensen et al. 2003). Nitrogen from legumes is more sustainable in terms of ecological integrity, energy flows and food security than nitrogen from industrial sources (Crews and Peoples 2004).

Conventional stockless arable farms are dependent on the input of synthetic nitrogen fertilizers, and manure and slurry from livestock farms become an environmental issue. In these livestock operations, nutrients are available in excess and over-fertilization may occur. The concept of either mixed farms or close cooperation between crop and livestock operations – a common practice in sustainable farming, especially organic ones – can contribute considerably to mitigation and adaptation. N-application rates in organic agriculture are usually 60–70% lower than in conventional agriculture because of the recycling of organic residues and manures. In addition, the limited availability of N in organic systems requires careful, efficient management (Kramer et al. 2006).

In organic agriculture, diversifying crop rotations with green manure improves soil structure and diminishes N_2O emissions. Soils managed organically are more aerated and have significantly lower mobile nitrogen concentrations, which further reduces N_2O emissions as higher soil carbon levels may lead to N_2 emission rather than N_2O (Flessa et al. 2002; Mathieu et al. 2006). A reduction of the Global Warming Potential has been observed on Dutch organic dairy farms and in organic pea production areas compared with conventional (Bos et al. 2007). Studies revealed that organic potatoes, tomatoes, and various other vegetables had less GHG emissions than conventional crops (Öko-Institut 2007). Most of the organic farms have lower input of nutrients by farmyard manure on grassland and pastures as well as to fewer environmental problems such as phosphorous run-off, nitrogen leaching into deeper soil layers and emission of N_2O (Weiske et al. 2006; Olesen. 2006).

The application of sustainable management techniques that build up soil organic matter have the potential to balance a large part of the agricultural emissions, although their effects over time may be reduced as soils are built up. Long-term comparison field trials in temperate climate zones have shown no slowing of sequestration for more than 30 years. Modelling of sequestration potentials of a shift from conventional to organic agriculture in Scandinavia gives a time span of 50 to 100 years (Foereid and Høgh-Jensen 2004). Shift to organic farming may help to reduce approximately 20% of the greenhouse gases emissions from agriculture sector.

In the context of subsistence agriculture and in regions with periodic disruptions of water supply like in Pakistan, organic agriculture is competitive to conventional agriculture and often superior with respect to cop yields. Numerous case studies showed that crop yields in organic farming were 20% higher than conventional subsistence farming (Halberg et al. 2006; Badgley et al. 2007; Sanders 2007).

Organic agriculture has huge potential, both in terms of the recommendations of the IPCC (IPCC 2007) and for future food security. Organic agriculture helps to reduce erosion caused by wind and water along with improving soil organic matter and fertility, conserves biodiversity, reduces environmental degradation impacts and integrates farmers in to high value food chains (Niggli et al. 2008).

Organic agriculture, is based on practical farming skills, observation, personal experience and intuition – traditional systems that function without reliance on modern inputs compared with conventional intensive agriculture which has neglected the basic skill and knowledge of farming. This practical adaptation "reservoir" of knowledge (Tengö and Belfrage 2004) is important for manipulating complex agro-ecosystems, for breeding locally adapted seeds and livestock, and for producing on-farm fertilizers (compost, manure, green manure) and inexpensive nature-derived pesticides.

Farming practices that conserve the soil fertility are important for the future of agriculture and food security. Soil organic matter may help mitigate the effects of extreme weather events, while increasing primary crop productivity. Soils under organic management have been reported to retain more rainwater. In different long-term field experiments in the USA, organic matter was considerably higher in organically managed than in conventional soils (Marriott and Wander 2006). In addition, higher organic matter content and more biomass in soils make organic fields less prone to soil erosion (Reganold et al. 1987; Siegrist et al. 1998).

In the Rodale farming system trial, the amount of water percolating through the top 36 cm of soil was 15–20% higher in the organic systems than in the conventional ones. Moreover, the organic soils held 816,000 l per ha in the top 15 cm of soil. This water reservoir significantly improved the yields of corn and soybean in dry years (Lotter et al. 2003). Under limiting water conditions during the growing period, yields of organic farms are equal or higher than those of conventional agriculture. A meta-analysis of 133 scientific papers by Badgley et al. (2007) showed that organic agriculture was particularly competitive in the low yield environments in developing countries. These findings underline that the technique inherent to organic farming of investing in soil fertility by means of green manure, leguminous intercropping, composting and recycling of livestock manure could contribute considerably to reducing greenhouse gases while also increasing global food productivity.

Water capture during torrential rains was two folds higher in organic settings than those in conventional practices (Lotter et al. 2003). This extra water holding capacity of fields subjected to organic farming significantly reduced the risk of floods, an effect that could be very important if organic agriculture were practiced on larger scale. Observations of biodynamic systems in India found decreased irrigation needs of 30–50%. Better soil structure, aeration and drainage, lower bulk density, higher organic matter content, soil respiration, more earthworms and a deeper topsoil layer are all associated with the lower irrigation need (Proctor and Cole 2002). In Tigray Province, one of the most degraded parts of Ethiopia, agricultural productivity was doubled by soil fertility techniques such as compost application and introduction of leguminous plants into the crop sequence.

The diversity of landscapes, farming activities, fields and agro biodiversity is greatly enhanced in organic agriculture (Niggli et al. 2008), which makes these farms more resilient to unpredictable weather patterns resulting from climate change (Bengtsson et al. 2005; Hole et al. 2005). Enhanced biodiversity reduces pest outbreaks (Zehnder et al. 2007; Pfiffner and Luka 2003; Pfiffner et al. 2003). Diversified agro-ecosystems reduce the severity of plant and animal diseases, while improving utilization of soil nutrients and water (Altieri et al. 2005). A brief account of comparison of organic and conventional intensive farming is presented in Table 10.5.

Sustainable and organic agriculture offer multiple opportunities to reduce greenhouse gases and counteract global warming. For example, organic agriculture reduces energy requirements for production systems by 25–50% compared to conventional chemical-based agriculture. Reducing greenhouse gases through their sequestration in soil has even greater potential to mitigate climate change. Improving soil sequestration of carbon is desirable in both low and high yielding crop and animal systems. However, soil improvement is particularly important for agriculture in developing countries where crop inputs such as chemical fertilizers and pesticides are not readily available along with high costs, and the knowledge needed for their proper application is not widespread.

In order to reduce trade-offs among food security, climate change and ecosystem degradation, productive and ecologically sustainable agriculture is crucial. In that context, organic agriculture represents a multi-targeted and multifunctional strategy. It offers a proven alternative concept that is being implemented quite successfully by a growing number of farms and food chains. Many of the organic agriculture's practices can be implemented within other sustainable farming systems. The system oriented and participative concept of organic agriculture, combined with new sustainable technologies offer greatly needed solutions in the changing climatic scenarios.

Table 10.5 Comparison between organic and conventional farming

Organic farming	Conventional farming
Limited reliance on external inputs	Higher reliance on external inputs
Improved efficiency of nutrients through use of manures, composts produced at farm	Low fertilizer use efficiency of externally applied chemical fertilizers
Low levels of emissions	Higher levels of emissions
Low energy requirements for fertilizer production	Higher energy requirements for industrial scale production of chemical based fertilizers
Long term improvement of yield and yield stability	Long term decline in yield and yield stability
Improves soil water retention capacity	Decreases soil water retention capacity
Improves soil fertility and soil stability through buildup of organic matter over time	Decreases long term soil fertility and soil stability in most of the cases
Farming practices help to overcome adversaries of climate change	Farming practices enhance vulnerabilities to climate induced events

In Pakistan, organic farming can be practiced along with conventional agriculture. Low input responsive crop varieties need to be developed along with improving water and nutrient use efficiency at field level. Organic agriculture can help mitigate climate change by either reducing greenhouse gases emissions or by sequestering CO_2 from the atmosphere in the soil. By practicing organic farming, the country may be able to convert losses from climate change into gains particularly in terms of carbon sequestration.

The competitive advantages of organic farming systems over conventional farming as indicated in Table 10.4 reveal the tremendous potential that organic farming practices offer to farming community in the current scenario of climate change. The climate induced risks further hamper the capacity of farmers to invest in farming business, so switching over to organic farming may offer benefits in terms of saving of fertilizer costs, and once organic farming systems become self-sufficient in meeting farm N requirements through use of legumes, manures, etc. the yield and net returns to farmers becomes at par with conventional farming systems.

Fertilizer and water use efficiency is low in Pakistan thereby organic farming offers solution to improve the same to uplift the socio-economic conditions of farmers. The current energy crisis can also be reduced to some extent if demands of chemical fertilizers are reduced by switching over to organic farming systems. In the section on impacts of climate change on water resources, we have already highlighted that Pakistan is going to be a water scarce country by 2035. This scarcity can be managed by efficient utilization of available water resources and this efficient utilization can be achieved through adoption of organic farming practices that help to conserve water and enhance water retention capacity of soil.

The current farming practices in Pakistan are not sustainable as they are leading to nutrient mining and soil deterioration. The intensive cultivation is contributing towards decrease in soil stability and making them prone to wind and water erosion. The adoption of organic farming practices will help to improve soil organic matter contents and enable the farmers to fulfil the crop nutrient requirements from indigenous resources like compost, manures, etc. The soil building capacity of organic farming systems will help the Pakistani farmers to improve long term soil fertility and stability to mitigate potential damages caused by increasing frequency of floods and droughts.

The buildup of organic matter through organic farming helps to reduce the impact of drought and floods on account of potential benefits that soil organic matter provide in terms of improving water and nutrient holding capacity of soils. Organic matter helps the soils to hold and release water and nutrients over a longer period of time. Most of the Pakistani soils are inherently low in organic matter and current farming practices are further lowering its content, so it is highly recommended to adopt the practices that can naturally build up soil organic matter contents. It is pertinent to indicate that organic farming is climate neutral while conventional farming systems have no potential to neutralize the impacts of weather extremes. The shift from conventional to organic farming in Pakistan may be slow and initially less profitable but the potential long term benefits that organic farming offers in mitigating climate change may outweigh the short term gains under current unsustainable farming practices.

10.5 Conclusion

Climate change is posing a real threat to the agro-based economy of Pakistan despite the fact that country is not making a major contribution towards the global greenhouse gas emissions. Country has limited technical knowhow and poor infrastructure to withstand the adversaries of changing climate. The main impacts of climate change would be on the agriculture sector with indirect impact on national water and food security that has a nexus with energy sector as about one third of the national power generation is from hydro power resources. Living with changing climate has emerged as a challenge for developing country under the scenarios of rapidly increasing population and deteriorating natural resources. Situation demands to have a massive and quick change in policy as well as at farm level to identify and implement farming practices that are environment friendly and increase soil resilience to climate change. We analyzed the impact of changing climate on land, water, crops and livestock sector and presented a synthesis on how organic farming practices can help to achieve food security under changing climate in the context of Pakistan.

Climate change may affect land resources negatively under both irrigated and rain-fed management systems. The magnitude of damage may vary with duration and intensity of climate induced events. Lands under rain-fed conditions are liable to severe damage from droughts with consequences of losses of productivity. Farmers under such conditions have limited capacity to cope with the damage caused by weather extremes on account of their poor financial conditions. Floods may render irrigated lands unfit for production on account of damage by erosion, loss of organic matter and top productive soil. Climate change poses serious threats to national water and food security through increased intensity of droughts and floods. The limited water storage capacity and low water use efficiency particularly in the agriculture sector further magnify the issue of water scarcity and necessitate to switch over to efficient methods of using water.

Crop productivity is also negatively affected by changing climatic conditions. Climate change is expected to reduce yield of major crops. High temperatures may hasten the crop maturity and reduce portioning of assimilates for grain formation and increase the crop evaporative demands. Water scarcity may reduce area under cultivation of crops that need more water like rice and sugarcane and farmers will have to modify cropping pattern accordingly. Due to floods and droughts, area under fodder and minor crops may reduce and it may put pressure on food supplies for both humans as well as livestock. Disease outbreak in flood hit areas may lead to mortality of animals. High temperatures may negatively affect growth, milk and meat production from animals. Fisheries sector is vulnerable to changing climate and national fish production may drop if immediate attention is not given by policy makers and researchers.

We have demonstrated that climate change is going to negatively impact agriculture and its sub-sectors. Our synthesis in Sect. 10.4 clearly indicates that organic farming practices have globally proven potential to mitigate adversaries of climate change. Farming community can take advantage of potential benefits of organic farming in mitigating climate change by adopting sustainable farming practices. As farming in Pakistan is mainly for subsistence, so farmers need education and awareness on principles and positive aspects of such sustainable farming solutions available to them. Commercial farms need to recognize the fact that although yields are initially low under organic farming but net long term returns are reasonably high under these systems of farming. Pakistan needs to make a partial shift from conventional to organic farming on account of multiple advantages of organic farming. These benefits include improvements in long term soil fertility, increase in soil resilience, improved water holding capacity and increase in biodiversity.

References

Abid M, Schilling J, Scheffran J, Zulfiqar F (2016) Climate change vulnerability, adaptation and risk perceptions at farm level in Punjab, Pakistan. Sci Total Environ 547:447–460. https://doi.org/10.1016/j.scitotenv.2015.11.125. Epub 2016 Feb 4

Ainsworth EA, Ort DR (2010) How do we improve crop production in a warming world? Plant Physiol 154:526–530. https://doi.org/10.1104/pp.110.161349

Allison EH, Perry AL, Badjeck MC, Adger WN, Brown K, Conway D, Halls AS, Pilling GM, Reynolds JD, Andrew NL, Dully NK (2009) Vulnerability of national economies to the impacts of climate change on fisheries. Fish Fisher 10:173–196. https://doi.org/10.1111/j.1467-2979.2008.00310.x

Altieri MA, Ponti L, Nicholls C (2005) Enhanced pest management through soil health: toward a belowground habitat management strategy. Biodynamics (summer):33–40

Amir P (2008) Glaciers melt and agriculture: are we running against time? Need for a science-based response. In Planning Commission of Pakistan, "Glaciers behaviour under climate change and its impacts on agriculture in Pakistan". Seminar proceedings held on June 4

Archer DR, Forsythe N, Fowler HJ, Shah SM (2010) Sustainability of water resources management in the Indus Basin under changing climatic and socio economic conditions. Hydrol Earth Syst Sci 14:1669–1680. https://doi.org/10.5194/hess-14-1669-2010

Asif M (2013) Climate change, irrigation water crisis and food security in Pakistan. Dissertation, Uppsala University

Aslam M (2016) Agricultural productivity current scenario, constraints and future prospects in Pakistan. Sarhad J Agric 32(4):289–303. https://doi.org/10.17582/journal.sja/2016.32.4.289.303

Badgley C, Moghtader J, Quintero E, Zakem E, Chappell MJ, Avilés-Vàzquez K, Samulon A, Perfecto I (2007) Organic agriculture and the global food supply. Ren Agric Food Syst 22:86–108. https://doi.org/10.1017/S1742170507001640

Battisti DS, Naylor RL (2009) Historical warnings of future food insecurity with unprecedented seasonal heat. Sci 323:240–244. https://doi.org/10.1126/science.1164363

Baumgard LH, Rhoads RP (2012) Ruminant production and metabolic responses to heat stress. J Anim Sci 90:1855–1865. https://doi.org/10.2527/jas.2011-4675. Epub 2011 Dec 28

Baumgard LH, Rhoads RP (2013) Effects of heat stress on post absorptive metabolism and energetics. Ann Rev Anim Biosci 1:311–337. https://doi.org/10.1146/annurev-animal-031412-103644. Epub 2012 Dec 13

Bengtsson J, Ahnström J, Weibull A-C (2005) The effects of organic agriculture on biodiversity and abundance: a meta-analysis. J Appl Ecol 42:261–269. https://doi.org/10.1111/j.1365-2664.2005.01005.x

Boardman J, Favis-Mortlock DT (1993) Climate change and soil erosion in Britain. Geogr J 159:179–183. https://doi.org/10.2307/3451408

Bos JFFP, de Haan JJ, Sukkel W, Schils RLM (2007) Comparing energy use and greenhouse gas emissions in organic and conventional farming systems in the Netherlands. Paper presented at the 3rd QLIF congress: improving sustainability in organic and low input food production systems, University of Hohenheim, Germany, March 20–23, 2007

Bouwman AF (1990) Soil and the greenhouse effect. Wiley, New York, USA

Boyce DG, Tittensor DP, Worm B (2008) Effect of temperature on global patterns of tuna and billfish richness. Marine Ecol Prog Ser 355:267–276. https://doi.org/10.3354/meps07237

Brierley AS, Kingsford MJ (2009) Impacts of climate change on marine organisms and ecosystems. Curr Biol 19:602–614. https://doi.org/10.1016/j.cub.2009.05.046

Brodin G (2016) Organic farming's role in adaptation to and mitigation of climate change – an overview of ecological resilience and a model case study. Student thesis series INES nr 379. Department of Physical Geography and Ecosystem Science, Lund University, Sweden, p 48

Caldeira K, Jain AK, Hoffert MI (2003) Climate sensitivity uncertainty and the need for energy without CO2 emissions. Science 299:2052–2054. https://doi.org/10.1126/science.1078938

Chandiposha M (2013) Potential impact of climate change in sugarcane and mitigation strategies in Zimbabwe. Afr J Agric Res 8: 2814–2818. DOI: https://doi.org/10.5897/AJAR2013 .7083

Cochrane K, De Young C, Soto D, Bahri T (2009) Climate change implications for fisheries and aquaculture: overview of current scientific knowledge. FAO fisheries and aquaculture technical paper no. 530

Crews TE, Peoples MB (2004) Legume versus fertilizer sources of nitrogen: ecological tradeoffs and human needs. Agric Ecosyst Environ 102:279–297. https://doi.org/10.1016/j.agee.2003.09.018

de Carvalho AL, Menezes RSC, N'obrega RS, Pinto AS, JPH O, Rando C, Giarolla A (2015) Impact of climate changes on potential sugarcane yield in Pernambuco, northeastern region of Brazil. Renew Energy 78:26–34. https://doi.org/10.1016/j.renene.2014.12.023

de Moraes Sá JC, Lal R, Cerri CC, Lorenz K, Hungria M, de Faccio Carvalho PC (2017) Low-carbon agriculture in South America to mitigate global climate change and advance food security. Environ Int 98:102–112. https://doi.org/10.1016/j.envint.2016.10.020

De Souza K, Kituyi E, Harvey B, Leone M, Murali K, Ford J (2015) Vulnerability to climate change in three hot spots in Africa and Asia: key issues for policy-relevant adaptation and resilience building research. Reg Environ Chang 15:747–753. https://doi.org/10.1007/s10113-015-0755-8

Devendra C (2012) Climate change threats and effects: challenges for agriculture and food security. Akademi Sains, Kuala Lumpur, 66 pp

ECF (2004) What is dangerous climate change? Initial results of a symposium on key vulnerable regions, Climate Change and Article 2 of the UNFCCC, Buenos Aires, COP 10 Event at the EU pavilion, 27–30 October 2004, Beijing, China

Farooqi BA, Khan HA, Mir H (2005) Climate change perspective in Pakistan. Pak J Met 2:3

Fischer G, Shah M, Tubiello FN, van Velhuizen H (2005) Socio-economic and climate change impacts on agriculture: an integrated assessment, 1990–2080. Philos Trans R Soc 29:2067–2083. https://doi.org/10.1098/rstb.2005.1744

Flessa H, Ruser R, Dörsch P, Kamp T, Jimenez MA, Munch JC, Beese F (2002) Integrated evaluation of greenhouse gas emissions (CO_2, CH_4, N_2O) from two farming systems in southern Germany. Agric Ecosyst Environ 91(1–3):175–189. https://doi.org/10.1016/S0167-8809(01)00234-1

Foereid B, Høgh-Jensen H (2004) Carbon sequestration potential of organic agriculture in northern Europe – a modelling approach. Nutr Cycl Agroecosyst 68(1):13–24

Food and Agriculture Organization (FAO) (2009) High level expert forum-how to feed the world in 2050. Economic and social development department. Retrieved from http://www.fao.org/fileadmin/templates/wsfs/docs/Issues_papers/HLEF2050_Global_Agriculture.pdf

Food and Agriculture Organization (FAO) (2011) Indus basin. Retrieved from http://www.fao.org/nr/water/aquastat/basins/indus/index.stm

Food and Agriculture Organization of the United Nations (2007) Adaptation to climate change in agriculture, forestry and fisheries: perspective, framework and priorities. Interdepartmental

Working Group on Climate Change, Rome: United Nations Report. Available from ftp://ftp.fao.org/docrep/fao/009/j9271e/j9271e.pdf

Garcia SM, Grainger RJR (2005) Gloom and doom? The future of marine capture fisheries. Philos Trans R Soc 360:21–46. https://doi.org/10.1098/rstb.2004.1580

Gawander J (2007) Impact of climate change on sugar-cane production in Fiji. WMO Bull 56:34–39

Gebre SL, Ludwig F (2014) Spatial and temporal variation of impacts of climate change on the hydrometeorology of Indus River Basin using RCPs scenarios, South East Asia. J Earth Sci Clim Chang 5:241. https://doi.org/10.4172/2157-7617.1000241

Global Change Impact Study Centre (GCISC), Islamabad (2009) In Iqbal M., Sultana H and others, Vulnerability and adaptability of wheat production in different climatic zones of Pakistan under climate change scenarios. No GCISC-RR-14

Government of Pakistan (2013) Pakistan economic survey 2013–2014. Finance Division, Economic Advisors Wing, Islamabad

Halberg N, Sulser TB, Høgh-Jensen H, Rosegrant MW, Knudsen MT (2006) Chapter 10: The impact of organic farming on food security in a regional and global perspective. In: Halberg N, Alrøe HF, Knudsen MT, Kristensen ES (eds) Global development of organic agriculture: challenges and prospects. CABI Publishing, Wallingford, pp 277–322

Hansen JH (2007) Sea level rises are coming-unless we act now. New Sci 2614:30–33

Hay RKM, Walker AJ (1989) An introduction to the physiology of crop yield. Longman Scientific and Technical, New York, p 292

Hole DG, Perkins AJ, Wilson JD, Alexander IH, Grice PV, Evans AD (2005) Does organic farming benefit biodiversity? Biol Conserv 122:113–130. https://doi.org/10.1016/j.biocon.2004.07.018

Hussain SS, Mudasser M (2007) Prospects for wheat production under changing climate in mountain areas of Pakistan-an econometric analysis. Agric Syst 94:494–501. https://doi.org/10.1016/j.agsy.2006.12.001

Immerzeel WW, Van Beek LP, Bierkens MF (2010) Climate change will affect the Asian water towers. Sci 328:1382–1385. https://doi.org/10.1126/science.1183188

Imran A, Zaman Q, Rasul G, Mahmood A (2014) An analytical study of variations in the monsoon patterns over Pakistan. Pak J Met 10:20

Intergovernmental Panel on Climate Change (2007) The physical science basis. IPCC Secretariat, Geneva

International Monetary Fund (2015) .Issues in managing water challenges and policy instruments: Regional perspectives and case studies. Retrieved from http://www.imf.org/external/pubs/ft/sdn/2015/sdn1511tn.pdf

IPCC (2014) Climate Change 2014: Synthesis report, contribution of working groups I, II and III to the fifth assessment report of the Intergovernmental Panel on Climate Change (Core Writing Team, Pachauri RK, Meyer LA (eds)). IPCC, Geneva, Switzerland, p 151

IRSA (2013) At http://www.pakirsa.gov.pk/irsa.html. Accessed 20 Mar 2013

Schmidhuber J, Tubiello FN (2007) Global food security under climate change. Proc Natl Acad Sci 104(50):19703–19708

Jacobi J, Bottazzi P, Pillco MI, Schneider M, Rist S (2017) Building farm resilience in a changing climate: challenges, potentials, and ways forward for smallholder cocoa production in Bolivia. In: Identifying emerging issues in disaster risk reduction, migration, climate change and sustainable development. Springer, Cham, pp 231–247

Jaggard KW, Qi A, Semenov MA (2007) The impact of climate change on sugarbeet yield in the UK: 1976–2004. J Agric Sci 145:367–375. https://doi.org/10.1017/S0021859607006922

Janjua PZ, Samad G, Khan NU (2010) Impact of climate change on wheat production: a case study of Pakistan. Pak Dev Rev 49:799–822

Johnson JE, Welch DJ (2010) Marine fisheries Management in a Changing Climate: a review of vulnerability and future options. Rev Fish Sci 18:106–124. https://doi.org/10.1080/10641260903434557

Khaliq T (2008) Modeling the impact of climate change on maize (*Zea mays* L.) productivity in the Punjab. Dissertation, University of Agriculture, Faisalabad, Pakistan

Kleypas JA, Feely RA, Fabry VJ, Langdon C, Sabine CL, Robbins LL (2006) Impacts of ocean acidification on coral reefs and other marine calcifiers: a guide for future research. Report of the ocean acidification workshop-the 11th international coral reef symposium in Ft. Lauderdale, Florida

Knox JW, Rodríguez Díaz JA, Nixon DJ, Mkhwanazi M (2010) A preliminary assessment of climate change impacts on sugarcane in Swaziland. Agric Syst 103:63–72. https://doi.org/10.1016/j.agsy.2009.09.002

Kramer SB, Reganold JP, Glover JD, Bohannan BJM, Mooney HA (2006) Reduced nitrate leaching and enhanced denitrifier activity and efficiency in organically fertilized soils. Proc Natl Acad Sci U S A 103:4522–4527

Kreft S, Eckstein D, Junghans L, Kerestan C, Hagen U (2015) Global climate risk index 2015. Who suffers most from extreme weather events? pp 1–31

Lotter D, Seidel R, Liebhardt W (2003) The performance of organic and conventional cropping systems in an extreme climate year. Am J Altern Agric 18:146–154. https://doi.org/10.1079/AJAA200345

Lough J (2007) Climate and climate change on the great barrier reef. In: Johnson JE, Marshall PA (eds) Climate change and the great barrier reef: a vulnerability assessment. Great Barrier Reef Marine Park Authority, Townsville

Lu X, Wang L, McCabe MF (2016) Elevated CO2 as a driver of global dryland greening. Sci Rep 6:20716. https://doi.org/10.1038/srep20716

Malik W, Hussain MS, Abid S, Mudasser M, Hussain SS (2005) Impact of climate change on wheat productivity in selected cropping systems in Punjab. In: Muhammed A, Hussain SS (eds) Global change perspective in Pakistan. Proceedings of National Workshop, April 28–30, Islamabad, Pakistan. Asianics Agro Dev International, Islamabad, pp 55–61

Marriott EE, Wander MM (2006) Total and labile soil organic matter in organic and conventional farming systems. Soil Sci Soc Am J 70:950–959

Mathieu O, Lévêque J, Hénault C, Milloux MJ, Bizouard F, Andreux F (2006) Emissions and spatial variability of N2O, N2 and nitrous oxide mole fraction at the field scale, revealed with 15N isotopic techniques. Soil Biol Biochem 38:941–951. https://doi.org/10.1016/j.soilbio.2005.08.010

McNeil BI, Matear RJ (2007) Climate change feedbacks on future oceanic acidification. Tellus Ser. B Chem Phy Meteor 59:191–198. https://doi.org/10.1111/j.1600-0889.2006.00241.x

Misra AK (2013) Climate change impact, mitigation and adaptation strategies for agricultural and water resources, in ganga plain (India). Mitig Adapt Strateg Glob Chang 18:673–689. https://doi.org/10.1007/s11027-012-9381-7

Moreki J, Tsopito C (2013) Effect of climate change on dairy production in Botswana and its suitable mitigation strategies. J Animal Feed Res 3:216–221

Munday PL, Jones GP, Pratchett MS, Williams AJ (2008) Climate change and the future of coral reef fishes. Fish Fish 9:1–25. https://doi.org/10.1111/j.1467-2979.2008.00281.x

Musemwa L, Muchenje V, Mushunje A, Zhou L (2012) The impact of climate change on livestock production amongst the resource-poor farmers of third world countries: a review. Asian J Agric Rural Dev 2:621–631

Nardone A, Ronchi B, Lacetera N, Ranieri MS, Bernabucci U (2010) Effects of climate changes on animal production and sustainability of livestock systems. Livestock Sci 130:57–69. https://doi.org/10.1016/j.livsci.2010.02.011

Niggli U, Hepperly P, Fließbach A, and Mäder P (2008) Does organic farming have greater potential to adapt to climate change? In: Neuhoff D et al (eds), Cultivating the future based on science. Volume 2 – Livestock, Socio-economy and Cross disciplinary Research in Organic Agriculture. Proceedings of the Second Scientific Conference of the International Society of Organic Agriculture Research (ISOFAR), held at the 16th IFOAM Organic World Congress in Cooperation with the International Federation of Organic Agriculture Movements (IFOAM) and the Consorzio ModenaBio, 18–20 June 2008 in Modena, Italy, 586–589

Niggli U, Fließbach A, Hepperly P, Scialabba N (2009) Low greenhouse gas agriculture: mitigation and adaptation potential of sustainable farming systems. FAO, Rome. Rev. 2 – 2009

Öko-Institut (2007) Arbeitspapier: Treibhausgasemissionen durch Erzeugung und Verarbeitung von Lebensmitteln. Authors: Fritsche U, Eberle U Öko-Institut Darmstadt. Download at the Öko-Institut Homepage at http://www.oeko.de/oekodoc/328/2007-011-de

Olesen JE (2006) Climate change as a driver for European agriculture. SCAR-foresight in the field of agricultural research in Europe, expert paper

Olesen JE, Bindi M (2002) Consequences of climate change for European agricultural productivity, land use and policy. Eur J Agron 16:239–262. https://doi.org/10.1016/S1161-0301(02)00004-7

Parry ML, Carter TR, Konjin NT (eds) (1988) The impact of climatic variations on agriculture, Assessments in Semi-Arid Regions. Kluwer Academic Publishers, Dordrecht

Parry ML, Canziani OF, Palutikof JP, van der Linden PJ, Hanson CE (Ed.) (2007) Climate change: Impacts, adaptation and vulnerability. Contribution of Working Group II to the Fourth Assessment Report of the Intergovernmental Panel on Climate Change, 2007. Cambridge University Press, Cambridge/New York

Pfiffner L, Luka H (2003) Effects of low-input farming systems on carabids and epigeal spiders – a paired farm approach. Basic Appl Ecol 4:117–127. https://doi.org/10.1078/1439-1791-00121

Pfiffner L, Merkelbach L, Luka H (2003) Do sown wildflower strips enhance the parasitism of lepidopteran pests in cabbage crops? International Organisation for Biological and Integrated Control of Noxious Animals and Plants/West Palaearctic Regional Section Bulletin 26:111–116

Proctor P, Cole G (2002) Grasp the nettle: making biodynamic farming and gardening work. Random House Publishing, Auckland

Rasul G (2010) The role of the Himalayan mountain systems in food security and agricultural sustainability in South Asia. Int J Rural Manage 6(1):95–116

Rasul G, Dahe Q, Chaudhry QZ (2008) Global warming and melting glaciers along southern slopes of HKH ranges. Pak J Met 5(9):63–76

Rasul G, Chaudhary QZ, Mahmood A, Hyder KW, Dahe Q (2011) Glaciers and Glacial Lakes under changing climate in Pakistan. Pak J Met 8(15):1–8

Raza A, Ahmad M (2015) Analyzing the impact of climate change on cotton productivity in Punjab and Sindh, Pakistan. Climate change working paper series (no.9). Retrieved from http://www.pide.org.pk/pdf/Climate Change 9.pdf

Raza A, Friedel JK, Bodner G (2012a) Increasing water use efficiency for sustainable agriculture. In: E. Lichtfouse (ed.), Agroecology and strategies for climate change, Sust Agric Rev 8. https://doi.org/10.1007/978-94-007-1905-7_8, © Springer Science+Business Media B.V., pp 167–211

Raza A, Khanzada SD, Ahmad S, Afzal M (2012b) Improving water use efficiency for wheat production in Pakistan. Pak J Agric Agric Engg Vet Sci 28(1):27–39

Reganold JP, Elliot LF, Unger YL (1987) Long-term effects of organic and conventional farming on soil erosion. Nature 330:370–372. https://doi.org/10.1038/330370a0

Rhoads RP, Baumgard LH, Suagee JK (2013) Metabolic priorities during heat stress with an emphasis on skeletal muscle. J Anim Sci 91:2492–2503. https://doi.org/10.2527/jas.2012-6120

Rounsevell MDA, Evans SP, Bullick P (1999) Climate change and agricultural soils—impacts and adaptation. Climatic Change. In: IPCC, 2001 (ed), Climate Change 2001: impacts, adaptation and vulnerability, vol. 43. Cambridge University Press, Cambridge, MA, pp 683–709

Sanders J (2007) Economic impact of agricultural liberalisation policies on organic farming in Switzerland. PhD thesis, Aberystwyth University

Sautier M, Duru M, Martin-Clouaire R (2013) Use of productivity-defined indicators to assess exposure of grassland-based livestock systems to climate change and variability. Crop Pasture Sci 64:641–651. https://doi.org/10.1071/CP13076

Sayed AS (2011) Climate change and its realities for Pakistan. Symposium on "changing environmental pattern and its impact with special focus on Pakistan". Pakistan Engineering Congress

Schneider SH (2009) The worst-case scenario. Nature 458:1104–1105. https://doi.org/10.1038/4581104a

SDPI (2009) Food insecurity in Pakistan, 129 pp. http://www.sdpi.org/publications/files/food%20insecurity%20in%20Pakistan%202009%20.pdf

Shrestha A, Agrawal N, Alfthan B, Bajracharya S, Maréchal J, van Oort B (eds) (2015) The Himalayan climate and water atlas: impact of climate change on water resources in five of Asia's Major River basins. ICIMOD, GRID-Arendal and CICERO, Kathmandu

Siegrist S, Staub D, Pfiffner L, Mäder P (1998) Does organic agriculture reduce soil erodibility? The results of a long-term field study on loess in Switzerland. Agric Ecosyst Environ 69:253–264. https://doi.org/10.1016/S0167-8809(98)00113-3

Silanikove N, Koluman N (2015) Impact of climate change on the dairy industry in temperate zones: predications on the overall negative impact and on the positive role of dairy goats in adaptation to earth warming. Small Ruminant Res 123:27–34. https://doi.org/10.1016/j.smallrumres.2014.11.005

Sohag MA, Mahessar AA (2004) Telemetry system in irrigation system. In National Conference on Emerging Technologies

Tariq A, Tabasam N, Bakhsh K, Ashfaq M, Hassan S (2014) Food security in the context of climate change in Pakistan. Pak J Commerce Soc Sci 8:540–550

Tengö M, Belfrage K (2004) Local management practices for dealing with change and uncertainty: a cross-scale comparison of cases in Sweden and Tanzania. Ecol Soc, 9. Online: www.ecologyandsociety.org/vol9/iss3/art4

Thorup-Kristensen K, Magid J, Jensen LS (2003) Catch crops and green manures as biological tools in nitrogen management in temperate zones. Adv Agron 79:227–302

Tubiello FN, Soussana JFO, Howden SM (2007) Crop and pasture response to climate change. Proc Natl Acad Sci 104:19686–19690. https://doi.org/10.1073/pnas.0701728104

UNFCCC (2006) Technologies for adaptation to climate change. Bonn, Germany. Retrieved from http://unfccc.int/resource/docs/publications/tech_for_adaptation_06.pdf

United Nations Environment Programme (2007) Global outlook for snow and ice. United Nations Report

United Nations Environment Programme (2008) In dead water; merging of climate change with pollution, over-harvest and infestations in the world's fishing grounds. United Nations Report

Untenecker J, Tiemeyer B, Freibauer A, Laggner A, Luterbacher J (2017) Tracking changes in the land use, management and drainage status of organic soils as indicators of the effectiveness of mitigation strategies for climate change. Ecol Indic 72:459–472. https://doi.org/10.1016/j.ecolind.2016.08.004

Vu JCV, Allen LH Jr (2009) Growth at elevated CO2 delays the adverse effects of drought stress on leaf photosynthesis of the C4 sugarcane. J Plant Physiol 166:107–116. https://doi.org/10.1016/j.jplph.2008.02.009

Weiske A, Vabitsch A, Olesen JE, Schelde K, Michel J, Friedrich R, Kaltschmitt M (2006) Mitigation of greenhouse gas emission in European conventional and organic dairy farming. Agric Ecosyst Environ 112:221–232

World Wide Fund for Nature – Pakistan (2009): Pakistan Forest Programme

Yu W, Yang YC, Savitsky A, Alford D, Brown C, Wescoat J, Debowicz D (2013) The Indus basin of Pakistan: the impacts of climate risks on water and agriculture. World Bank Publications, Washington, DC

Zehnder G, Gurr GM, Kühne S, Wade MR, Wratten SD, Wyss E (2007) Arthropod pest management in organic crops. Ann Rev Entomol 52:57–80. https://doi.org/10.1146/annurev.ento.52.110405.091337

Zewdie A (2014) Impacts of climate change on food security: a literature review in sub Saharan Africa. J Earth Sci Clim Chang 5:225. https://doi.org/10.4172/2157-7617.1000225

Chapter 11
Impact of Recombinant DNA Technology and Nanotechnology on Agriculture

Praveen Guleria and Vineet Kumar

Abstract Agriculture has direct impact on the food status and economy of any country. For the last two decades, attempts have been made to improve agricultural production by using recombinant DNA technology and nanotechnology. Application of recombinant DNA technology and nanotechnology induces direct interaction of transgene and nanoparticles with the components of the agroecosystem. Escape of transgene from transgenic plants invades wild plant types, leading to the generation of superweeds with enhanced invasiveness. Further, the transgene occurs within soil particles in suspended form in the soil microbiome. In this form, the transgene interacts with soil microbial communities and enters the food chain via bioaccumulation and biomagnification. Likewise, interaction of nanoparticles with soil components may enhance nanoparticle toxicity. Such alterations may modify the growth and survival of microbes and plants. This chapter presents the toxic effects of recombinant DNA technology and nanotechnology on the various components of agroecosystem.

Keywords Plants · Soil · Microbes · Ecosystem · Impact · Recombinant DNA technology · Nanotechnology

11.1 Introduction

Agricultural ecosystem or agroecosystem is a conventional unit of the well known biological ecosystems. It is an artificial ecosystem under direct human impact. Agroecosytem constitutes of soil, soil dwelling microorganisms, plant types, their

P. Guleria (✉)
Plant Biotechnology & Genetic Engineering Laboratory, Department of Biotechnology, School of Life Sciences, DAV University, Jalandhar, Punjab, India

V. Kumar
Department of Biotechnology, School of Biotechnology and Biosciences, Faculty of Technology and Sciences, Lovely Professional University (LPU), Phagwara, Punjab, India

E. Lichtfouse (ed.), *Sustainable Agriculture Reviews 32*, Sustainable Agriculture Reviews 32, https://doi.org/10.1007/978-3-319-98914-3_11

pollinators, predators of agricultural pests, prevailing genetic diversity of food crops and the dependent animals (Swift and Anderson 1994). Agricultural ecosystem not only supports the food crops, but also serves as an efficient habitat for wild plant types. Wild plant species are bank of valuable genetic resource, hence, their maintenance in agroecosystem offers gene bank for supporting plant breeding to sustain the productivity and produce in global climate change.

Agroecosystem affects the environmental, societal and economic development of any area. So, enhancement in the agricultural land and produce will not only feed the poor/hungry but also help in maintenance of rural development and urbanization. However to sustain the food requirements, agricultural ecosystem management is highly required. Presently, population and urbanization is increasing day by day. So, in order to meet the qualitative and quantitative requirements of food, agriculture ecosystem needs to be enhanced (Fischer et al. 2007).

Recombinant DNA (rDNA) technology and nanotechnology are the two emerging technologies that have revolutionized various aspects of agriculture. rDNA technology has undoubtedly being documented to improve the growth, development, yield and plant-environmental interactions (Dunwell 2000; Sharma et al. 2002; Wan 2015). Likewise, nanotechnology is an interdisciplinary field that is presently being widely explored for assessing their role in plant growth, development and environmental interactions (Mukhopadhyay 2014; Sekhon 2014). Present review thus discusses the impact of rDNA and nanotechnology on agricultural ecosystem and its components.

11.2 Recombinant DNA Technology: Impact on Agroecosystem

Recombinant DNA Technology is collection of molecular genetic techniques that allow isolation, cloning and expression of gene from one organism in same or another organism (Firidin 2010). The new DNA thus expressed or formed is known as recombinant DNA (rDNA). This process is also known as genetic engineering (Firidin 2010). The rDNA approaches has been widely implemented in agriculture with the aim of improvement in plant growth, development and yield. Hence collectively these techniques are termed as agricultural biotechnology when implemented for agricultural ecosystem. The plants developed using these techniques are termed as transgenic plants (Herdt 2006).

Transgenic crops like canola, maize, rice, brinjal and cotton have already been developed. Various private organizations have already generated and commercialized such transgenic crops (James 2003). Most of the transgenic plants developed till date is genetically modified for single trait. Only 20% of the total developed transgenic plants are engineered for multiple traits (James 2004). Primarily, herbicide and pesticide transgenic plants are developed that have remarkably improved the status of overall agricultural produce. Later, transgenic plants for improved

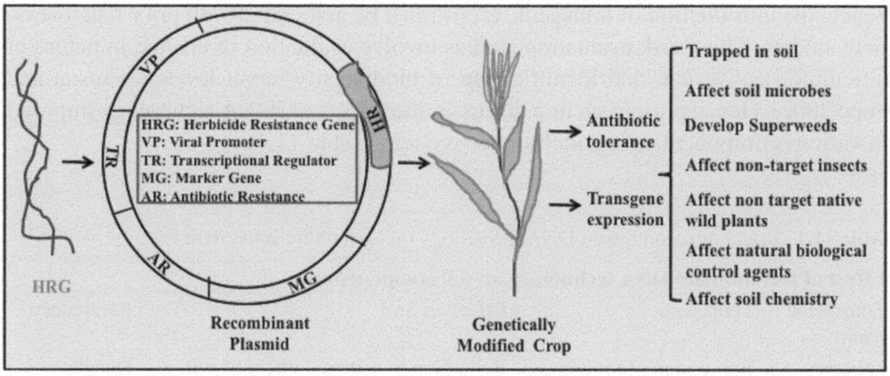

Fig. 11.1 The figure depicts the development of genetically modified plants using recombinant DNA technology. A plasmid carrying viral promoters, transcriptional regulator, marker genes is modified with herbicide resistance transgene (HRG) to form a recombinant plasmid. This recombinant plasmid is used to develop transgenic plants having resistance against herbicides. Since the transgene interacts with complete host system, it can undergo rearrangements and can lead to certain predictable as well as unpredictable responses like influence on soil functionality, microbial populations and wild plants

nutrition and stress tolerance have also been developed (Datta 2013). Various reports documenting the impact of rDNA technology on agriculture improvement are already present (Dunwell 2000; Sharma et al. 2002; Wan 2015).

For making transgenic plants, the gene of interest is inserted into the test plant (Fig. 11.1). The transgene/gene works in any cell by intercommunication and reciprocity. Hence, a single gene having a particular expression can respond variously inside the cell (Ho 2000). Further for inserting the gene into plant, vector vehicles are also required. The vectors possess viral promoters, antibiotic resistance and marker genes and certain transcriptional regulators.

Hence, the transgene can simply integrate in the plant genome, induce mutation to the host genome or itself undergo rearrangements. Therefore, the transgenic crop might produce the functional required protein or can synthesize toxic/ allergic proteins (Maghari and Ardekani 2011).

As a result inspite of the success stories of rDNA technology in agriculture, various articles/reports stating the rDNA technology as a danger to environment/biodiversity has been documented. It has been documented that genetically developed plants should undergo more stringent pre-marketing risk assessments than conventionally breeding developed plants. So, the toxicity evaluation of generation of genetically modified organisms has now become a product based approach rather than process based approach (Johnson and Hope 2009).

The agricultural ecosystem is generally related with the farmed lands. The biodiversity of agricultural ecosystem keeps on changing due to various known or unknown reasons like pollution, climate change, impact of human activities or due to variation in predator prey relationship. In the presence of such unstable ecosystem, the introduction of transgenic plants can further violate the conditions of agricultural ecosystem.

Hence, the introduction of transgenic crops must be associated with prior risk assessment studies. These risk evaluation studies involve evaluation of change in nature of agricultural ecosystem and identification of biodiversity target levels (Johnson and Hope 2009). Hence present section discusses the effects of rDNA technology imposed on various components of agricultural ecosystem (Table 11.1).

Table 11.1 Effect of recombinant DNA technology on agriculture ecosystem

Effect of recombinant DNA technology on soil components			
Transgenic plant	Transgene	Effect on Soil	References
Tobacco	Expression of proteinase inhibitor I having insecticidal activity	Reduction in carbon content of soil, reduction in decomposition process, shift in trophic group composition	Donegan et al. (1997)
Tobacco	Gene encoding neomycin phosphotransferase II	Adsorption of the recombinant DNA on the soil particles	Widmer et al. (1996); Widmer et al. (1997)
Cotton	*Bacillus thuringiensis* Cry gene	Variation in enzymatic activities of urease, phosphatase, dehydrogenase, phenol oxidase and protease	Shen et al. (2006)
Maize	*Bacillus thuringiensis* Berliner Cry1Fa2 protein (Bt) and glyphosate herbicide tolerance	Enhanced nitrogen content of soil but reduced phosphorus levels	Liu et al. (2010a, b)
Effect of recombinant DNA technology on non-target soil organisms			
Transgenic plant	Transgenic trait	Response	References
Cotton	*Bacillus thuringiensis* var. kurstaki endotoxin	Enhanced population of soil bacteria and fungi	Donegan et al. (1995)
Corn	*Bacillus thuringiensis* (Bt) Cry1Ab endotoxin	Reduced population of lacewing larvae, biological control agents of insects and mite pests in the agricultural fields	Hilbeck et al. (1998)
Alfalfa	Genes overepxressing α-amylase and lipid peroxidase genes	Variation in soil microbial populations, plant growth parameters and soil chemistry, increment in population of culturable, cellulose utilizing and aerobic spore forming bacteria, pH of soil reduced and enzyme activity of alkaline phosphatase and dehydrogenase enzymes reduced	Donegan et al. (1999)
Corn	*Bacillus thuringiensis* (Bt) Cry1Ab endotoxin	Dispersed pollens to nearby milkweed plants, infected larvae of monarch butterfly, reduced growth and increased mortality rates of monarch larvae	Losey, et al. (1999)
Sugarbeet	*Kanamycin resistance*	Horizontal gene transfer from transgenic sugarbeet to *Acinetobacter* sps, bacterial strain expressed kanamycin resistance	Nielsen et al. (2000)

(continued)

Table 11.1 (continued)

Brassica napus	Herbicide tolerance	Increased fatty acid component of root associated microbial communities	Dunfield and Germida (2001)
Tobacco	*aadA* gene expressing spectinomycin and streptomycin resistance	*In situ* transfer of *aadA* gene to *Acinetobacter* sp. strain BD413	Kay et al. (2002)
Potato	Gene expressing cysteine proteinase inhibitors	Suppressed population growth of bacterial and fungal microbial communities	Cowgill et al. (2002)
Corn	*Bacillus thuringiensis* (Bt) Cry1Ab endotoxin	Colonization of non-target mychorrhizal symbiont *Glomus mosseae* lowered	Castaldini et al. (2005)
Maize	*Bacillus thuringiensis* (Bt) Cry1Ab endotoxin	Reduced body mass and growth parameters and hatchability of snails	Kramarz et al. (2009)
Cotton	*Bacillus thuringiensis* (Bt) Cry1Ab endotoxin	Enhanced population sizes of mirid bugs, non-Bt toxin target plant pests	Lu et al. (2010)
Effect of recombinant DNA technology on non-target plants			
Transgenic plant	Transgenic trait	Response	References
Oilseed rape	Glufosinate tolerance	Gene flow to *B. campestris*	Mikkelsen et al. (1996)
Oilseed rape	Glufosinate tolerance	Gene flow to wild mustard	Lefol et al. (1996)
Arabidopsis thaliana	Csr1-1 gene for herbicide resistance	20 times higher affinity to pollinate wild plant types and contaminate wild genetic base	Bergelson et al. (1998)
B. napus	Glyphosate resistance	Introgression of glyphosate resistance transgene wild *B. rapa,* stable trasgene expression in wild plants in absence of glyphosate	Snow et al. (1999); Warwick et al. (2008)

11.2.1 Effect of rDNA on Soil Chemistry

Soil is the important component of agricultural ecosystem as it supports plant germination, growth, survival and nutrient cycling. The solid matter, liquid and gas fractions of soil determine its structure. The varying aggregation of these components determines the chemical composition and structure of soil. The formation of capillary complexes is quite important as they regulate the water exchange between soil and air, thus regulating the growth of plants (Atlas and Bartha 1982). The presence of pores in soil particles determine the soil atmosphere also known as atmosphere lithosphere interface. These pores if filled with water, displaces the soil atmosphere. Soil atmosphere containing oxygen is termed as oxic and if lacks oxygen, then it is termed as anoxic (Sexstone et al. 1985). The presence of various organic and inorganic components like carbon, nitrogen determines the chemical composition of soil (Brady 1984). Hence, genetically modified plants if sown in fields could have possible access to affect the discussed components of soil.

However, reports documenting the impact of genetically modified crops on soil and its components are few. Some of these are discussed as under.

In 1997, the effect of planned introduction of genetically modified tobacco on soil parameters was evaluated. The tobacco plants were modified for the expression of proteinase inhibitor I having insecticidal activity. The effect of decomposition of transgenic and parent tobacco plants on soil composition and related microbial parameters was assessed. A significant variation in the decomposition of both samples was observed. The carbon content of transgenic litter was considerably reduced than control litter. Reduction in carbon has further delayed the process of litter decomposition. Moreover, increase in the population of nematodes and reduction in population of colembolla surrounding transgenic litter was observed. Increase in nematode population reflected shift in trophic group composition. Further the changes were associated with variation in the feeding habits of bacteriovorous and fungal nematodes. Hence, overall the field exposure of transgenic tobacco had induced negative alterations in the agroecosystem (Donegan et al. 1997).

The soil parameters on germination of transgenic Bt cotton (Sukang-103) and its non-Bt cotton counterpart (Sumian-12) were evaluated. Few significant variations in the enzymatic activities of urease, phosphatase, dehydrogenase, phenol oxidase and protease were observed (Shen et al. 2006). However, it has been demonstrated that Bt plants secrete transgenic bacterial crystal protein into soil from root exudates and decomposed plant parts. This altered the composition of soil and persists for longer durations in soil. Increased persistence of bacterial crystal proteins in soil can possibly become part of food chains and food webs (Saxena et al. 1999; Zwahlen et al. 2003). Further, persistence of insecticidal toxin in clay and humic acid components of soil can pose threat to the life of non target insect species as well (Saxena and Stotzky 2000).

Likewise, the transgenic tobacco overexpressing gene encoding neomycin phosphotransferase II when sown in fields lead to adsorption of the recombinant DNA on the soil particles (Widmer et al. 1996; Widmer et al. 1997). Similarly, introduction of genetically modified maize expressing the *Bacillus thuringiensis* Berliner Cry1Fa2 protein (Bt) and glyphosate herbicide tolerance to fields enhanced the nitrogen content of soil but reduced the phosphorus levels. Further, the study presumed that the soil management practices during cropping transgenic crops have additional negative effects on the soil chemistry (Liu et al. 2010a, b).

11.2.2 Effect of rDNA on Non-target Soil Organisms

Transgenic cotton modified to produce *Bacillus thuringiensis* var. kurstaki endotoxin in soil was reported to enhance the total population of soil bacteria and fungi. However, the production of endotoxin by transgenics was not responsible for the variation in microbial populations. The alteration in transgenic plants due to genetic manipulations was considered responsible for the increment in soil microbial density (Donegan et al. 1995).

The control and two genotypes of genetically modified alfalfa plants individually overexpressing α-amylase and lipid peroxidase genes were treated with wild PC strain, RMB7201 strain having antibiotic resistance for spectinomycin and streptomycin and RMBPC-2 strain expressing gene for antibiotic resistance and an additional nifA gene for nitrogen fixation. Post microbial treatments, the three genotypes of alfalfa were evaluated variation in the soil microbial populations, plant growth parameters and soil chemistry. Among the three plant types, the lipid peroxidase expressing plants showed significant reduction in shoot biomass with an increment in the accumulation of nitrogen and phosphorus content. Further, the field soil around lipid peroxidase transgenics showed increment in the population of culturable, cellulose utilizing and aerobic spore forming bacteria. Moreover, the pH of the surrounding soil was reduced and the enzyme activity of alkaline phosphatase and dehydrogenase enzymes was highly reduced. Hence the planned introduction of genetically modified plants has induced unintentional changes in the soil, plant and microbial populations (Donegan et al. 1999).

Bt, non-Bt corn plants and their residues were reported to be associated with different rhizobial eubacterial colonies. The colonization of non-target mychorrhizal symbiont *Glomus mosseae* was significantly lowered in Bt corn in comparison to non-Bt corn plants. Further, an associated reduction in the soil respiration in Bt corn soil samples compared to non-Bt corn soil samples was observed (Castaldini et al. 2005). However, introduction of Bt and non-Bt cotton varieties were not observed to affect the population and functional diversity of rhizoshpere microorganisms (Shen et al. 2006).

Nematodes are plant parasites significantly affecting crop growth and yield. The potato plant overexpressing cysteine proteinase inhibitors (cystatins) were raised for resistance against potato-cyst nematode. The field introduction of these transgenics was observed to alter the non-target microbial soil populations but without any adverse affect on the soil functionality. In the first year of transgenic introduction, one of the transgenic lines enhanced the fungal community than bacterial population. Whereas another transgenic line reduced fungal growth by degrading the fungal fatty acid 18:2ω6. Subsequently the second year, the transgenics suppressed the population growth of both bacterial and fungal microbial communities (Cowgill et al. 2002).

The herbicide tolerant transgenic *Brassica napus* were observed to increase the fatty acid component of the root associated microbial communities than conventional *B. napus* plants. A significant increment in the levels of 10:02OH, 12:02OH, 12:03OH, a15:0, 15:1ω5c, cy17:0, 18:3ω6, 9, 12c and 19:0ω8c has been reported (Dunfield and Germida 2001). Bt-maize overexpressing active *Bacillus thuringiensis* (Bt) Cry1Ab endotoxin has also been reported to affect the growth efficiency of land snails. Snails exposed to Bt-toxin via food and soil were found to show reduced body mass and growth parameters compared to snails exposed to control conditions. Further, negative effect on the hatchability of snails due to Bt toxin exposure was also evident (Kramarz et al. 2009).

The non-Bt toxin target plant pests have became progressively evident due to introduction of Bt crops in the fields. Since in Bt crop fed fields insecticide pest management has been altered that promoted the growth of Bt toxin non-targeted pests. As in China, secondary pest population was significantly enhanced post Bt cotton introduction to the fields (Ho and Xue 2008; Zhao et al. 2011). Likewise, a 10 year field investigation of Bt cotton in Northern China showed enhanced population sizes of mirid bugs and simultaneously they emerged as significant pest affecting cotton growth and production (Lu et al. 2010).

The monarch butterfly on the milkweed plants of surrounding Bt corn plants was observed to show detrimental effects. Bt corn was observed to disperse their pollens to nearby milkweed plants. The larvae of butterfly thus infected by Bt corn pollens showed less growth and increased mortality rates (Losey et al. 1999). Further, it has also been demonstrated that the transgene of genetically modified plant can possibly transmitted to and taken up by the native soil microbes. The homologous recombination has induced horizontal transfer of gene from transgenic sugarbeet to *Acinetobacter* sps. The bacterial strain was observed to express kanamycin resistance like the transgenic plants (Nielsen et al. 2000). Likewise, a study has also reported the *in situ* transfer of *aadA* gene expressing protein for spectinomycin and streptomycin resistance from transgenic tobacco to *Acinetobacter* sp. strain BD413 (Kay et al. 2002).

Trangenic crops have also been reported to negatively affect the survival of pest predators. The lacewing bugs are biological control agents of insects and mite pests in the agricultural fields The transgenic corn overexpressing Cry1Ab were reported to reduce the population of lacewing larvae. (Hilbeck et al. 1998). Further, widespread use of Bt crops has induced intense pressure on pests for selection. Reports have documented the evolution of Bt resistant pests post Bt crop cultivation. Like, the pest diamond black moth was reported to develop resistance against Bt toxin in the fields (Tabashnik 1994). The reports of diamond black moth resistance for Bt toxin has been documented from Malaysia, China, Japan, Thailand and Philippines (Iqbal et al. 1996; Liu and Tabashnik 1997). Few other insects have been reported to develop Bt resistance in lab conditions but not in the open fields. Some of these are tobacco budworm, cottonwood leaf beetle, mosquito, pink bollworm moth, almond moth, yellow fever mosquito, cabbage looper moth and Tiger moth (Huang et al. 1999; Gould et al. 1997; Liu et al. 1999; Tabashnik et al. 1994; Wirth et al. 1997; Frutos et al. 1999; Whalon and McGaughey 1998). Further, it has been documented that change/ absence in the receptor and proteinase required for binding/ activation of Bt toxin has lead to the development to resistance for Bt toxin in the insects (Tabashnik et al. 1997; Oppert et al. 1997). Later, development of tolerance for Bt toxin mediated osmotic lysis in the insect midgut cells was also reported as a mechanism of insect resistance against Cry genes of *Bacillus thuringiensis* (Liu et al. 2005).

11.2.3 Effect of rDNA on Non-target Plant Populations

The *Arabidopsis thaliana* transgenics overexpressing Csr1–1 gene for herbicide resistance were observed to possess 20 times higher affinity to pollinate wild plant types (Bergelson et al. 1998). Hence, transfer of pollens from genetically modified plants to wild types can potentially contaminate the wild genetic base.

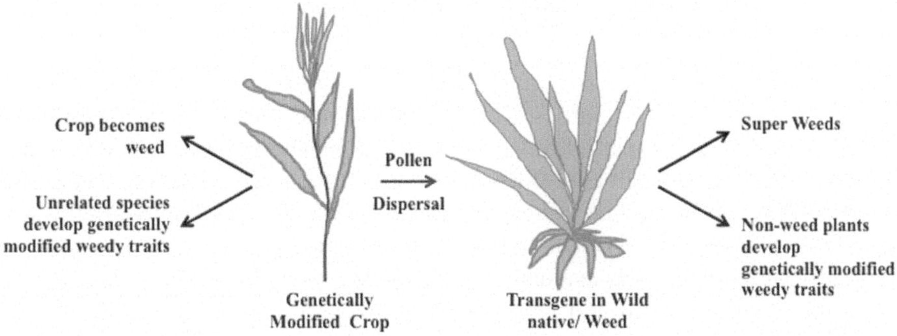

Fig. 11.2 Potential ways and modifications induced by transgene to wild weed populations. The transgenic crops can themselves behave as weedy plants after long term exposure of transgene. This may lead to displacement of native crop or weed or natural vegetation. Free pollen dispersal from transgenics can cause horizontal gene transfer or introgression of transgene to weeds or natural vegetative plants, thus developing herbicide resistant super weeds

Transgenics are also thought to potentially change the weed communities in various ways (Fig. 11.2). Transgenic crops can themselves behave as weedy plants and in that case they displace the native crop or weed or natural vegetation of agroecosystem. The process of hybridization or horizontal gene transfer from transgenics to weeds or natural vegetative plants can develop a generation of super weeds. Further, introgression of transgene in weeds can give rise to herbicide tolerant weeds. Development of such super weeds will affect the agricultural crops and ultimately their overall yield (Warwick et al. 1999). Detailed description of transfer/ introgression of transgenes from transgenic plants to wild natives has been reviewed by Warwick et al. (1999).

Introgression of glyphosate resistance transgene from *B. napus* to wild *B. rapa* has been reported. Further, the transgene showed stable expression in wild plants even in the absence of glyphosate (Snow et al. 1999). The introgressed wild relative *B. rapa* was observed for stable transgene persistence for 6 years. One of the introgressed plant showed reduced pollen viability and produced as total of 480 seeds. 22 plants from next progeny showed high pollen viability and stable expression of herbicide resistance gene. Thus, the wild *B. rapa* plant showed transgene expression continuously for 6 years without any herbicide selection pressure (Warwick et al. 2008).

Genetic modifications have been reported to induce invasiveness in the transgenics. The expression of transgenic trait accompanies variation in the various other traits of plants producing a novel organism to the existing agroecosystem. This invasiveness of genetically modified plants induces transgene flow from transgenics to wild natives leading to introgression of genes. These variations ultimately affect the existing ecosystem (Wolfenbarger and Phifer 2000). The thirteen agricultural crops namely, sunflower, rice, sorghum, soybean, wheat, beans and millets have been reported to hybridize with their native wild relatives. This hybridization has induced evolution of weed species. Further, the event of hybridization has lead to high level of introgression in native plants. These variations have ultimately reduced the genetic base of wild species leading to their extinction (Ellstrand et al. 1999; Lefol et al. 1996).

11.3 Nanotechnology Influence on Agroecosystem

Nanotechnology is the science of synthesis, characterization and application of nanometric particles. The particles having atleast one dimension in nanometric range are known as nanoparticles. Various types of nanoparticles have been synthesized and characterized till date (Siegel et al. 2012; Iravani et al. 2014; Ingale and Chaudhari 2013). Researchers are presently working on the various applicatory aspects of nanoparticles. But along with these areas, presently the concern of nanoparticles toxicological evaluations is also underway. However, nanoparticles toxicity evaluation in soil and agroecosystem are very limited. Less information regarding the use of appropriate nanoparticles dose limits the agricultural ecosystem-nanoparticle interaction studies. Further, soil is multilayered and can act as huge reactive sink for nanoparticles. Further, it can also not be estimated that how nanoparticles will transform when mixed with soil (Klaine et al. 2008).

Various reports documenting the growth promotory as well as growth inhibiting roles of nanoparticles on limited plant types are present. Independent reports on the influence of nanoparticles on soil microbial communities are also present. Hence, present article discusses the effect of nanoparticles interaction on agroecosystem and its components including plants, soil microbial population and soil biochemistry (Table 11.2).

11.3.1 Nanoparticles Interactions with Soil Components

Nanoparticles if present in agricultural fields are going to interact with plants, microbes as well as soil particles. Nanoparticles applications are presently limiting due to lack of reports regarding their toxicological studies. Therefore, exposure of nanoparticles to soil may or may not affect the soil components and functions. However, the related studies are still lacking. Few of these are discussed as follows. Exposure of TiO_2 nanoparticles was observed to reduce the nitrogen flow of soil. A reduction in nitrogen cycle and expression of nitrification-denitrification genes was reported. Likewise, the activities of nitrification and denitrification enzymes were also downregulated. These variations were further associated with shift in the diversity of bacteria, archaea and the ammonia-oxidizing clades. The study thus concluded that nanoparticles can pollute soil health parameters and related ecosystem which require further evaluations (Simonin et al. 2016).

In another study, the colloidal behavior of Al_2O_3 nanoparticles in the presence of natural organic matter during varying pH was evaluated. The physiochemical properties of natural organic matter can potentially affect the colloidal nature of nanoparticles. The natural organic matter could stabilize the colloidal form of nanoparticles during neutral or alkaline pH. However, the stability of colloidal Al_2O_3 nanoparticles decreased in acidic pH in natural organic matter via charge neutralization (Ghosh et al. 2008). The variation in the colloidal stability of nanoparticles causes

Table 11.2 Effect of nanoparticles on the agriculture ecosystem

Effect of nanoparticles on soil microbial communities			
Type of Nanoparticles	Affected Organism	Response	References
TiO_2, SiO_2 and ZnO nanoparticles	Gram-positive *Bacillus subtilis* and Gram-negative *E. coli*	Inhibition of bacterial growth	Adams et al. (2006)
Ultrafine ZnO nanoparticles	*E. Coli*	Increased bacterial membrane permeability leading to membrane disorganization and bacterial growth inhibition	Brayner et al. (2006)
Fullrenes	*E. Coli* and *Bacillus subtilis*	Reduced growth and respiration rate	Tong et al. (2007)
Silver nanoparticles	Nitrifying bacteria *E. coli PHL628-gfp*	Attachment to bacterial wall and cell wall pitting	Choi et al. (2008)
Quantum dots	*E. Coli*	Nanoparticles crossed trophic levels in food chain	Holbrook et al. (2008)
CdSe Quantum Dots	*Pseudomonas aeruginosa*	Damage bacterial membrane and generate reactive oxygen species, cross the trophic levels in food chain	Priester et al. (2009), Werlin et al. (2011)
Polyvinyl pyrolidone coated Silver nanoparticles, Oleic acid coated Silver nanoparticles	Earthworm, *Eisenia fetida*	Bioaccumulatoin and reduction in reproduction rate	Shoults-Wilson et al. (2011a, b)
TiO_2 and ZnO nanoparticles	Soil microbial populations	Reduced diversity and biomass	Ge et al. (2011)
Silver nanoparticles	Nitrogen-fixing bacteria *Bradyrhizobium canariense*	Reduced growth	Kumar et al. (2011)
Nano-sized zerovalent iron	Two species of earthworms, *Eisenia fetida* and *Lumbricus rubellus*	Reduced growth, weight and reproduction, enhanced mortality	El-Temsah and Joner (2012)
Copper oxide and magnetite nanoparticles	Microbial communities on Bet-Dagan sandy loam and Yatir sandy clay loam soil	Reduced abundance, composition, hydrolytic activity and oxidative potential of microbial communities	Frenk et al. (2013)
TiO_2 nanoparticles	*E. Coli* and *Aeromonas hydrophila*	Reactive oxygen species generation and photoinactivation of bacteria	Tong et al. (2013)
Effect of nanoparticles on soil components			
Type of Nanoparticles	Effect on soil		References
Al_2O_3 nanoparticles	Decreased stability of colloidal Al_2O_3 nanoparticles in acidic pH in natural organic matter via charge neutralization		Ghosh et al. (2008)

(continued)

Table 11.2 (continued)

Citrate coated and non-coated silver nanoparticles	Non-coated silver nanoparticles showed increased bioavailability with increased time of persistence in soil		Coutris et al. (2012)
CuO and ZnO nanoparticles	Inhibit methane oxidation activity of soil, reduced microbial abundance of soil		Mohanty et al. (2014)
TiO_2 nanoparticles	Reduced carbon mineralization process of silty clay soil		Simonin et al. (2015)
TiO_2 nanoparticles	Reduction in nitrogen cycle, downregulation of nitrification-denitrification genes and enzymes		Simonin et al. (2016)
ZnO nanoparticles	Reduce process of carbon and nitrogen mineralization of date palm leaf decomposition in arid sandy soils		Rashid et al. (2016)
Effect of nanoparticles on plants			
Type of Nanoparticles	Affected Plants	Alteration Induced	References
C_{70} fullrenes	Rice	Fullrenes bioaccumulated, transmitted to next progeny	Lin et al. (2009)
Water-soluble fullrenes	*Arabidopsis thaliana*	Retarded root length, abnormalities in cell division, mitochondrial activity and microtubular organization, inhibition of root gravitropism responses	Liu et al. (2010a, b)
Ceria nanoparticles	Cucumber	Two dimensional distribution of NPs in leaves, roots and shoots	Zhang et al. (2011)
CuO nanoparticles	Maize	Growth inhibition	Wang et al. (2012)
Silver nanoparticles	Wheat	Reduction in shoot- root length	Dimkpa et al. (2013)
Fe_3O_4 and TiO_2 nanoparticles	Soybean	Reduction in growth and accumulation of leaf carbon content on TiO_2 exposure. Reduction in leaf phosphorus content on Fe_3O_4 exposure	Burke et al. (2015)
ZnO nanoparticles	*Phaseolus vulgaris*	Reduced shoot-root length and root ferric reductase activity	Dimkpa et al. (2015)

variation in its interaction with soil/ cell culture medium. These interactions can induce toxic/ non-toxic effects of nanoparticles in soil or cell culture medium (Moore et al. 2015). Hence, detailed evaluations in this respect are still required. However, the organic matter present in soil can affect the stability of nanoparticles that may induce nanoparticles mediated toxicity to agroecosystem.

Dispersion of manufactured nanoparticles in soil can affect their bioavailability, suspension ability, transport and aggregation ability. Dissolved organic matter of soil can change the physical and chemical properties of nanoparticles by coating them. This leads to variation in their surface chemistry, pore size and most importantly mechanism of their toxicity behavior (Wang et al. 2011). The behavior of $AgNO_3$, citrate coated and non-coated silver nanoparticles was evaluated in mineral soil and soil rich in organic matter content. Nanoparticles as well as $AgNO_3$ showed mobility in both soil types. Among these three types of silver, the uncoated silver nanoparticles showed increased bioavailability with their increased time of persistence in soil. So, in this form they can act as a consistent source of bioaccessible silver. Hence, the soil components can variously interact with silver nanoparticles for longer duration that may or may not lead to toxic effects (Coutris et al. 2012).

The CuO and ZnO nanoparticles have been reported to inhibit the methane oxidation activity of soil. The nanoparticles have induced cell mediated damage to soil microbial populations thus reducing the microbial abundance of soil. Further, the moisture content of soil was also related to the toxicity of nanoparticles. Lesser was the moisture content of soil, more was the nanoparticles mediated toxicity (Mohanty et al. 2014). Exposure of TiO_2 nanoparticles was observed to reduce the carbon mineralization process of silty clay soil having high content of organic matter. The aggregation of TiO_2 nanoparticles was responsible for the observed alterations in soil functionality (Simonin et al. 2015). Similarly, the exposure of ZnO nanoparticles has been documented to reduce the process of carbon and nitrogen mineralization of date palm leaf decomposition in arid sandy soils. Nanoparticles mediated bacterial growth and colony formation inhibition was responsible for the observed changes in soil activity (Rashid et al. 2016).

11.3.2 Effect of Nanoparticles on Soil Microbes and Associated Organisms

The epigeic earthworm, *Eisenia fetida* (Lumbricidae, Savigny, 1826) is a model soil organism representing the biota of agricultural ecosystem. Evaluation of nanoparticles exposure on earthworm can be extrapolated to assess their influence on soil ecology. Earthworms exposed to polyvinylpyrolidone or oleic acid coated silver nanoparticles via artificial soil composed of quartz, limestone and sphagnum were accumulated in earthworms and reduced their reproduction rate at 773.3 mg kg-1 and 727.6 mg kg-1 doses, respectively (Shoults-Wilson et al. 2011a). In a similar study, exposure of silver nanoparticles to earthworm by artificial soil was also reported to reduce their reproductive ability. Release of Ag^+ ions from silver nanoparticles and the type of soil was responsible for the nanoparticles mediated toxicity (Shoults-Wilson et al. 2011b). They have also shown avoidance of silver nanoparticles in the contaminated soils for over 48 hrs. Thus, nanoparticles interaction with organisms can lead to ecologically significant but unpredictable responses that may be toxic/ non-toxic (Shoults-Wilson et al. 2011c).

Similarly, nano-sized zerovalent iron was found to negatively affect the growth, weight, mortality and reproduction of two species of earthworms, *Eisenia fetida* and *Lumbricus rubellus* (El-Temsah and Joner 2012). Copper oxide and magnetite nanoparticles were evaluated for their toxicity on microbial communities on Bet-Dagan sandy loam and Yatir sandy clay loam soil. The bacterial communities of Bet-Dagan soil were more severely affected by both types of nanoparticles than those of Yatir soil. A significant variation in abundance, composition, hydrolytic activity and oxidative potential of bacterial communities on exposure of these nanoparticles was evident (Frenk et al. 2013). TiO_2 and ZnO nanoparticles have also been reported to negatively affect the soil microbial populations. The both type of nanoparticles were observed to reduce the diversity and biomass of microbial populations (Ge et al. 2011).

Nanoparticles have also been reported to transmit from one trophic level to another trophic level in the food web/ chain. Quantum dots present on the membrane of *E. Coli* were observed to cross the trophic levels in the food chain. The membrane modified *E. Coli* was consumed by its primary consumer ciliate bacterivores that were lately consumed by rotifers (Holbrook et al. 2008). Interaction of CdSe quantum dots with *Pseudomonas aeruginosa* was found to damage their membrane and generate reactive oxygen species (Priester et al. 2009). In another study, these CdSe quantum dots were observed to transfer and bioaccumulate in protozoan *Tetrahymena thermophila* that is the predator of *P. aeruginosa*. The concentration of cadmium of predator was reportedly five times higher than the bacterial prey. The bacteria were not digested in protozoan food vacuoles. Hence, the quantum dots present in bacteria were sustained inside the protozoa for higher plants of agroecosystem (Werlin et al. 2011).

In another study, a three component based approach for toxicity of nano-TiO_2 on bacteria *E. Coli* and *Aeromonas hydrophila* was deciphered (Fig. 11.3). On exposure of TiO_2 to bacteria, intrinsic photoactivity of TiO_2 nanoparticles, their aggregation and the nano-TiO_2/bacteria surface interface was considered responsible for their bacterial nanotoxicity (Tong et al. 2013).

The photoactivity as well as aggregation of nano-TiO_2 was responsible for reactive oxygen species generation and photoinactivation of bacteria. Further, component III stated that alignment of nanoparticles at bacterial surface was essentially inducing reactive oxygen generation and inactivation of bacteria (Tong et al. 2013). Silver nanoparticles have been reported to induce detrimental effects on the nitrifying bacteria *E. Coli PHL628-gfp*. The silver nanoparticles were found to attach to the microbial cells and induced cell wall pitting (Choi et al. 2008). Likewise, plant associated nitrogen-fixing bacteria *Bradyrhizobium canariense* belonging to arctic soil was reported sensitive for silver nanoparticles (Kumar et al. 2011). Fullrenes have also been reported to inhibit the growth of soil microbial populations (Fortner et al. 2005). Further, treatment of soil with C_{60} fullrenes has enhanced the proportion of gram-negative bacteria than gram-positive bacteria. Inhibition in the growth and respiration of *E. Coli* and *Bacillus subtilis* in response to fullrenes in minimal soil media was also evident (Tong et al. 2007).

11.3.3 Influence of Nanoparticles on Plants

Nanoparticles have also been reported to interact with plants and induce toxic/ non-toxic effects. In agroecosystem, interaction of plants, microbes and soil components is essential to maintain a healthy ecosystem. Hence, understanding the interaction

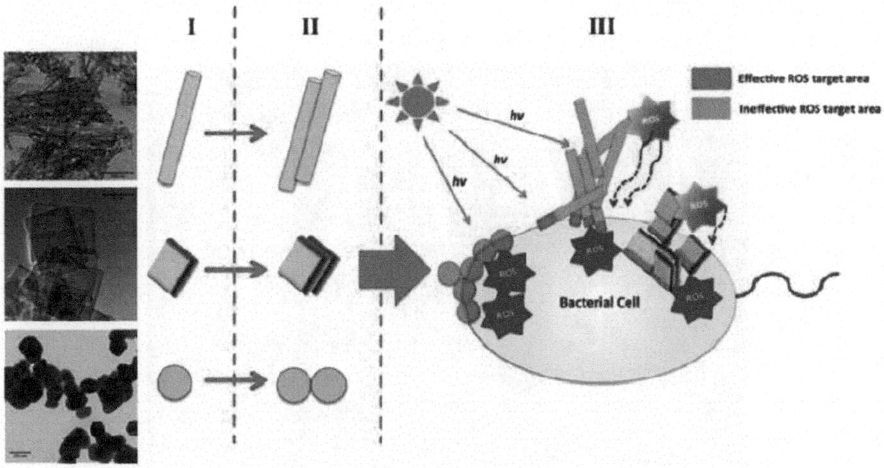

Fig. 11.3 The figure depicts the three component based approach of phototoxicity induced by TiO_2 nanoparticles on bacteria *E. Coli* and *Aeromonas hydrophila.* The nano-TiO_2 are photoactive (component I), as a result the aggregation of nano-TiO_2 (component II) and alignment on bacterial surface (component III) induces generation of reactive oxygen species. These collectively lead to the degradation of bacterial cells. (Reprinted with permission from (Tong et al. 2013). Copyright (2013) American Chemical Society)

and level of interaction between nanoparticles and plants is important. Various articles have discussed the phytotoxicity of varying nanoparticles types on plants (Ma et al. 2010; Miralles et al. 2012; Thul et al. 2013; Phogat et al. 2016). Few of these are discussed here.

Maize plants exposed to CuO nanoparticles showed phytotoxic responses in terms of growth inhibition. The nanoparticles were observed in the xylem sap of plants showing their movement from roots to shoot via xylem and back from shoot to root via phloem. Further, bioaccumulation of CuO nanoparticles in maize was evident that posed a risk of nanoparticles to food safety (Wang et al. 2012). Likewise, uptake, distribution and transformation of fullrenes C_{70} in rice plants were reported. Further, these fullrenes were transmitted to the next progeny seeds (Lin et al. 2009). Exposure of 7 and 25 nm ceria nanoparticles to cucumber has reported presence of 7 nm nanoparticles in the roots and shoots of cucumber plants. Further, the two dimensional distribution of nanoparticles in cucumber leaves was observed (Zhang et al. 2011).

Arabidopsis thaliana plants exposed to water-soluble fullrenes showed retarded root growth in terms of reduced length and abnormalities in cell division, mitochondrial activity and microtubular organization. Further, the roots showed inhibition of root gravitropism responses (Liu et al. 2010a, 2010b). Fe_3O_4 and TiO_2 nanoparticles were exposed to soybean plants for 6 weeks under green house conditions. The plants showed significant reduction in growth and accumulation of leaf carbon content on TiO_2 exposure than Fe_3O_4 nanoparticles. Further, the soybean showed reduction in leaf phosphorus content on Fe_3O_4 exposure (Burke et al. 2015).

ZnO nanoparticles have been documented to reduce the shoot-root length and root ferric reductase activity of beans (*Phaseolus vulgaris*). The nano form of zinc

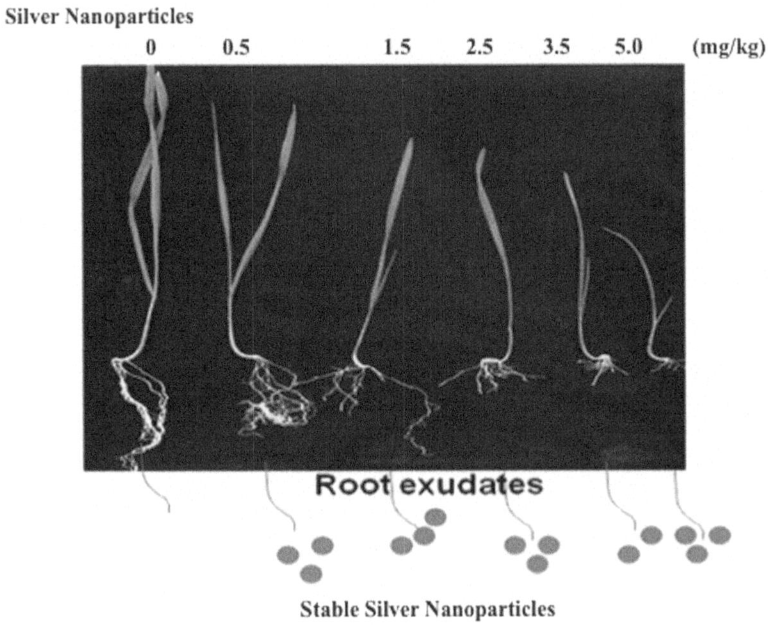

Fig. 11.4 Dose dependent phytotoxicity of silver nanoparticles on wheat. The nanoparticles phytotoxicity was evaluated in sandy soil matrix. The nanoparticles were taken up by the plant roots and transported to shoot leading to their accumulated in shoots and roots. The accumulation of nanoparticles was responsible for the reduction of shoot- root length of wheat plants. (Reprinted with permission from (Dimkpa et al. 2013). Copyright (2013) American Chemical Society)

was responsible for the observed alterations. The enzymatic activity was recovered on colonization of *Pseudomonas chlororaphis* O6. However, a simultaneous reduction in the uptake of essential elements, Zn and Fe was reported (Dimkpa et al. 2015). Likewise, silver nanoparticles were observed to reduce the growth of wheat in dose dependent manner as shown in Fig. 11.4 (Dimkpa et al. 2013).

11.4 Conclusion

Recombinant DNA technology and nanotechnology are two most recent technologies that are looked upon as novel strategies to counter the worldwide trouble of food insecurity. However, application of these techniques in agricultural ecosystems needs to understand the level of interactions that rDNA and nanoparticles can possess with soil, soil microbial communities and plants. As discussed, rDNA technology as well as nanoparticles can negatively affect these components of agroecosystem. Hence, while stepping head to take-up these technologies for improvement in agriculture, their various toxic/ non-toxic effects needs to be considered.

References

Adams LK, Lyon DY, Alvarez PJJ (2006) Comparative eco-toxicity of nanoscale TiO2, SiO2, and ZnO water suspensions. Water Res 40:3527–3532. https://doi.org/10.1016/j.watres.2006.08.004

Atlas RM, Bartha R (1982) Microbial ecology: Fundamentals and Applications. J Ecol 70:686. https://doi.org/10.2307/2259932

Bergelson J, Purrington CB, Wichmann G (1998) Promiscuity in transgenic plants. Nature 395:25. https://doi.org/10.1038/25626

Brady NC (1984) The nature and properties of soils. Macmillan, New York

Brayner R, Ferrari-Iliou R, Brivois N, Djediat S, Benedetti MF, Fievet F (2006) Toxicological impact studies based on *Escherichia coli* bacteria in ultrafine ZnO nanoparticles colloidal medium. Nano Lett 6:866–870. https://doi.org/10.1021/nl052326h

Burke DJ, Pietrasiak N, Situ SF, Abenojar EC, Porche M, Kraj P, Lakliang Y, Samia ACS (2015) Iron oxide and titanium dioxide nanoparticle effects on plant performance and root associated microbes. Int J Mol Sci 16:23630–23650. https://doi.org/10.3390/ijms161023630

Castaldini M, Turrini A, Sbrana C, Benedetti A, Marchionni M, Mocali S, Fabiani A, Landi S, Santomassimo F, Pietrangeli B, Nuti MP, Miclaus N, Giovannetti M (2005) Impact of Bt corn on rhizospheric and soil eubacterial communities and on beneficial mycorrhizal symbiosis in experimental microcosms. Appl Environ Microbiol 71:6719–6729. https://doi.org/10.1128/AEM.71.11.6719-6729.2005

Choi O, Deng KK, Kim NJ, Ross L, Surampalli RY, Hu Z (2008) The inhibitory effects of silver nanoparticles, silver ions, and silver chloride colloids on microbial growth. Water Res 42:3066–3074. https://doi.org/10.1016/j.watres.2008.02.021

Coutris C, Joner EJ, Oughton DH (2012) Aging and soil organic matter content affect the fate of silver nanoparticles in soil. Sci Total Environ 420:327–333. https://doi.org/10.1016/j.scitotenv.2012.01.027

Cowgill SE, Bardgett RD, Kiezebrink DT, Atkinson HJ (2002) The effect of transgenic nematode resistance on non-target organisms in the potato rhizosphere. J Appl Ecol 39:915–923. https://doi.org/10.1046/j.1365-2664.2002.00774.x

Datta A (2013) Genetic engineering for improving quality and productivity of crops. Agric Food Secur 2:15. https://doi.org/10.1186/2048-7010-2-15

Dimkpa CO, McLean JE, Martineau N, Britt DW, Haverkamp R, Anderson AJ (2013) Silver nanoparticles disrupt wheat (*Triticum aestivum L.*) growth in a sand matrix. Environ Sci Technol 47:1082–1090. https://doi.org/10.1021/es302973y

Dimkpa CO, Hansen T, Stewart J, McLean JE, Britt DW, Anderson AJ (2015) ZnO nanoparticles and root colonization by a beneficial pseudomonad influence essential metal responses in bean (*Phaseolus vulgaris*). Nanotoxicology 9:271–278. https://doi.org/10.3109/17435390.2014.900583

Donegan KK, Palm CJ, Fieland VJ, Porteous LA, Ganio LM, Schaller DL, Bucao LQ, Seidler RJ (1995) Changes in levels, species and DNA fingerprints of soil microorganisms associated with cotton expressing the *Bacillus thuringiensis* var. kurstaki endotoxin. Appl Soil Ecol 2:111–124. https://doi.org/10.1016/0929-1393(94)00043-7

Donegan KK, Seidler RJ, Fieland VJ, Schaller DL, Palm CJ, Ganio LM, Cardwell DM, Steinberger Y (1997) Decomposition of genetically engineered tobacco under field conditions: Persistence of the proteinase inhibitor I product and effects on soil microbial respiration and protozoa, nematode and microarthropod populations. J Appl Ecol 34:767–777. https://doi.org/10.2307/2404921

Donegan KK, Seidler RJ, Doyle JD, Porteous LA, Digiovanni G, Widmer F, Watrud LS (1999) A field study with genetically engineered alfalfa inoculated with recombinant *Sinorhizobium meliloti* effects on the soil ecosystem. J Appl Ecol 36:920–936. https://doi.org/10.1046/j.1365-2664.1999.00448.x

Dunfield KE, Germida JJ (2001) Diversity of bacterial communities in the rhizosphere and root interior of field-grown genetically modified *Brassica napus*. FEMS Microbiol Ecol 38:1–9. https://doi.org/10.1111/j.1574-6941.2001.tb00876.x

Dunwell JM (2000) Transgenic approaches to crop improvement. J Exp Bot 51:487–496. https://doi.org/10.1093/jexbot/51.suppl_1.487

Ellstrand NC, Prentice HC, Hancock JF (1999) Gene flow and introgression from domesticated plants into their wild relatives. Ann Rev Ecol Syst 30:539–563. https://doi.org/10.1146/annurev.ecolsys.30.1.539

El-Temsah YS, Joner EJ (2012) Ecotoxicological effects on earthworms of fresh and aged nano-sized zero-valent iron (nZVI) in soil. Chemosphere 89:76–82. https://doi.org/10.1016/j.chemosphere.2012.04.020

Firidin S (2010) Recombinant DNA technology. Yunus Res Bull 10:16–19

Fischer, G., Huang, J., Keyzer, M.A., Qiu, H., Sun, L., van Veen, W.C.M. (2007) China's agricultural prospects and challenges. Report on scenario simulations until 2030 with the Chinagro welfare model covering national, regional and county level

Fortner JD, Lyon DY, Sayes CM, Boyd AM, Falkner JC, Hotze EM, Alemany LB, Tao YJ, Guo W, Ausman KD, Colvin VL, Hughes JB (2005) C_{60} in water: Nanocrystal formation and microbial response. Environ Sci Technol 39:4307–4316. https://doi.org/10.1021/es048099n

Frenk S, Ben-Moshe T, Dror I, Berkowitz B, Minz D (2013) Effect of metal oxide nanoparticles on microbial community structure and function in two different soil types. PLoS One 8:e84441. https://doi.org/10.1371/journal.pone.0084441

Frutos R, Rang C, Royer M (1999) Managing insect resistance to plants producing *Bacillus thuringiensis* toxins. Crit Rev Biotechnol 19:227–276. https://doi.org/10.1080/0738-859991229251

Ge Y, Schimel JP, Holden PA (2011) Evidence for negative effects of TiO2 and ZnO nanoparticles on soil bacterial communities. Environ Sci Technol 45:1659–1664. https://doi.org/10.1021/es103040t

Ghosh S, Mashayekhi H, Pan B, Bhowmik P, Xing B (2008) Colloidal behavior of aluminum oxide nanoparticles as affected by ph and natural organic matter. Langmuir 24:12385–12391. https://doi.org/10.1021/la802015f

Gould F, Anderson A, Jones A, Sumerford D, Heckel DG, Lopez J, Micinski S, Leonard R, Laster M (1997) Initial frequency of alleles for resistance to *Bacillus thuringiensis* toxins in field populations of *Heliothis virescens*. Proc Natl Acad Sci U S A 94:3519–3523

Herdt RW (2006) Biotechnology in Agriculture. Annu Rev Environ Resour 34:265–295

Hilbeck A, Baumgartner M, Fried PM, Bigler F (1998) Effects of transgenic *Bacillus thuringiensis* corn-fed prey on mortality and development time of immature *Chrysoperla carnea* (Neuroptera: Chrysopidae). Environ Entomol 27:480–487. https://doi.org/10.1093/ee/27.2.480480-487

Ho MW (2000) Genetic engineering dream or nightmare?: Turning the tide on the brave new world of bad science and big business. 2 Rev Updated edition. International Publishing Group Continuum. https://doi.org/10.1604/9781858600512

Ho P, Xue D (2008) Farmers' perceptions and risks of agro-biotechnological innovations in China: ecological change in Bt cotton? Int J Environ Sustain Dev 7:396–417. https://doi.org/10.1504/IJESD.2008.022386

Holbrook RD, Murphy KE, Morrow JB, Cole KD (2008) Trophic transfer of nanoparticles in a simplified invertebrate food web. Nat Nanotechnol 3:352–355. https://doi.org/10.1038/nnano.2008.110

Huang F, Buschman LL, Higgins RA, McGaughey WH (1999) Inheritance of resistance to *Bacillus thuringiensis* toxin (Dipel ES) in the European corn borer. Science 284:965–967. https://doi.org/10.1126/science.284.5416.965

Ingale AG, Chaudhari AN (2013) Biogenic synthesis of nanoparticles and potential applications: an eco-friendly approach. J Nanomed Nanotechol 4:165. https://doi.org/10.4172/2157-7439.1000165

Iqbal M, Verkerk RHJ, Furlong MJ, Ong PC, Rahman SA, Wright DJ (1996) Evidence for resistance to *Bacillus thuringiensis* (Bt) subsp. kurstaki HD-1, Bt subsp. aizawai and Abamectin in field populations of *Plutella xylostella* from Malaysia. Pestic Sci 48:89–97. https://doi.org/10.1002/(SICI)1096-9063(199609)48:1<89::AID-PS450>3.0.CO;2-B

Iravani S, Korbekandi H, Mirmohammadi SV, Zolfaghari B (2014) Synthesis of silver nanoparticles: chemical, physical and biological methods. Res Pharm Sci 9:385–406

James C (2003) Global review of commercialized transgenic crops. Curr Sci 84:303–309

James C (2004) Preview: global status of commercialized biotech/GM crops: 2004

Johnson BR, Hope AR (2009) Biotechnology in the environment: potential effects on biodiversity. In: Doelle HW, Rokem JS, Berovic M (eds) Biotechnology: fundamentals in biotechnology, vol X. Encyclopedia of Life Support Systems (EOLSS), Abu Dhabi, U.A.E, pp 1–11

Kay E, Vogel TM, Bertolla F, Nalin R, Simonet P (2002) *In Situ* transfer of antibiotic resistance genes from transgenic (transplastomic) tobacco plants to bacteria. Appl Environ Microbiol 68:3345–3351. https://doi.org/10.1128/AEM.68.7.3345-3351.2002

Klaine SJ, Alvarez PJJ, Batley GE, Fernandes TF, Handy RD, Lyon DY, Mahendra S, McLaughlin MJ, Lead JR (2008) Nanomaterials in the environment: Behavior, fate, bioavailability, and effects. Environ Toxicol Chem 27:1825–1851. https://doi.org/10.1897/08-090.1

Kramarz P, Vaufleury A, Gimbert F, Cortet J, Tabone E, Andersen MN, Krogh PH (2009) Effects of Bt-maize material on the life cycle of the land snail *Cantareus asperses*. Appl Soil Ecol 42:236–242. https://doi.org/10.1016/j.apsoil.2009.04.007

Kumar N, Shah V, Walker VK (2011) Perturbation of an arctic soil microbial community by metal nanoparticles. J Hazard Mater 190:816–822. https://doi.org/10.1016/j.jhazmat.2011.04.005

Lefol E, Danielou V, Darmency H (1996) Predicting hybridization between transgenic oilseed rape and wild mustard. Field Crop Res 45:153–161. https://doi.org/10.1016/0378-4290(95)00067-4

Lin S, Reppert J, Hu Q, Hudson JS, Reid ML, Ratnikova TA, Rao AM, Luo H, Ke PC (2009) Uptake, translocation, and transmission of carbon nanomaterials in rice plants. Small 5:1128–1132. https://doi.org/10.1002/smll.200801556

Liu YB, Tabashnik BE (1997) Experimental evidence that refuges delay insect adaptation to *Bacillus thuringiensis*. Proc R Soc Lond B 264:605–610. https://doi.org/10.1098/rspb.1997.0086

Liu YB, Tabashnik BE, Dennehy TJ, Patin AL, Bartlett AC (1999) Development time and resistance to Bt crops. Nature 400:519. https://doi.org/10.1038/22919

Liu K, Zheng J, Hong H, Peng J, Yang H, Peng R (2005) Mechanisms for Bt toxin resistance and increased chemical pesticide susceptibility in Cry1Ac10-resistant cultured insect cells. Cytotechnology 49:153–160. https://doi.org/10.1007/s10616-006-6880-y

Liu N, Zhu P, Peng C, Kang L, Gao H, Clarke NJ, Clarke JL (2010a) Effect on soil chemistry of genetically modified (GM) vs. non-GM maize. GM Crop 1:1–5. https://doi.org/10.4161/gmcr.1.3.12810

Liu Q, Zhao Y, Wan Y, Zheng J, Zhang X, Wang C, Fang X, Lin J (2010b) Study of the inhibitory effect of watersoluble fullerenes on plant growth at the cellular level. ACS Nano 4:5743–5748. https://doi.org/10.1021/nn101430g

Losey JE, Raynor LS, Carter ME (1999) Transgenic pollen harms monarch larvae. Nature 399:214. https://doi.org/10.1038/20338

Lu Y, Wu K, Jiang Y, Xia B, Li P, Feng H, Wyckhuys KA, Guo Y (2010) Mirid bug outbreaks in multiple crops correlated with wide-scale adoption of Bt cotton in China. Science 328:1151–1154. https://doi.org/10.1126/science.1187881

Ma X, Geisler-Lee J, Deng Y, Kolmakov A (2010) Interactions between engineered nanoparticles (ENPs) and plants: Phytotoxicity, uptake and accumulation. Sci Total Environ 408:3053–3061. https://doi.org/10.1016/j.scitotenv.2010.03.031

Maghari BM, Ardekani AM (2011) Genetically modified foods and social concerns. Avicenna J Med Biotech 3:109–117

Mikkelsen TR, Jensen J, Jorgensen RB (1996) Inheritance of oilseed rape (Brassica napus) RAPD markers in a backcross progeny with *Brassica campestris*. Theoret Appl Genetics 92:492–497. https://doi.org/10.1007/BF00223698

Miralles P, Church TL, Harris AT (2012) Toxicity, uptake, and translocation of engineered nanomaterials in vascular plants. Environ Sci Technol 46:9224–9239. https://doi.org/10.1021/es202995d

Mohanty SR, Rajput P, Kollah B, Chourasiya D, Tiwari A, Singh M, Rao AS (2014) Methane oxidation and abundance of methane oxidizers in tropical agricultural soil (vertisol) in response to CuO and ZnO nanoparticles contamination. Environ Monit Assess 186:3743–3753. https://doi.org/10.1007/s10661-014-3654-4

Moore TL, Rodriguez-Lorenzo L, Hirsch V, Balog S, Urban D, Jud C, Rothen-Rutishauser B, Lattuada B, Petri-Fink A (2015) Nanoparticle colloidal stability in cell culture media and impact on cellular interactions. Chem Soc Rev 44:6287–6305. https://doi.org/10.1039/C4CS00487F

Mukhopadhyay SS (2014) Nanotechnology in agriculture: prospects and constraints. Nanotechnol Sci Appl 7:63–71. https://doi.org/10.2147/NSA.S39409

Nielsen KM, Elsas JD, Smalla K (2000) Transformation of *Acinetobacter sp. Strain BD413*(pFG4ΔnptII) with transgenic plant DNA in soil Microcosms and effects of kanamycin on selection of transformants. Appl Environ Microbiol 66:1237–1242. https://doi.org/10.1128/AEM.66.3.1237-1242.2000

Oppert B, Kramer KJ, Beeman RW, Johnson D, McGaughey WH (1997) Proteinase-mediated insect resistance to *Bacillus thuringiensis* toxins. J Biol Chem 272:23473–23476. https://doi.org/10.1074/jbc.272.38.23473

Phogat N, Khan SA, Shankar S, Ansary AA, Uddin I (2016) Fate of inorganic nanoparticles in agriculture. Adv Mater Lett 7:03–12. https://doi.org/10.5185/amlett.2016.6048

Priester JH, Stoimenov PK, Mielke RE, Webb SM, Ehrhardt C, Zhang JP, Stucky GD, Holden PA (2009) Effects of soluble cadmium salts versus CdSe quantum dots on the growth of planktonic *Pseudomonas aeruginosa*. Environ Sci Technol 43:2589–2594. https://doi.org/10.1021/es802806n

Rashid MI, Shahzad T, Shahid M, Ismail IMI, Shah GM, Almeelbi T (2016) Zinc oxide nanoparticles affect carbon and nitrogen mineralization of Phoenix dactylifera leaf litter in a sandy soil. J Hazard Mater 324:298–305. https://doi.org/10.1016/j.jhazmat.2016.10.063

Saxena D, Stotzky G (2000) Insecticidal toxin from *Bacillus thuringiensis* is released from roots of transgenic Bt corn in vitro and in situ. FEMS Microbiol Ecol 33:35–39. https://doi.org/10.1111/j.1574-6941.2000.tb00724.x

Saxena D, Flores S, Stotzky G (1999) Insecticidal toxin in root exudates from Bt corn. Nature 402:480. https://doi.org/10.1038/44997

Sekhon BS (2014) Nanotechnology in agri-food production: an overview. Nanotechnol Sci Appl 7:31–53. https://doi.org/10.2147/NSA.S39406

Sexstone AJ, Revsbech NP, Parkin TB, Tiedje JM (1985) Direct measurement of oxygen profiles and denitrification rates in soil aggregates. Soil Sci Soc Am J 49:645–651. https://doi.org/10.2136/sssaj1985.03615995004900030024x

Sharma HC, Crouch JH, Sharma KK, Seetharama N, Hash CT (2002) Applications of biotechnology for crop improvement: prospects and constraints. Plant Sci 163:381–395. https://doi.org/10.1016/S0168-9452(02)00133-4

Shen RF, Cai H, Gong WH (2006) Transgenic Bt cotton has no apparent effect on enzymatic activities or functional diversity of microbial communities in rhizosphere soil. Plant Soil 285:149–159. https://doi.org/10.1007/s11104-006-9000-z

Shoults-Wilson WA, Reinsch BC, Tsyusko OV, Bertsch PM, Lowry GV, Unrine JM (2011a) Effect of silver nanoparticle surface coating on bioaccumulation and reproductive toxicity in earthworms (*Eisenia fetida*). Nanotoxicology 5:432–444. https://doi.org/10.3109/17435390.2010.537382

Shoults-Wilson WA, Reinsch BC, Tsyusko OV, Bertsch PM, Lowry GV, Unrine JM (2011b) Role of particle size and soil type in toxicity of silver nanoparticles to earthworms. Soil Sci Soc Am J 75:1–13. https://doi.org/10.2136/sssaj2010.0127nps

Shoults-Wilson WA, Zhurbich OI, McNear DH, Tsyusko OV, Bertsch PM, Unrine JM (2011c) Evidence for avoidance of Ag nanoparticles by earthworms (*Eisenia fetida*). Ecotoxicology 20:385–396. https://doi.org/10.1007/s10646-010-0590-0

Siegel J, Ondrej K, Pavel U, Kolska Z, Slepicka P, Svorcik V (2012) Progressive approach for metal nanoparticle synthesis. Mater Lett 89:47–50. https://doi.org/10.1016/j.matlet.2012.08.048

Simonin M, Guyonnet JP, Martins JMF, Ginot M, Richaume A (2015) Influence of soil properties on the toxicity of TiO_2 nanoparticles on carbon mineralization and bacterial abundance. J Hazard Mater 283:529–535. https://doi.org/10.1016/j.jhazmat.2014.10.004

Simonin M, Richaume A, Guyonnet JP, Dubost A, Martins JMF, Pommier T (2016) Titanium dioxide nanoparticles strongly impact soil microbial function by affecting archaeal nitrifiers. Sci Rep 6. https://doi.org/10.1038/srep33643

Snow AA, Andersen B, Jorgensen RB (1999) Costs of transgenic herbicide resistance introgressed from *Brassica napus* into weedy *B. rapa*. Mol Ecol 8:605–615. https://doi.org/10.1046/j.1365-294x.1999.00596.x

Swift MJ, Anderson JM (1994) Biodiversity and ecosystem function in agricultural systems. In: Schulze ED, Mooney AM (eds) Biodiversity and ecosystem function, vol 99, pp 15–41. https://doi.org/10.1007/978-3-642-58001-7_2

Tabashnik BE (1994) Evolution of resistance to *Bacillus thuringiensis*. Annu Rev Entomol 39:47–79. https://doi.org/10.1146/annurev.en.39.010194.000403

Tabashnik BE, Finson N, Groeters FR, Moar WJ, Johnson MW, Luo K, Adang MJ (1994) Reversal of resistance to *Bacillus thuringiensis* in *Plutella xylostella*. Proc Natl Acad Sci U S A 91:4120–4124. https://doi.org/10.1073/pnas.91.10.4120

Tabashnik BE, Liu YB, Malvar T, Heckel DG, Masson L, Ballester V, Granero F, Mensua JL, Ferre J (1997) Global variation in the genetic and biochemical basis of diamond back moth resistance to *Bacillus thuringiensis*. Proc Natl Acad Sci U S A 94:12780–12785. https://doi.org/10.1073/pnas.94.24.12780

Thul ST, Sarangi BK, Pandey RA (2013) Nanotechnology in agroecosystem: Implications on plant productivity and its soil environment. Expert Opin Environ Biol 2:1. https://doi.org/10.4172/2325-9655.1000101

Tong Z, Bischoff M, Nies L, Applegate B, Turco RF (2007) Impact of fullerene (C_{60}) on a soil microbial community. Environ Sci Technol 41:2985–2991. https://doi.org/10.1021/es061953l

Tong T, Shereef A, Wu J, Binh CTT, Kelly JJ, Gaillard JF, Gray KA (2013) Effects of material morphology on the phototoxicity of nano-TiO_2 to bacteria. Environ Sci Technol 47:12486–12495. https://doi.org/10.1021/es403079h

Wan B (2015) Transgenic pyramiding for crop improvement. In: Al-Khayri JM, Jain SM, Johnson DV (eds) Advances in plant breeding strategies: breeding, biotechnology and molecular tools, pp 369–396. https://doi.org/10.1007/978-3-319-22521-0

Wang Z, Li J, Zhao J, Xing B (2011) Toxicity and internalization of CuO nanoparticles to prokaryotic alga *Microcystis aeruginosa* as affected by dissolved organic matter. Environ Sci Technol 45:6032–6040. https://doi.org/10.1021/es2010573

Wang Z, Xie X, Zhao J, Liu X, Feng W, White JC, Xing B (2012) Xylem- and phloem-based transport of CuO nanoparticles in maize (*Zea mays* L.). Environ Sci Technol 46:4434–4441. https://doi.org/10.1021/es204212z

Warwick SI, Beckie HJ, Small E (1999) Transgenic crops : new weed problems for Canada? Phytoprotection 80:71–84. https://doi.org/10.7202/706182ar

Warwick SI, Legere A, Simard MJ, James T (2008) Do escaped transgenes persist in nature? The case of an herbicide resistance transgene in a weedy *Brassica rapa* population. Mol Ecol 17:1387–1395. https://doi.org/10.1111/j.1365-294X.2007.03567.x

Werlin R, Priester JH, Mielke RE, Kramer S, Jackson S, Stoimenov PK, Stucky GD, Cherr GN, Orias E, Holden PA (2011) Biomagnification of cadmium selenide quantum dots in a simple experimental microbial food chain. Nat Nanotech 6:65–71. https://doi.org/10.1038/nnano.2010.251

Whalon ME, McGaughey WH (1998) Bacillus thuringiensis: use and resistance management. In: Ishaaya I, Degheele D (eds) Insecticides with novel modes of action: mechanism and application. Springer, Berlin, pp 106–137. https://doi.org/10.1007/978-3-662-03565-8_7

Widmer F, Seidler RJ, Watrud LS (1996) Sensitive detection of transgenic plant marker gene persistence in soil microcosms. Mol Ecol 5:603–613. https://doi.org/10.1111/j.1365-294X.1996.tb00356.x

Widmer F, Seidler RJ, Donegan KK, Reed GL (1997) Quantification of transgenic plant marker gene persistence in the field. Mol Ecol 6:1–7. https://doi.org/10.1046/j.1365-294X.1997.00145.x

Wirth MC, Georghiou GP, Federici BA (1997) CytA enables CryIV endotoxins of *Bacillus thuringiensis* to overcome high levels of CryIV resistance in the mosquito, Culex quinquefasciatus. Proc Natl Acad Sci U S A 94:10536–10540. https://doi.org/10.1073/pnas.94.20.10536

Wolfenbarger LL, Phifer PR (2000) The ecological risks and benefits of genetically engineered plants. Science 290:2088–2093. https://doi.org/10.1126/science.290.5499.2088

Zhang Z, He X, Zhang H, Ma Y, Zhang P, Ding Y, Zhao Y (2011) Uptake and distribution of ceria nanoparticles in cucumber plants. Metallomics 3:816–822. https://doi.org/10.1039/c1mt00049g

Zhao JH, Ho P, Azadi H (2011) Benefits of Bt cotton counterbalanced by secondary pests? Perceptions of ecological change in China. Environ Monit Assess 173:985–994. https://doi.org/10.1007/s10661-010-1439-y

Zwahlen C, Hilbeck A, Gugerl P, Nientwig W (2003) Degradation of the Cry1Ab protein within transgenic *Bacillus thuringiensis* corn tissue in the field. Mol Ecol 12:765–775. https://doi.org/10.1046/j.1365-294X.2003.01767.x

Index

E. Lichtfouse (ed.), *Sustainable Agriculture Reviews 32*, Sustainable Agriculture Reviews 32, https://doi.org/10.1007/978-3-319-98914-3

The manufacturer's authorised representative in the EU is Springer
Nature Customer Service Centre GmbH, Europaplatz 3, 69115 Heidelberg,
Germany. If you have any concerns regarding our products, please
contact ProductSafety@springernature.com

Printed and bound by CPI Group (UK) Ltd, Croydon, CR0 4YY
12/08/2026
02194519-0002